Praise for *Your Money and Your Brain*

"This short and entertaining book packs a vast amount of serious information about your brain, about your mind and about your money. You will learn a lot when you read it for the first time, and you will probably want to read it again to learn some more."

—Daniel Kahneman, professor of psychology at Princeton University and winner of the 2002 Nobel Prize in Economics

"Jason Zweig has written a pioneering work. His findings challenge many of our conventional beliefs about investor behavior. Zweig goes an important step further by laying down a series of rules that, if followed, will prevent the reader from making many of the emotional decisions that have cost investors dearly over time. *Your Money and Your Brain* is a book that stands head and shoulders above the conventional pablum served up in most stock market books."

—David Dreman, author of *Contrarian Investment Strategies: The Next Generation*

"Backed by stellar research and written in an entertaining, informal style that makes a complex subject accessible to the layperson, Zweig makes clear how we can understand what our brains are doing and how to use that knowledge to get out of our own way and invest wisely."

—*Publishers Weekly* (starred review)

"Jason Zweig is one the world's experts on the investing process. He has written the best book yet on the emerging science of neuroeconomics. Buy it, read it, and become a more thoughtful, and a better, investor."

—Bill Miller, chairman and chief investment officer, Legg Mason Capital Management

"Jason Zweig knows your financial demons, where they live, why they're making you poor, and how you can beat them. You owe it to yourself, your future, and your heirs to read *Your Money and Your Brain*."

—William Bernstein, associate clinical professor of neurology, Oregon Health and Science University; cofounder, Efficient Frontier Advisors; and author of *The Four Pillars of Investing*

"I am not sure that neuroeconomics will make you rich, but I am certain Jason Zweig's thoughtful review will make you wiser. Here is an informative book from a pro of scientific reporting."

—Antonio Damasio, author of *Descartes' Error* and *Looking for Spinoza,* and director, Brain and Creativity Institute, University of Southern California

"As advertised, this book is about your brain, but yours is not the only brain in this book. Lucky for you, that other brain is Jason Zweig's, and what a brilliant, fascinating, illuminating, powerful, and unique device that is. Listen to Zweig carefully. I have read a zillion books on investing, and none of them comes close to what he has so generously bestowed upon us here."

—Peter L. Bernstein, author of *Against the Gods: The Remarkable Story of Risk*

"Money and the brain: strange bedfellows resting nicely together in Jason Zweig's enjoyable book."

—Joseph LeDoux, university professor, New York University, author of *The Emotional Brain* and *Synaptic Self*

"The most remarkable breakthrough in personal finance this century is identifying the reasons we are so poorly programmed for investing. . . . Of the recent books about investor psychology and neuroeconomics, journalist Jason Zweig's is the first comprehensive treatment for the lay investor."

—Robert Frick, senior editor, *Kiplinger's Personal Finance* (one of *Kiplinger's* "Best Investing Reads of 2007")

"By exploring why investors behave as they do, Zweig's book puts you, the buyer of investment advice and products, on a more equal footing with the sellers. It belongs on every investor's bookshelf."

—Bill Barnhart, *Chicago Tribune*

"This is a book every financial professional should read—and reread at least annually. It will give them insight into the investment decisions they make for themselves and for clients."

—Mike Clowes, *Investment News*

"A fascinating and entertaining read about the new field of neuroeconomics—the study of psychology, economics, and the brain."

—Kara McGuire, *Minneapolis-St. Paul Star Tribune*

"Compelling. . . . Go and get a copy of *Your Money and Your Brain*. It is likely to be one of your better investments."

—Tom Stevenson, *The Daily Telegraph* (U.K.)

Your Money and Your Brain

Simon & Schuster Paperbacks

How the New

Science of

Neuroeconomics

Can Help

Make You Rich

Jason Zweig

NEW YORK LONDON TORONTO SYDNEY

SIMON & SCHUSTER PAPERBACKS
A Division of Simon & Schuster, Inc.
1230 Avenue of the Americas
New York, NY 10020

First Simon & Schuster trade paperback edition September 2008

SIMON & SCHUSTER PAPERBACKS and colophon are registered trademarks of Simon & Schuster, Inc.

For information about special discounts for bulk purchases, please contact Simon & Schuster Special Sales at 1-800-456-6798 or business@simonandschuster.com.

Designed by Paul Dippolito

Frontispiece art by Katie Ris

Manufactured in the United States of America

10 9 8 7 6 5 4 3 2 1

The Library of Congress has cataloged the hardcover as follows:

Zweig, Jason.
Your money and your brain : how the new science of neuroeconomics can help make you rich / Jason Zweig.
p. cm.
Includes bibliographic references and index.
1. Investments—Psychological aspects. 2. Finance—Decision making. 3. Neuroeconomics.
HG4515.15 .Z84 2007
332.601'9—dc22 2006100986

ISBN-13: 978-0-7432-7668-9
ISBN-10: 0-7432-7668-X
ISBN-13: 978-0-7432-7669-6 (pbk)
ISBN-10: 0-7432-7669-8 (pbk)

For my wife, who did the real work with love and grace

Contents

To view the color illustrations please visit www.jasonzweig.com.

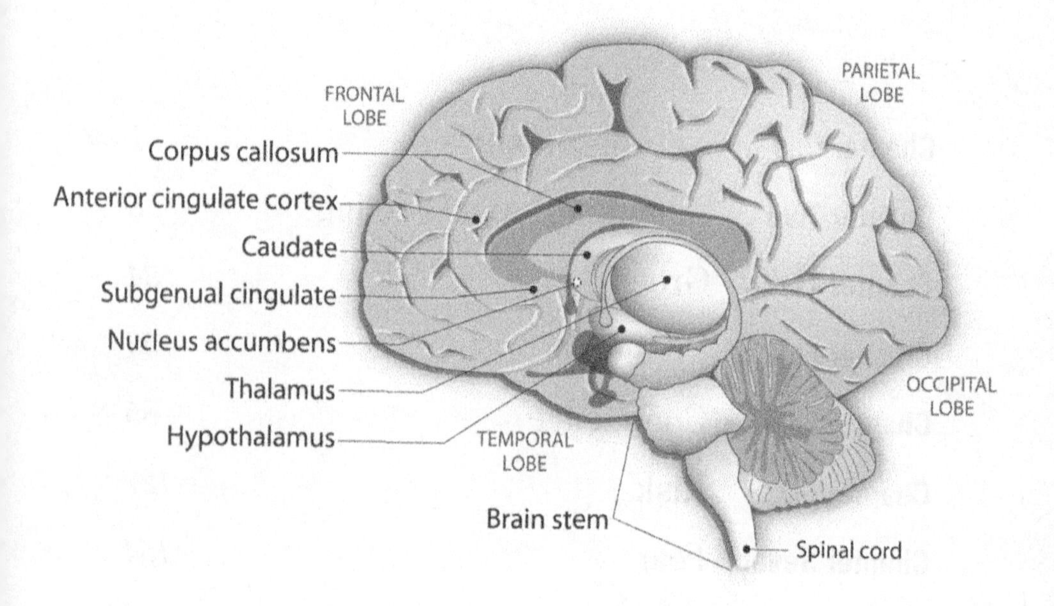
FRONTAL LOBE
PARIETAL LOBE
Corpus callosum
Anterior cingulate cortex
Caudate
Subgenual cingulate
Nucleus accumbens
Thalamus
Hypothalamus
OCCIPITAL LOBE
TEMPORAL LOBE
Brain stem
Spinal cord

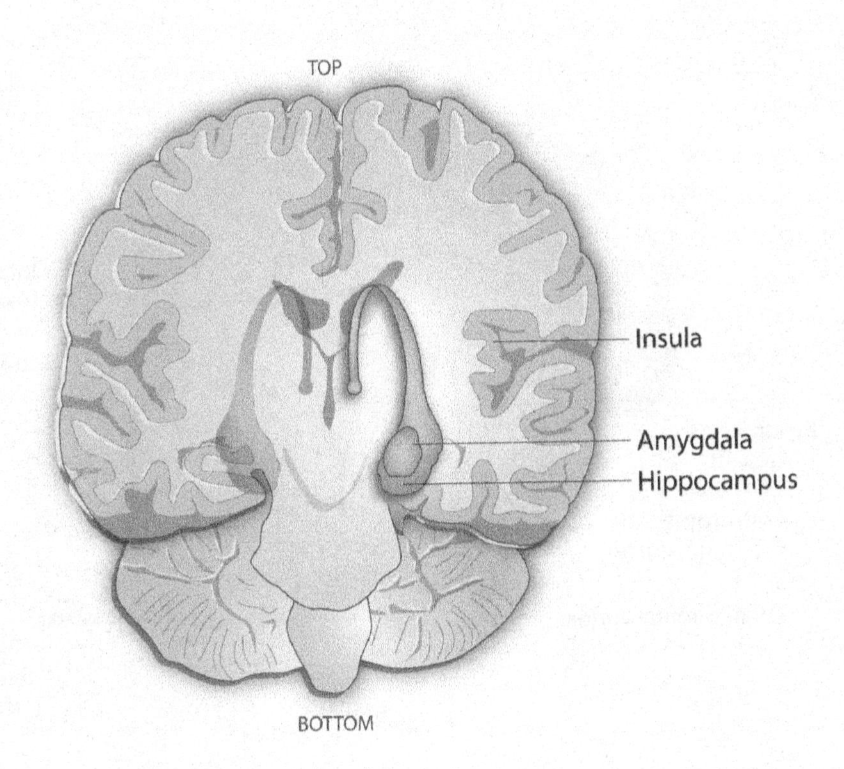
TOP
Insula
Amygdala
Hippocampus
BOTTOM

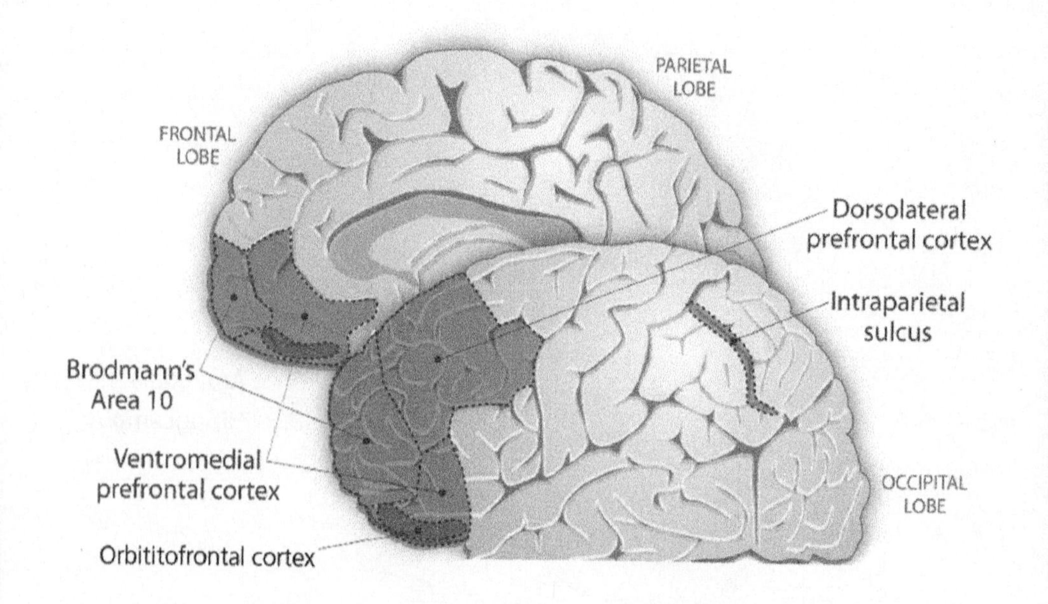
PARIETAL
LOBE
FRONTAL
LOBE
Dorsolateral
prefrontal cortex
Intraparietal
sulcus
Brodmann's
Area 10
Ventromedial
prefrontal cortex
OCCIPITAL
LOBE
Orbititofrontal cortex

CHAPTER ONE

Neuroeconomics

BRAIN, n. An apparatus with which we think that we think.

—*Ambrose Bierce*

"HOW COULD I HAVE BEEN SUCH AN IDIOT?" IF YOU'VE never yelled that sentence at yourself in a fury, you're not an investor. There may be nothing across the entire spectrum of human endeavor that makes so many smart people feel so stupid as investing does. That's why I've set out to explain, in terms any investor can understand, what goes on inside your brain when you make decisions about money. To get the best use out of any tool or machine, it helps to know at least a little about how it works; you will never maximize your wealth unless you can optimize your mind. Fortunately, over the past few years, scientists have made stunning discoveries about the ways the human brain evaluates rewards, sizes up risks, and calculates probabilities. With the wonders of imaging technology, we can now observe the precise neural circuitry that switches on and off in your brain when you invest.

I've been a financial journalist since 1987, and nothing I've ever learned about investing has excited me more than the spectacular findings emerging from the study of "neuroeconomics." Thanks to this newborn field—a hybrid of neuroscience, economics, and psychology—we can begin to understand what drives investing behavior not only on the theoretical or practical level, but as a basic biological function. These flashes of fundamental insight will enable you to see as never before what makes you tick as an investor.

On this ultimate quest for financial self-knowledge, I'll take you inside laboratories run by some of the world's leading neuroeconomists and describe their fascinating experiments firsthand, since I've had my own brain studied again and again by these researchers. (The scientific consensus on my cranium is simple: It's a mess in there.)

The newest findings in neuroeconomics suggest that much of what we've been told about investing is wrong. In theory, the more we learn about our investments, and the harder we work at understanding them, the more money we will make. Economists have long insisted that investors know what they want, understand the tradeoff between risk and reward, and use information logically to pursue their goals.

In practice, however, those assumptions often turn out to be dead wrong. Which side of this table sounds more like you?

IN THEORY	IN PRACTICE
You have clear and consistent financial goals.	You're not sure what your goals are. Last time you thought you knew, you had to change them.
You carefully calculate the odds of success and failure.	That stock your cousin recommended was "a sure thing"—until it stunned you both by going to zero.
You know exactly how much risk you're comfortable taking.	When the market was going up, you said you had a high tolerance for risk. When it went down, you became intolerant in a hurry.
You efficiently process all the available information to maximize your future wealth.	You owned stock in Enron and WorldCom, but you never read the fine print in their financial statements—missing the signs of trouble to come.
The smarter you are, the more money you'll make.	In 1720, Sir Isaac Newton was wiped out in a stock-market crash, blazing a trail of financial failure that geniuses have been following ever since.
The more closely you follow your investments, the more money you'll make.	People who keep up with the news about their stocks earn lower returns than those who pay almost no attention.
The more work you put into investing, the more money you'll make.	"Professional" investors, on average, do not outperform "amateurs."

You're not alone. Like dieters lurching from Pritikin to Atkins to South Beach and ending up at least as heavy as they started, investors habitually are their own worst enemies, even when they know better.

- Everyone knows that you should buy low and sell high—and yet, all too often, we buy high and sell low.

- Everyone knows that beating the market is nearly impossible—but just about everyone thinks he can do it.

- Everyone knows that panic selling is a bad idea—but a company that announces it earned 23 cents per share instead of 24 cents can lose $5 billion of market value in a minute-and-a-half.

- Everyone knows that Wall Street strategists can't predict what the market is about to do—but investors still hang on every word from the financial pundits who prognosticate on TV.

- Everyone knows that chasing hot stocks or mutual funds is a sure way to get burned—yet millions of investors flock back to the flame every year. Many do so even though they swore, just a year or two before, never to get burned again.

One of the themes of this book is that our investing brains often drive us to do things that make no logical sense—but make perfect emotional sense. That does not make us irrational. It makes us human. Our brains were originally designed to get more of whatever would improve our odds of survival and to avoid whatever would worsen the odds. Emotional circuits deep in our brains make us instinctively crave whatever feels likely to be rewarding—and shun whatever seems liable to be risky.

To counteract these impulses from cells that originally developed tens of millions of years ago, your brain has only a thin veneer of relatively modern, analytical circuits that are often no match for the blunt emotional power of the most ancient parts of your mind. That's why knowing the right answer, and doing the right thing, are very different.

- An investor I'll call "Ed," a real estate executive in Greensboro, North Carolina, has rolled the dice on one high-tech and biotech company after another. At last count, Ed had lost more than 90% of his investment on at least four of these stocks. After Ed had lost 50% of his money, "I swore I'd sell if they fell another 10%," he recalls. "When they still kept dropping, I kept dropping my selling point instead of getting out. It felt like the only thing worse than losing all that money on paper would be selling and losing it for real." His accountant has reminded him that if he sells, he can write off the losses and cut his income tax bill—but Ed still can't bear to do it. "What if they go up from here?" he asks plaintively. "Then I'd feel stupid *twice*—once for buying them and once for selling them."

- In the 1950s, a young researcher at the RAND Corporation was pondering how much of his retirement fund to allocate to stocks and how much to bonds. An expert in linear programming, he knew that "I should have computed the historical co-variances of the asset classes and drawn an efficient frontier. Instead, I visualized my grief if the stock market went way up and I wasn't in it—or if it went way down and I was completely in it. My intention was to minimize my future regret. So I split my contributions 50/50 between bonds and equities." The researcher's name was Harry M. Markowitz. Several years earlier, he had written an article called "Portfolio Selection" for the *Journal of Finance* showing exactly how to calculate the tradeoff between risk and return. In 1990, Markowitz shared the Nobel Prize in economics, largely for the mathematical breakthrough that he had been incapable of applying to his own portfolio.

- Jack and Anna Hurst, a retired military officer and his wife who live near Atlanta, seem like very conservative investors. They have no credit card debt and keep almost all of their money in dividend-paying, blue-chip stocks. But Hurst also has what he calls a "play" account, in which he takes big gambles with small amounts of money. Betting on a few long shots in the stock market is Hurst's way of trying to fund what he calls his "lottery dreams." Those dreams are important to Hurst, because he has amyotrophic lateral sclerosis (ALS, or Lou Gehrig's disease); he's been completely paralyzed since 1989. Hurst can invest only by operating a laptop computer with a special switch that reads the electrical signals in his facial muscles. In 2004, one of his "lottery" picks was Sirius Satellite Radio, one of the most volatile stocks in America. Hurst's dreams are to buy a Winnebago customized for quadriplegics and to finance an "ALS house" where patients and their families can get special care. He is both a conservative *and* aggressive investor.

In short, the investing brain is far from the consistent, efficient, logical device that we like to pretend it is. Even Nobel Prize winners fail to behave as their own economic theories say they should. When you invest—whether you are a professional portfolio manager overseeing billions of dollars or a regular Joe with $60,000 in a retirement account—you combine cold calculations about probabilities with instinctive reactions to the thrill of gain and the anguish of loss.

The 100 billion neurons that are packed into that three-pound clump of tissue between your ears can generate an emotional tornado when you think about money. Your investing brain does not just add and multiply and estimate and evaluate. When you win, lose, or risk money, you stir up some of the most profound emotions a human being can ever feel. "Financial decision-making is not necessarily about money," says psychologist Daniel Kahneman of Princeton University. "It's also about intangible motives like avoiding regret or achieving pride." Investing requires you to make decisions using data from the past and hunches in the present about risks and rewards you will harvest in the future—filling you with feelings like hope, greed, cockiness, surprise, fear, panic, regret, and happiness. That's why I've organized this book around the succession of emotions that most people pass through on the psychological roller coaster of investing.

For most purposes in daily life, your brain is a superbly functioning machine, instantly steering you away from danger while reliably guiding you toward basic rewards like food, shelter, and love. But that same intuitively brilliant machine can lead you astray when you face the far more challenging choices that the financial markets throw at you every day. In all its messy, miraculous complexity, your brain is at its best and worst—its most profoundly human—when you make decisions about money.

And it's not as if emotion is the enemy and reason is the ally of good financial decisions. People who have suffered head injuries that prevent them from engaging the emotional circuitry in their brains can be terrible investors. Pure rationality with no feelings can be as bad for your portfolio as sheer emotion unchecked by reason. Neuroeconomics shows that you will get the best results when you harness your emotions, not when you strangle them. This book will help you strike the right balance between emotion and reason.

Most of all, this book should help you understand your investing self better than you ever have before. You may think you already know what kind of investor you are, but you are probably wrong. "If you don't know who you are," quipped the investment writer "Adam Smith" in his classic book *The Money Game,* then Wall Street "is an expensive place to find out." (The people who bought Internet stocks in 1999 because they thought they had a high tolerance for risk—and then lost 95% over the next three years—know just how expensive it can be.) Over the years, I've grown convinced that there are only three kinds of investors: those who think they are geniuses, those who think they are idiots, and those who aren't sure. As a general rule, the ones who aren't sure are the only ones who are right. If you think you're a financial genius, you're almost certainly dumber than you think—and you need to chain your brain so you can control your futile attempts to outsmart everyone else. If you think you're a financial idiot, you're probably smarter than you realize—and you need to train your brain so you can understand how to triumph as an investor.

Knowing more about who you are as an investor can make you a fortune—or save you one. That's why it's so important to learn the basic lessons that have emerged from neuroeconomics:

- a monetary loss or gain is not just a financial or psychological outcome, but a biological change that has profound physical effects on the brain and body;
- the neural activity of someone whose investments are making money is indistinguishable from that of someone who is high on cocaine or morphine;
- after two repetitions of a stimulus—like, say, a stock price that goes up one penny twice in a row—the human brain automatically, unconsciously, and uncontrollably expects a third repetition;
- once people conclude that an investment's returns are "predictable," their brains respond with alarm if that apparent pattern is broken;
- financial losses are processed in the same areas of the brain that respond to mortal danger;

- anticipating a gain, and actually receiving it, are expressed in entirely different ways in the brain, helping to explain why "money does not buy happiness";
- *expecting* both good and bad events is often more intense than *experiencing* them.

We all know that it's hard to solve a problem until you truly understand what caused it. Many investors have told me over the years that their biggest frustration is their inability to learn from their mistakes. Like hamsters in a spinning wheel, the faster they chase their financial dreams, the faster they go absolutely nowhere. The newest findings in neuroeconomics offer a real opportunity to jump off the hamster-wheel of frustration and find financial peace of mind. By enabling you to understand your investing brain better than ever before, this book should help you:

- set realistic and achievable objectives;
- earn higher returns with greater safety;
- become a calmer, more patient investor;
- use the news and tune out the noise in the market;
- measure the limits of your own expertise;
- minimize the number and severity of your mistakes;
- stop kicking yourself when you do make a mistake;
- control what you can and let go of everything else.

Again and again as I researched this book, I was struck by the overwhelming evidence that most of us do not understand our own behavior. There have been many books whose central message is "almost everything you ever thought you knew about investing is wrong." There have been very few that seek to make you a better investor by showing that everything you ever thought you knew about *yourself* is wrong. In the end, this book is about more than the inner workings of the investing brain. It is also about what it means to be human—with all our miraculous powers as well as embarrassing frailties. No matter how much or how little you may think you know about investing, there is always more to learn about the final financial frontier: yourself.

CHAPTER TWO

"Thinking" and "Feeling"

> It is necessary to know the power and the infirmity of our nature, before we can determine what reason can do in restraining the emotions, and what is beyond her power.
>
> —*Benedict de Spinoza*

The Colon Doctor with a Gut Feeling

Not long ago, Clark Harris, a gastroenterologist in New York City, bought stock in CNH Global N.V., a company that makes farming and construction equipment. When a friend asked why he thought the shares would go up, Dr. Harris—who normally does his homework on a stock before he buys—admitted that he knew next to nothing about the company (which is based in the Netherlands). Nor did the city-dwelling doctor have a clue about farm tractors, hay balers, bulldozers, or backhoes. But he loved the stock anyway. After all, explained Dr. Harris, his middle name is Nelson, so the ticker symbol for the company's shares (CNH) matches his initials. And that, he cheerfully admitted, was why he bought it. When his friend asked whether he had any other reasons to invest, Dr. Harris replied, "I just have a good feeling about it, that's all."

It's not just gastroenterologists who rely on gut feelings when they make financial decisions. In 1999, the stock of Computer Literacy Inc. shot up 33% in a single day, purely because the company changed its name to the more hip-sounding fatbrain.com. During 1998 and 1999, one group of stocks outperformed the rest of the technology industry by a scorching 63 percentage points—merely by changing their official corporate names to include *.com, .net,* or *Internet.*

During the years when shares in the Boston Celtics basketball team were publicly traded, they barely budged on news about important business factors like the construction of a new arena—but went jumping way up or down based on whether the team won or lost the previous evening's basketball game. At least in the short run, the stock price of the Celtics was not determined by such fundamental factors as revenues or net earnings. Instead, it was driven by the things that sports fans care about—like last night's score.

Other investors rely on their gut feelings even more literally than Dr. Harris or Celtics fans. Explaining why he had bought stock in Krispy Kreme Doughnuts Inc., one trader declared online in late 2002, "Incredible, my boss bought 30 dozen [doughnuts] for the whole office at $6.00 each . . . Hhhmmmmmmmmm fabulous, no need to ruin them with coffee. Buying more [stock] today." Another visitor to an online message board for Krispy Kreme stock proclaimed, "This stock will soar because these donuts are so good."

The first thing these judgments had in common is that they all were driven by intuition. The people who bought these stocks did not analyze the underlying business; instead, they went on a feeling, a sensation, a hunch. The second thing these judgments had in common is that they all were wrong. CNH has underperformed the stock market since Dr. Harris bought it, fatbrain.com no longer exists as an independent company, many "dot-com" stocks fell by more than 90% between 1999 and 2002, Boston Celtics shares generated higher returns during the off-season than while the team was playing, and Krispy Kreme stock has fallen by roughly three-fourths—even though its doughnuts are still as tasty as ever.

Nor is this kind of thinking unique to supposedly naïve individual investors: A survey of more than 250 financial analysts found that over 91% felt that the most important task in evaluating an investment is to arrange the facts into a compelling "story." Portfolio managers talk constantly about whether a stock "feels right," professional traders regularly move billions of dollars a day based on "what my gut is telling me," and George Soros, one of the world's leading hedge fund managers, reportedly considers dumping his holdings when he gets a backache.

In his book *Blink,* Malcolm Gladwell claims that "decisions made very quickly can be every bit as good as decisions made cautiously and deliberately." Gladwell is a superb writer, but when it comes to investing, his argument is downright dangerous. Intuition can yield wondrously fast and accurate judgments, but only under the right conditions—when the rules for reaching a good decision are simple and stable. Unfortunately, investment choices are seldom simple, and the keys to success (at least in the short run) can be very unstable. Bonds do well for a while and then, as soon as you buy them, they generate lousy returns; your emerging-market stock fund loses money for years and then, right after you bail out, it doubles in value. In the madhouse of the financial markets, the only rule that appears to apply is Murphy's Law. And even that guideline comes with a devilish twist: Whatever can go wrong will go wrong, but only when you least expect it to.

While Gladwell does concede that our intuitions can often mislead us, he fails to emphasize that our intuitions *about* our intuitions can be misleading. Among the most painful of the stock market's many ironies is this: One of the clearest signals that you are wrong about an investment is having a hunch that you're right about it. Often, the more convinced you are that your hunch will pay off big, the more money you are likely to lose.

In a game governed by rules like these, if all you do is "blink," your investing results will stink. Intuition has a legitimate role to play in investing—but it should be a subordinate, not a dominant, role. Fortunately, you can make your intuitions work better for you, and you do not have to invest by intuition alone. The best financial decisions draw on the dual strengths of your investing brain: intuition *and* analysis, feeling *and* thinking. This chapter will show you how to get the most out of both.

The Man with Two Brains

Quick: If John F. Kennedy had not been assassinated, how old would he be today?

Now, decide whether you'd like to reconsider your answer.

If you're like most people, your spontaneous first guess was that JFK would be 76 or 77 years old. After you gave it a second thought, you probably added about ten years to that estimate. (The precise answer: John F. Kennedy was born on May 29, 1917, so do the math.) It's not just amateurs whose first answer is wrong. In 2004, I sprang this little quiz on one of the world's leading experts on decision-making. His first guess was that Kennedy would be 75; when I gave him a few moments to think about it, he changed his answer to 86.

Why do we get this problem wrong at first—and then correct it so easily? When you first confront the question, your intuition instantly summons up a powerful visual memory of Kennedy as a vigorous, youthful leader. You then adjust the age of that young man upward, but not far enough—perhaps because the contrast with older presidents like Lyndon Johnson or Ronald Reagan makes Kennedy seem even younger than he actually was. Kennedy's boyish face is so vividly anchored in your memory that it overwhelms the other data you should consider, like how many years have passed since his death.

Psychologists call this process "anchoring and adjustment," and it gets us through most of daily life remarkably well. Once you were prompted to reconsider, the analytical part of your brain probably recognized and fixed your intuitive error somewhat like this: "Let's see, I guess Kennedy was in his mid-40s when he was shot, and that was around 1963, so he'd probably be around 90 if he were alive today."

But your intuition does not always give your rational side the chance to reconsider. In the early 1970s, psychologists Amos Tversky and Daniel Kahneman of the Hebrew University in Jerusalem asked people to spin a wheel of fortune that was numbered from 0 to 100—and then to estimate whether the percentage of total United Nations membership made up by African countries was higher or lower than the number they had just spun. The spins of the wheel made a big difference, even though such obviously random (and totally irrelevant) numbers should have had no influence on people. On average, people who spun a 10 guessed that only 25% of U.N. members were African nations, while people who spun a 65 guessed that 45% were African.

You can test your own tendency to anchor with this simple exercise. Take the last three digits of your telephone number, then add 400. (For example, if your phone number ends with the digits 237, adding 400 to it gives you 637.) Now answer these two questions: Was Attila the Hun defeated in Europe before or after that year? And what's your best guess of the exact year Attila the Hun was defeated?

Even though telephone numbers have nothing to do with battles against medieval barbarians, experiments on hundreds of people show that the average guess marches up in lockstep with the anchor:

WHEN THE PHONE NUMBER PLUS 400 IS BETWEEN:	THEN THE AVERAGE ESTIMATE OF WHEN ATTILA WAS DEFEATED IS:
400 and 599	629 A.D.
600 and 799	680 A.D.
800 and 999	789 A.D.
1,000 and 1,199	885 A.D.
1,200 and 1,399	988 A.D.

The correct answer, by the way, is 451 A.D.

As soon as your intuition seizes on a number—any number—it becomes stuck, as if it had been coated in glue. That's why real estate agents will usually show you the most expensive house on the market first, so the others will seem cheap by comparison—and why mutual fund companies nearly always launch new funds at $10.00 per share, enticing new investors with a "cheap" price at the beginning. In the financial world, anchoring is everywhere, and you can't be fully on guard against it until you understand why it works so powerfully.

Here's another thought experiment that shows the tug of war between intuition and analytical thinking:

A candy bar and a piece of gum together cost $1.10. The candy bar costs $1.00 more than the piece of gum. Quick: How much does the gum cost?

Now take thirty seconds or so to decide whether you would like to change your answer.

Almost everyone, at first blush, says that the gum costs 10 cents. Most people will never notice that answer is wrong unless they're explicitly asked to second-guess themselves. After thinking about it for a while, you probably realized you made a mistake: If the gum costs 10 cents, and the candy bar costs $1.00 more than the gum, then the candy bar would be $1.10. But $1.10 + 10 cents = $1.20, which can't be right. Scratch your head a bit and you'll arrive at the correct answer: The gum costs 5 cents and the candy costs $1.05.

You can correctly solve a problem like this if—but only if—the analytical part of your brain becomes aware that your intuition may have made a mistake. Using terms suggested by Matthew Lieberman, a psychologist at the University of California, Los Angeles, I call these two aspects of your investing brain the reflexive (or intuitive) system and the reflective (or analytical) system.

Most financial decisions are a tug of war between these two ways of thinking. To see how tough it can be for analysis to trump intuition, look at Figure 2.1. Even after you prove that your perceptions are playing a trick on you, it's hard to conquer the illusion. You know that what you're seeing has to be wrong, but you still *feel* that it's right. As Daniel Kahneman says, "You must learn to recognize that you need to use a ruler."

FIGURE 2.1 Which Line Is Longer?

In the famous Müller-Lyer illusion, the upper horizontal line looks shorter than the lower one. They are, in fact, the same length—which you can easily verify by measuring them with a ruler. But your intuition is so powerful that it will continue to tell you the bottom line is longer even after your analysis has proved otherwise.

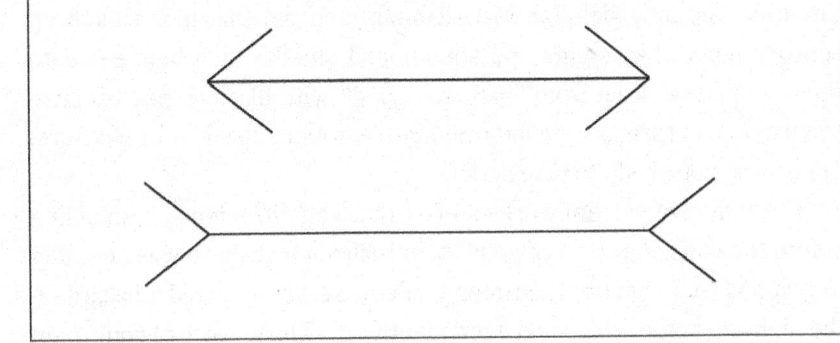

Nevertheless, it's far from true that your reflexive brain is powerful but dumb and your reflective brain is weak but smart. In fact, each system is good at some things and bad at others. Let's learn more about how the two systems work and how you can get them to work better for you as an investor.

The Reflexive Brain

The popular belief that emotional thoughts reside in the "right brain" while logical reasoning is based in the "left brain" isn't entirely wrong.

But the reality is more subtle. While the two kinds of thinking are largely carried out in different areas, right and left have less to do with it than above and below.

The reflexive system is primarily headquartered underneath the cerebral cortex that most of us visualize as the "thinking" part of the brain. Although the cerebral cortex is also a critical part of the emotional system, most reflexive processing goes on below it in the basal ganglia and the limbic areas. A knotty bundle of tissue in the core of the brain, the basal ganglia (also known as the "striatum" because of their striped or banded appearance) play a central role in identifying and seeking almost anything we recognize as rewarding: food, drink, social status, sex, money. They also act as a kind of relay station between the cortex, where complex thought is organized, and the limbic system, where many stimuli from the outside world are first processed.

All mammals have a limbic system, and ours works much like theirs—as a kind of flashpoint of the mind. If we are to survive, we need to pursue rewards and avoid risks as quickly as possible. Limbic structures like the amygdala (ah-MIG-dah-lah) and the thalamus snatch up sensory inputs like sights and sounds and smells, then help evaluate them on a basic scale from "bad" to "good" with blazing speed. Those evaluations, in turn, are transformed into emotions like fear or pleasure, motivating your body to take action.

The reflexive system works so fast that you often finish responding before the conscious part of your brain realizes that there was anything to respond to. (Think of the times you've swerved to avoid a hazard on the highway before you could even identify it.) These parts of your brain can activate an alarm in less than a tenth of a second.

The reflexive system (which some researchers call System 1) gets "first crack at making most judgments and decisions," says Matthew Lieberman of UCLA. We count on our intuition to make initial sense of the world around us—and we tap into our analytical system only when intuition can't figure something out. As Kahneman says, "We run mostly on System 1 software."

In truth, the reflexive brain is not a single integrated system, but a jumble of structures and processes that tackle different problems in varying ways, everything from the "startle reflex" and pattern recognition to the perception of risk or reward and the character judgments about the people we meet. What these processes have in common, however, is that they tend to run rapidly, automatically, and below the level of consciousness.

That enables us to ignore most of what goes on around us most of the time—unless it rises to the level of a risk or reward that we need to avoid or pursue. Reuven Dukas, a behavioral ecologist at McMaster University in Hamilton, Ontario, has shown that having to pay attention to more than one stimulus at once dramatically decreases the amount of food that animals like birds and fish can identify and capture. The human animal is no different: While "multitasking" is a fact of life, so is the falloff in awareness we can devote to each new task. Shifting your attention from one thing to the next causes what Dukas calls "a period of reduced efficiency." When you redirect your attention, your brain is like a bicyclist who stops pedaling for a moment and must then get back up to full speed. As Dukas says, we are designed to "focus attention on the stimuli most likely to be of importance."

After all, our brains can't possibly keep up with everything that's happening in our environment. When you are at rest, your brain—which accounts for roughly 2% of the typical person's body weight—consumes 20% of the oxygen you take in and the calories you burn. Because your brain operates at such a high "fixed cost," you need to ignore most of what is happening around you. The vast majority of it isn't meaningful, and if you had to pay separate, equal, and continual attention to everything, information overload would fry your brain in short order. "Thinking wears you out," says Lieberman of UCLA. "So the reflective system tends not to want to do anything unless it has to."

Therefore, our intuition acts as the first filter of experience, an instantaneous screen that enables us to conserve our vital mental energy for the things that are most likely to matter. Because of its phenomenal skill in recognizing similarities, the reflexive system sounds an instant alarm when it detects a difference. As you drive down the road, for example, hundreds of stimuli glide beneath the radar of your conscious attention every second: houses and trees and storefronts, exit signs and billboards and mileage markers, airplanes overhead, the makes and colors and license plates of passing vehicles, birds perching on lampposts, most of the music coming from your car's sound system, even much of what your kids are doing in the backseat. Everything flows along in a mercifully homogenized blur; because it is all part of a familiar pattern, you negotiate your way through it effortlessly.

But the moment anything is out of place—if a tire bursts on a truck ahead, a pedestrian steps into the road, or a sign announces a sale at your favorite store—your reflexive system will seize it out of the background and make you hit the brake. By glossing over whatever stays essentially the same in your environment, your reflexive system can rivet your attention onto anything unexpected, anything novel, anything that appears to change suddenly or significantly. You may think you've made a "conscious decision," but more often than not you have been driven by the same basic impulses that drove our ancestors to avoid risks and pursue rewards. As neuroscientist Arne Öhman puts it, evolution has designed our emotions "to make us want to do what our ancestors had to do."

Why should investors care? "The reflexive system," explains University of Oregon psychologist Paul Slovic, "is very sophisticated and served us well for millions of years. But in a modern world, where life is full of much more complicated problems than just immediate threats, it's not adequate and is likely to get us into trouble." Your reflexive system is so fixated on change that it makes it hard for you to focus on what remains constant. If the Dow Jones Industrial Average moves from 12,683.89 to 12,578.03, the newscaster will holler, "The Dow dropped 106 points today!" Your reflexive system reacts to the size of the change and ignores the base on which it is calculated. So 106 points feels like a big drop, raising your pulse and giving you sweaty palms—and maybe even scaring you out of the market entirely. Your emotions crowd out the fact that the level of the index has changed by less than 1%.

Likewise, your reflexive system will prompt you into paying more attention to a single stock rising like a rocket or sinking like a stone than the much more important (but less vivid) change in the overall value of your portfolio. And you'll always be tempted to invest in that mutual fund that went up 123% last year; that hot number grabs your attention and keeps you from noticing the fund's tepid performance over the longer term. (It's no accident that the fund's advertisements print "123%" in giant type and the long-term numbers in microscopic scale.)

Economist Colin Camerer of the California Institute of Technology sums up the reflexive system this way: "It's kind of like a guard dog. It makes rapid but sort of sloppy decisions. It will always attack the burglar, but sometimes it might attack the postman, too." That's why "blink" thinking can get investors into trouble.

The Reflective Brain

But there's more to your investing brain than intuition and emotion. There's a vital counterweight: the reflective system. This function resides largely in the prefrontal cortex, which lies behind your forehead and is part of the frontal lobe that curls like a cashew around the core of your brain. Neuroscientist Jordan Grafman of the National Institutes of Health calls the prefrontal cortex "the CEO of the brain." Here, neurons that are intricately connected with the rest of the brain draw general conclusions from scraps of information, organize your past experiences into recognizable categories, form theories about the causes of change around you, and plan for the future. Another hub in the reflective network is the parietal cortex, back above your ears, which processes numerical and verbal information.

While the reflective system does play a role in processing emotion, you use it largely for solving more complex problems like "Is my investment portfolio sufficiently well-diversified?" or "What should I get my wife for our anniversary?" The reflective brain can intervene when the reflexive brain encounters situations it cannot solve by itself. If the reflexive parts of your brain are the default system, the "go-to" circuits that use intuitive processing to tackle problems first, then the reflective areas are the backup machinery, the "uh-oh" circuits that engage analytical thinking. If someone asked you to count backwards by 17s from 6,853, your intuition would draw a blank. Then, a moment later, you would consciously be aware of thinking, *6,836 . . . 6,819 . . . 6,802 . . .* "It never feels like it's going on by itself," says Matthew Lieberman of UCLA. "You not only know about it, but you feel like you're responsible for what it's doing. You feel like you turned it on for a reason that you could put into words."

Jordan Grafman has shown that people with damage to the prefrontal cortex—from a stroke or tumor, for example—have a hard time evaluating advice and making long-term plans. Grafman presented prefrontal patients with business forecasts. These predictions came not from live experts but rather from images of advisors projected onto a computer screen; the objective was to figure out which advisor to trust. Over the course of forty trial runs, the participants in the experiment had ample opportunities to compare each forecaster's predictions against actual outcomes. A control group of people with undamaged brains readily learned to favor the forecaster whose predictions turned out to be most accurate. The prefrontal patients, however, made their judgments *perceptually* instead of *conceptually,* relying on what Grafman calls "cues that typically had nothing to do with a good choice." One patient, for example, preferred the advisor whose image was displayed on a green background, "since it is springtime." It seems that if the prefrontal cortex is impaired, the brain's internal checks-and-balances system breaks down—and the reflexive areas may take over unopposed.

At the University of Iowa, students were briefly shown numbers that they had to memorize. Then they were offered the choice of either a fruit salad or a chocolate cake. When the number the students memorized was seven digits long, 63% of them chose the cake. When the number they were asked to remember had just two digits, however, 59% opted for the fruit salad. Our reflective brains know that the fruit salad is better for our health, but our reflexive brains crave that gooey, fattening chocolate cake. If the reflective brain is busy figuring something else out—like trying to remember a seven-digit number—then impulse can easily prevail. On the other hand, if we're not thinking too hard about something else (with only a minor distraction like memorizing two digits), then the reflective system can overrule the emotional impulse of the reflexive side.

If You're So Smart, Why Do You Act So Stupid?

But the reflective brain is hardly infallible. Robin Hogarth, a psychologist at Pompeu Fabra University in Barcelona, Spain, suggests imagining yourself in a supermarket checkout line. Your shopping cart is heaped high. How much will all these groceries cost? For an intuitive estimate, you would make a quick-and-dirty comparison between how full your cart is this time and how much a cartful of groceries usually costs. If you have, say, 30% more stuff than usual, you reflexively multiply your typical grocery bill by 1.3. In a couple of seconds, your intuition tells you, "Looks like about $100 to me." You can do all this without even realizing you're doing it. But what if you try to figure out the total bill with the reflective part of your brain? Then you have to add up each of the dozens of items in the cart separately and keep a running total in your head until you've counted every purchase (including confusing ones like "1.8 pounds of grapes at $1.79 a pound, or was that $2.79?"). Chances are, after the exacting effort of adding up barely a handful of individual prices, you will lose track and give up.

Computational neuroscientists—who use the principles of computer design to study the function and design of the human brain—believe that the reflective system may rely on what they call "tree-search" processing. Nathaniel Daw, a researcher in computational neuroscience at University College London, explains that this processing method takes its name from the classic image of a decision tree: On a chessboard, for example, the set of potential future choices grows wider with each subsequent move, like the branches of a tree fanning outward as they get farther from the trunk. If Daw and his colleagues are right, your reflective system laboriously sorts through experiences, predictions, and consequences one at a time to arrive at a decision—much like an ant moving up and down, back and forth, along the branches and twigs of a tree to find what it wants. As our earlier example of the shopping cart shows, the success of the tree-search method is limited by the power of your memory and the complexity of what you are measuring.

In the financial markets, people who rely blindly on their reflective systems often end up losing the forest for the trees—and their shirts as well. Although doctors get a bad rap as investors, in my experience engineers are worse. That may be because they are trained to calculate and measure every possible variable. I've met engineers who spend two or three hours a day analyzing stocks. They are often convinced that they've discovered a unique statistical secret that will enable them to beat the market. Because they have squelched their intuition, their analysis fails to alert them to the most obvious fact of all: There's always something to measure on Wall Street, which spews out a torrent of statistics on everything under the sun. Unfortunately, at least 100 million other investors can view the same data, taking away most of its value—while, at any moment, an unforeseen event can blindside the market, rendering anyone's statistical analysis at least temporarily useless.

That's what happened in 1987, when the arcane computer programs called "portfolio insurance" did not fully protect giant investors from losses—and may, in fact, have contributed to the U.S. stock market's record plunge of 23% in a single day. It happened again in 1998, when the PhDs, Nobel Prize winners, and other geniuses who ran the Long-Term Capital Management hedge fund measured everything imaginable—except the risks of borrowing too much money and assuming that markets would remain "normal." When the markets went crazy, LTCM went under, and nearly took the global financial system with it.

When a problem is hard to solve, the reflective system may "hand back" the challenge and let the reflexive brain take over. In an experiment by Robin Hogarth and the late Hillel Einhorn of the University of Chicago, people were told that an expert claimed that the market always went up after he predicted that it would rise. They were told they could verify the expert's claim by observing outcomes from one or more of the following categories of evidence:

1. what the market did after he predicted that it would rise
2. what the market did after he predicted that it would fall
3. what he predicted before the market rose
4. what he predicted before the market fell

Then they were asked what was the *minimum* evidence they would need in order to establish for certain whether the expert's claim was true. Fully 48% of the people responded that No. 1 was all they would need. Only 22% gave the correct answer: The minimum evidence needed to see whether the expert's claim was true is No. 1 and No. 4. Even though he says the market always goes up when he predicts it will, you still need to know what he said before the market went down. (After all, it does *not* always go up.) Subjecting him to both these tests is the only way to be positive about the truth. Surprisingly, this study was conducted among professors and graduate students in the statistics department at the University of London, who worked with numbers all day long and certainly should have known better.

To answer Hogarth and Einhorn's question correctly, you need only to understand that the most reliable way of determining whether something is true is to try proving that it is false. That's the cornerstone of the scientific method, the critical mind-set that overturned old orthodox "truths" like *the world is flat* and *the earth is the center of the universe.* But that kind of critical thinking is anathema to your intuition, which is most comfortable when dealing with the concrete reality of "what is." To handle a conceptual abstraction like "what is not," you need your reflective system to kick in with the hard mental effort of comparing alternatives and evaluating evidence. That requires asking tough questions like "under what conditions would this no longer be true or fail to work?" And the human mind, which functions as what psychologists Susan Fiske of Princeton and Shelley Taylor of UCLA have called a "cognitive miser," tends to shy away from that kind of effort. If the reflective system can't readily find a solution, the reflexive brain will resume control, using sensory and emotional cues as shortcuts. That's why even professional statisticians failed to solve Hogarth and Einhorn's task correctly: Why go through the trouble of trying to test the logic of all four answers when answer No. 1 feels and sounds so right at first blush?

The Jellybean Syndrome

The conflict between "thinking" and "feeling" can lead to results that are downright bizarre. Psychologists at the University of Massachusetts filled a small bowl and a large bowl with jellybeans. The small bowl held 10 jellybeans, of which 9 were always white and 1 red. The large bowl con-

FIGURE 2.2 **Which Bowl Would You Pick From?**

In this experiment, researchers instructed people to try picking a colored jellybean from either of two bowls. In the one on the left, 10% of the jellybeans were colored; in the one on the right, only 9% were. But people still preferred to pick from the bowl that they "knew" had lower odds of success, because they "felt" it offered more ways to win.

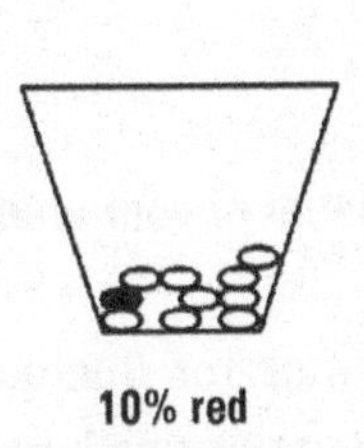

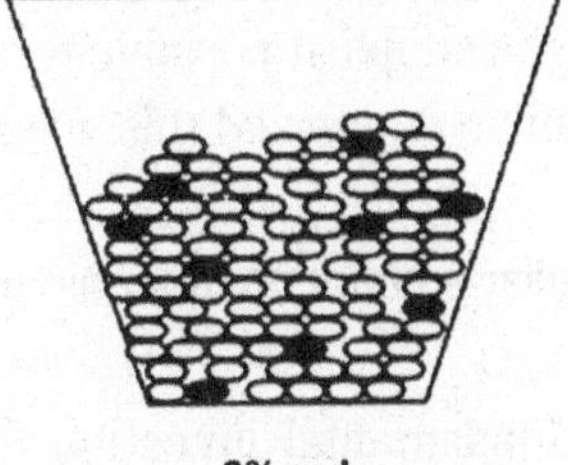

tained 100 jellybeans; on each run of the experiment, between 91 and 95 were white, and the rest were red. The people taking part in the experiment could earn one dollar if they were able to draw a red jellybean out of either of the two bowls. First, however, they were reminded that red jellybeans made up 10% of the total in the small bowl but no more than 9% of the total in the big one. (See Figure 2.2 for a schematic view of the experiment.) Right before each person tried to pluck out a red bean, the bowls were shaken up, then shielded from view to prevent cheating.

Which bowl did people pick from? Someone using the reflective system to think analytically would always choose from the small bowl, since it offered a constant 10% chance of success, while the odds of getting a red jellybean out of the big bowl could never go above 9%. Nevertheless, just under two-thirds of people preferred to pick from the big bowl when it contained 9% red jellybeans.

Even when it held only 5% red jellybeans, nearly one-quarter of the participants in the experiment chose to pick from the big bowl—regardless of what their reflective system was telling them about logic and probability. "I picked the [bowl] with the more red jellybeans," one person explained to the researchers, "because it looked like there were more ways to get a winner, even though I knew there were also more whites, and that the percents [sic] were against me." The participants, explain psychologists Seymour Epstein and Veronika Denes-Raj, "readily acknowledged the irrationality of their behavior. . . . Although they knew the probabilities were against them, they felt they had a better chance when there were more red beans."

The technical term for the jellybean syndrome is "denominator blindness." Every fraction, of course, looks like this:

numerator / denominator

And, in the simplest possible terms, the impact of every investment you make can be expressed this way:

dollar amount of your gain or loss / total amount of your wealth

In this fundamental investing fraction, the numerator fluctuates constantly, and often widely, while the denominator varies much more gradually over time. Let's say, for instance, that you have a total net worth of $200,000 and that yesterday the market value of your stocks did not change at all. If your stocks gain $1,000 today, then the numerator of your fraction instantly shoots from zero to $1,000. The denominator, however, nudges slightly upward from $200,000 to $201,000. The leap from zero to $1,000 is exciting and vivid, while the movement from $200,000 to $201,000 is hardly noticeable at all.

But it's the denominator that matters; that's where the real money is. After all, the sum total of your wealth is a much more important number than the amount by which it rose or fell on any given day. Even so, many investors fixate on the numbers that change the most, overlooking the much larger amounts of money that are at stake overall.

In the late 1980s, psychologist Paul Andreassen ran a series of remarkable studies in which he set up an artificial stock market in his laboratories at Columbia and Harvard. He showed the level of stock prices to one group of investors; another group could view only the change in stock prices. Depending on how much the stocks fluctuated, investors who focused on price levels earned between five and ten times higher profits than those who paid attention to price changes. That's because the investors who fixated on price *changes* traded too much, trying to shave profits off the interim fluctuations, while those who paid attention to price *levels* were more content to hold on for the long haul.

The jellybean syndrome hits home in other ways. The fees and expenses charged by mutual funds are a small number—typically less than 2% a year—while performance can be a big number, sometimes surpassing 20% a year. And the expense figures barely fluctuate at all, while the performance numbers are forever flashing up and down. No wonder individual investors consistently say that they consider past performance to be much more important than current expenses when they pick a fund.

Investment professionals, who are supposed to know better, are at least as prone to the jellybean syndrome: Financial advisors recently ranked expenses as the eighth most important factor when analyzing a mutual fund, after such things as performance, risk, how old the fund is, and how long the current manager has been in charge. Unfortunately, none of those factors will help these so-called experts to identify funds that will earn top returns. Decades of rigorous research have proven that the single most critical factor in the future performance of a mutual fund is that small, relatively static number: its fees and expenses. Hot performance comes and goes, but expenses never go away. The flashier factors like performance and reputation have almost no power to predict a fund's return—but they are more vivid and changeable than the fund's expenses, so they hijack our attention. And that prompts amateur and professional investors alike to pick their mutual funds from the wrong jellybean jar.

Getting the Best Out of Both

Add it all up, and it's clear that when you invest, acting like the coldly rational Mr. Spock from the old *Star Trek* TV series is no more practical than being a sputtering cauldron of emotion like Dr. McCoy. Since both systems have their strengths and weaknesses, the challenge for you as an investor is to make your reflective and reflexive sides work better together, so that you can strike the right balance between thinking and feeling. Here are a few suggestions that can help.

➢ **WHEN IT COMES TO TRUST, TRUST YOUR GUT.** A mutual fund manager named Fred Kobrick once attended a compelling presentation by the CEO of a fast-growing company. Afterward, Kobrick went up to the CEO to tell him how impressed he was with the company and that he would probably buy the stock. When the CEO reached out to shake his hand, Kobrick noticed that the executive's shirt cuff was monogrammed in a distinctive style. Then Kobrick saw that several of the company's other managers had their shirtsleeves monogrammed the same way. "In a flash, I knew I no longer wanted to buy the stock," he recalls. "How could these guys ever bring bad news to the boss if they couldn't even think for themselves when they were buying shirts?"

Most investors, of course, are unlikely to meet with CEOs face to face. But you should read the two documents that may reveal the character of company bosses—the annual proxy statement and the chairman's letter to shareholders in the annual report—with your emotional eyes open. The proxy will give you a feel for how much the managers get paid and whether they have conflicts of interest that make you queasy. The chairman's letter will show whether the boss unfairly takes credit for good markets (which are not within his control) or dodges blame for bad decisions (which are). If the chairman's letter brags about how the company will thrive in the future but ignores how badly it is doing right now, that's another sign that should trouble you. "If you start getting a fishy feeling," says Robin Hogarth of Pompeu Fabra University, "treat your emotions as data. A suspicion is a signal that you should consider delaying your decision."

Whenever you do make character judgments face to face, drawing on your intuition can keep you from being too coldly rational. If, for instance, you are picking a stockbroker or a financial planner to help manage your finances, it can be a mistake to make your choice based primarily on how impressive the person's professional credentials are. (Many investors are too easily dazzled by the alphabet soup of financial resumes, which can brim with MBAs, CPAs, CFAs, CFPs, JDs, and PhDs.) Selecting a financial advisor based on educational or professional credentials alone could leave you paired up with someone who is technically competent but doesn't "click" with you and can't help manage your emotions when the markets go to extremes. Therefore, you should first research each candidate's background at www.nasaa.org, www.napfa.org, www.advisorinfo.sec.gov, and http://pdpi.nasdr.com/pdpi/. That will establish whether he or she has ever been disciplined by the regulatory authorities for unfair dealings with other investors. Once you have at least two candidates with an equally clean record, then you can factor in each person's education and other qualifications—and pick the one your intuition tells you is more empathetic and compatible with your own personality.

➢ **KNOW WHEN REFLEX WILL RULE.** It's no surprise that mutual fund investors tend to lose their shirts whenever they buy and sell industry-specific "sector funds." Analyzing an entire industry requires a great deal of reflective study; with dozens of companies offering hundreds of competing products and services, it's hard to put together an objective picture of how profitable the entire sector will be in the future. But your reflexive system will pick up much simpler messages—"Oil prices are booming!" or "The Internet is changing the world!"—that can easily distract you from a more detailed analysis. Whenever this kind of excitement is in the air, warns psychologist Paul Slovic, "it's hard to engage the analytical system. It's much easier to go with your feelings about what's hot and whatever generates the most vivid images—like buying nanotech stocks because someone says, 'You know, those tiny machines can turn out giant profits.'" Instead of giant profits, however, most people end up with big losses when they rely on this kind of reflexive thinking. In 1999 and 2000, investors lost at least $30 billion by piling into technology-sector mutual funds just before the hot returns of the tech industry turned cold as ashes.

When the financial markets are quietly muddling along, it's easier for reflective judgment to prevail over reflexive intuitions. But when bull markets are pumping out sensationally high returns—or when bear markets are generating demoralizing losses—the reflexive system gets the upper hand and it becomes urgently important to think twice.

➢ **ASK ANOTHER QUESTION.** One way to think twice is to have a procedure for making sure you ask the right questions. As Daniel Kahneman says, "People who are confronted with a difficult question sometimes answer an easier one instead." That's because the reflexive system hates uncertainty, and will quickly reframe problems into terms it can understand and answer with ease. Faced, for example, with a difficult problem to solve—"Will this stock keep going up?"—many investors consult a chart of recent price performance. If the trend line slopes upward, then they immediately answer "Yes," without realizing that their reflexive system has tricked them into answering an entirely different question. All the chart really shows is the answer to a much easier problem: "Has this stock been going up?" People in this kind of situation "are not confused about the question they are trying to answer," says Kahneman. "They simply fail to notice that they are answering a different one."

Asking a follow-up query—for example, "How do I know?" or "What is the evidence?" or "Do I need more information than this?"—can force you to notice that your reflexive system was answering the wrong question. Christopher Hsee, a psychologist at the University of Chicago, has another suggestion: "If this happened to someone else, and they asked for your advice, what would you tell them to do? I oftentimes try to make decisions that way, by putting myself in someone else's shoes." Hsee's tip is particularly helpful because once you imagine giving advice to someone else, you can also imagine that person pressing you: "Are you sure? How do you know?"

➢ **DON'T JUST PROVE; TRY TO DISPROVE.** As we've seen, the reflexive brain believes that the best way to prove an assertion is to keep looking for more proof that it is true. But the only way to be more certain it's true is to search harder for proof that it is *false.* Money managers often say things like, "We add value by selling any stock that drops 15% from our purchase price." As proof, they show the performance of the stocks they kept. Instead, you should ask to see the subsequent returns of the stocks that got sold; that's the only way to tell whether the firm really should have sold them. Likewise, when an investment-consulting firm boasts that it achieves superior returns by firing underperforming fund managers, ask for data on the performance of the fired managers *after* they are fired. Only by looking at such "unobserved outcomes" can you truly test these people's claims. (Embarrassingly often, these experts have never analyzed this evidence themselves!)

➢ **CONQUER YOUR SENSES WITH COMMON SENSE.** In general, sights and sounds engage your reflexive system, while words and numbers activate your reflective system. That's why brokerage firms and insurance companies produce advertisements showing golden people strolling along a golden beach with their golden retriever; that image sets off powerful feelings of comfort and security in your reflexive system. That's also why mutual fund companies display the performance of their portfolios in the form of "mountain charts," those graphics that show an initial investment growing over time until it has heaped up into a vivid, Himalayan peak of wealth.

Motion imparts a power all its own. A half-century ago, the neurobiologist Jerome Lettvin showed that specialized cells in the optic nerve of a frog will send signals to the frog's brain at the sight of abstract displays that mimic the movement of a fly—even if the color and shape of the displays are not very bug-like. When the fake fly does not move, or its trajectory does not resemble that of a real fly, the cells do not respond. Lettvin concluded that frogs are designed to perk up whenever an object "moves about intermittently," in a pattern characteristic of their insect prey. It is not only the prey itself, but also its motion, that triggers the frogs' reaction.

Like frogs, people are inherently excited by motion. Investors are much more inclined to expect the stock market to keep rising when its activity is described with an action verb like "climbing" or "leaping" than when it is summed up in neutral terms like "posting a gain." Figures of speech that evoke rapid movement fire up our minds to expect that the market must be "trying to do something."

The emotional power of metaphors shows why you should never passively accept information in its original packaging. Instead, make sure you unwrap it in several different ways. When a broker or financial planner pushes a brightly colored chart across the table, ask questions like: What does this investment look like when it's measured over different—and longer—time periods? How does this stock or fund stack up against other comparable investments and against an objective benchmark like a market index? Based on its past record, when might an investment like this tend to perform poorly? Considering all the evidence that past performance does not predict future results, how does this investment rate on other criteria that are at least as important, like annual expenses and after-tax return?

➢ **ONLY FOOLS INVEST WITHOUT RULES.** When the great investment analyst Benjamin Graham was asked what it takes to be a successful investor, he replied: "People don't need extraordinary insight or intelligence. What they need most is the character to adopt simple rules and stick to them." I've set out ten basic investment rules in Appendix 1. The first letters of these ten commandments form the words THINK TWICE. Whenever an emotional moment in the market threatens to sweep you away, check your first impulse. By relying on your THINK TWICE rules before you make any investing decision, you can prevent yourself from being governed by guesswork and whipsawed by the momentary whims of the market.

➢ **COUNT TO TEN.** When your emotions are running high, take a timeout before you make a hasty decision you might regret later. Psychologists Kent Berridge of the University of Michigan and Piotr Winkielman of the University of California, San Diego, have shown that we can be swept away by emotion without any awareness whatsoever of what's going on inside us. Berridge and Winkielman call this phenomenon "unconscious emotion." To see how it works, consider one of their experiments in which thirsty people had to decide how much they would pay for a drink. On average, one group would pay only 10 cents; the other was willing to pay 38 cents. The only difference between the two groups: The cheapskates had been shown a photo of an angry face for less than one-fiftieth of a second—a visual exposure so brief that no one was conscious of seeing it—while the big spenders had seen a photo of a happy face for the same split second. None of the participants was aware of feeling happier or more anxious. But, for a period lasting approximately one minute, their behavior was governed by the unconscious emotions generated by these subliminal images. "Stimuli that are presented very briefly often have a greater effect than those that are presented for a longer time," says Winkielman. "Because you're unaware of what caused your mood or belief, you may be more inclined to just go with it."

The reflexive system "is tuned to respond to the current situation," says Norbert Schwarz, a psychologist at the University of Michigan. "Your mood affects your momentary behavior, but the consequences of your decision may extend far beyond this moment." When you're feeling unusually optimistic, you might take a financial risk you would normally shy away from; on the flip side, if you're feeling anxious or shaky, you might avoid a risk that you would readily accept at another time.

Nearly all of us feel more chipper when the sun is shining than when the sky is gray. Sure enough, stocks earn slightly higher returns on days when the sun is bright than on days when the sky is overcast—even though daily cloud cover has no rational economic significance. Some studies have even documented a werewolf effect, in which stocks earn half the returns when the moon is full that they earn during a new moon. And stock markets in countries whose national soccer teams lose World Cup elimination matches underperform the global stock index by an average of 0.4 percentage points the day after the defeat.

Companies can exploit your reflexive system just by giving their stocks catchy ticker symbols. Stocks whose trading names resemble familiar words (like MOO or GEEK or KAR) tend to outperform those with unpronounceable symbols like LXK or CINF or PHM—at least in the short run. In the long run, however, stocks with cute ticker symbols have a disconcerting tendency to go bust.

So unless you guard against investing under the influence of your own momentary mood swings, you may never achieve financial stability. When Norbert Schwarz warns, "Never make an important decision before you've slept on it," he is not repeating a cliché. He is stating fundamental wisdom freshly confirmed by the latest scientific research. You will almost always make a better investing decision if you sleep on it rather than acting on your first impulse.

Another option, says Matthew Lieberman, is to get a "second opinion from someone who isn't invested in your point of view." Try running your investment ideas past someone who knows you well but is accustomed to disagreeing with you. While your spouse might sound ideal for this purpose, the best person is someone who would be a good business—rather than romantic—partner for you. It's worth noting that many of America's most innovative and successful companies are led by duos who check and counterbalance each other's ideas: Berkshire Hathaway has Warren Buffett and Charles Munger; Yahoo! has David Filo and Jerry Yang; Google has Larry Page and Sergey Brin. If you have someone who is both trustworthy and tough on you, make a habit of testing your investing ideas out on him or her before you do anything else.

A final approach is to try "embodying" your thoughts. "You do your thinking in a body," explains Schwarz, "and your bodily responses interact with your neural activity." Odd as it sounds, pushing away from a hard surface with your forearm can be a helpful step in thinking more reflectively and less emotionally. By shoving back against your computer or your desk, for example, you can literally distance yourself from the emotional aspects of the decision. Next, use this stiff-arming motion as a reminder. As you "push back," remember to "think twice" by referring to the ten investing commandments in Appendix 1. You could even use a Palm Pilot or BlackBerry to send yourself automatic prompts—"Did I push back and think twice?"—before you make any moves in the market.

All these mental tricks provide what Santa Clara University finance professor Meir Statman calls "a cold shower" that can help keep you from getting caught up in the heat of the moment.

➢ **WHEN THE MARKET "BLINKS," BLINK BACK.** If a stock crashes on bad news, it might be permanently damaged—or just suffering from a temporary overreaction. By doing your homework on an investment ahead of time, you can be ready to pounce. If you're a serious stock investor, you should always be braced to buy the stock of companies you understand if they suddenly become bargains. That's how Bill Miller, the renowned manager of the Legg Mason Value and Legg Mason Opportunity mutual funds, does it. In the summer of 2004, the stock of Career Education Corp. dropped from around $70 to $27 as investors panicked over reports that regulators were investigating the company's accounting and business practices. But Miller knew that the operator of technical schools was solidly profitable, and he felt it was likely to remain that way. So when the market "blinked," Miller blinked right back, snapping up 2 million shares of Career Education's stock from frightened sellers at bargain prices. (By year-end 2004, the stock had already risen almost 50% from its lows in the summer panic.)

Doing your homework ahead of time enables you to take advantage of the "blink" thinking of fair-weather investors who panic at the first sign of trouble. Warren Buffett read the annual report of Anheuser-Busch every year for twenty-five years, familiarizing himself with the company while he patiently waited for the stock to become cheap enough for him to want to own it. Finally, in early 2005, the stock dropped—and Buffett, who now knew the business intimately, snapped up a major stake in Anheuser-Busch.

➢ **STOCKS HAVE PRICES; BUSINESSES HAVE VALUES.** In the short run, a stock's price will change whenever someone wants to buy or sell it and whenever something happens that seems like news. Sometimes the news is nothing short of ridiculous. On October 1, 1997, for example, shares in Massmutual Corporate Investors jumped by 2.4% on 11 times their normal trading volume. That day, WorldCom announced a bid to acquire MCI Communications. Massmutual's ticker symbol on the New York Stock Exchange is MCI—and hundreds of investors evidently rushed to buy it, believing the stock would rise after WorldCom's takeover offer. But MCI Communications traded on NASDAQ under the ticker symbol of MCIC—so the price of Massmutual's stock had shot up in a farce of mistaken identity. Likewise, in early 1999, the stock of Mannatech Inc. shot up 368% in its first two days of trading when Internet-crazed traders mistakenly thought Mannatech was a technology stock; in fact, it is a marketer of laxatives and nutritional supplements.

In the long run a stock has no life of its own; it is only an exchangeable piece of an underlying business. If that business becomes more profitable over the long term, it will become more valuable, and the price of its stock will go up in turn. It's not uncommon for a stock's price to change as often as a thousand times in a single trading day, but in the world of real commerce, the value of a business hardly changes at all on any given day. Business value changes *over* time, not *all* the time. Stocks are like weather, altering almost continuously and without warning; businesses are like climate, changing much more gradually and predictably. In the short run, it's the weather that gets our notice and appears to determine the environment, but in the long run it's the climate that really counts.

All this motion can be so distracting that Warren Buffett has said, "I always like to look at investments without knowing the price—because if you see the price, it automatically has some influence on you." Conductors have likewise learned that they can evaluate a classical musician more objectively if the audition is played from behind a screen, where no preconceptions about how the musician looks can affect the perception of how he or she sounds.

Therefore, once you become interested in a company, it's a good idea to let two weeks go by without ever checking its share price. At the end of that period, now that you no longer know exactly where the shares are trading, do your own evaluation—ignoring stock price and focusing exclusively on business value. Start with questions like these: Do I understand this company's products or services? If the stock did not trade publicly, would I still want to own this kind of business? How have similar firms been valued in recent corporate acquisitions? What will make this enterprise more valuable in the future? Did I read the company's financial statements, including the "statement of risk factors" and the footnotes (where the weaknesses are often revealed)?

All this research, says Buffett, really points you back to one central issue: "My first question, and the last question, would be, 'Do I understand the business?' And by understand it, I mean have a reasonably good idea of what it will look like in five or ten years from an economic standpoint." If you aren't comfortable answering that basic question, you shouldn't buy the stock.

➢ **TAKE ACCOUNT OF WHAT MATTERS.** The account statements you receive from your brokerage or mutual fund company are often designed to set your reflexive system jangling: They emphasize the short-term price changes of each investment, not the level of your total wealth. That approach makes your account statements less boring to read, especially when the markets are racing up and down. But it makes you even more likely to go with your first gut instinct, which may often be to buy high and sell low. "The most natural way to think about a decision," warns Daniel Kahneman, "is not always the best way to make the decision."

If your financial adviser or fund company can't overhaul your financial statements, do it yourself: Set up a spreadsheet in Excel or similar software. At the end of each calendar quarter, enter the value of each of your investments; use small, plain type. Use Excel's "autosum" function to calculate the grand total of all your holdings. Highlight that total in big, boldface type. (If you can't use Excel, a lined piece of paper will do fine; be sure to record the sum total in larger, bolder numbers.) To see how your portfolio has done, don't read the entries row by row; instead, compare this quarter's sum total against the previous quarter and against one year, three years, and five years ago. You will now readily be able to see whether a sharp move by any of your holdings has a meaningful impact on your overall portfolio. This kind of deliberate comparison, which Kahneman calls "taking the global view," engages your reflective system—helping to keep losses or gains on single investments from goading you into a reflexive action you may regret later (like selling low or buying high).

Regardless of whether you invest in individual stocks or just have some money in a 401(k) or other retirement plan, taking the global view will help you focus on the long-term growth and stability of all your assets, rather than the short-term jitterbugging of a single asset. This way, you will end up both richer and calmer.

CHAPTER THREE

Greed

He that loveth silver shall not be satisfied with silver.

—*Ecclesiastes*

"I Know How Good It Would Feel"

Laurie Zink can't stop buying lottery tickets. She is well aware that the chances of winning California's SuperLotto Plus are 41,416,353 to one, but she doesn't care. "I know the odds are ridiculous," says Zink. "But there's a disconnect in my mind between the reality of the odds and the knowledge of how great it would feel to win." Zink, who majored in anthropology at Vanderbilt University, is a bright, hardworking TV producer. But she has been bitten by the long-shot bug, and she can't get its sweet toxin out of her system. In 2001, three months after her college graduation, Zink became a contestant on the NBC reality show *Lost.* She and a partner were blindfolded and dropped off on a remote mountaintop in Bolivia with $100, a few days of food and water, a first-aid kit, and a tent. Three weeks later, Zink and her teammate made it back to the Statue of Liberty in New York harbor—and were stunned to find out that they had won $100,000 each.

The thrill of winning that jackpot has been branded onto Laurie Zink's brain ever since. And now, whenever the California lottery jackpot gets "huge," she buys a ticket. Zink is hardly a lotto addict—she indulges the urge only a few times a year—but when the feeling hits her, it's like a compulsion. "I know that 'what if?' feeling is completely irrational," Zink says, "but it will drive me into 7-Eleven and make me get out my wallet and buy a lottery ticket. Rationally, I know I'm not going to win.

But, like they say, 'Hey, you never know.' And I do know how good it would feel if it happened." In coming back to the lottery like a moth to the flame, Laurie Zink is typical: A survey of people who had won at least $1 million in the Ohio state lottery found that 82% kept buying lottery tickets on a regular basis after their windfall.

Whether or not you've ever hit the jackpot like Laurie Zink, you already know it must feel good to make money. What you might not be aware of is that it can feel better to anticipate making money than it actually does to make it.

Of course, you don't have a money meter in your brain that is uniquely activated by a financial reward. Instead, your brain treats potential investing (or gambling) profits as part of a broad class of basic rewards like food, drink, shelter, safety, sex, drugs, music, pleasant aromas, beautiful faces, even social interactions like learning to trust someone or pleasing your mother. A financial gain is merely the most modern member of this ancient group of feel-good experiences. Because we find out early and often that money is essential to providing many other pleasures, the reflexive part of the human brain responds to potential financial gains at least as powerfully as it does to the prospects of capturing more fundamental rewards. Peter Kirsch, a neuroscientist at Justus-Liebig-University in Giessen, Germany, explains it this way: "Even though money cannot satisfy any primary needs—you can't eat it or mate with it—the association between money and reward is very strong."

Anticipating a financial gain puts the reflexive part of your investing brain on red alert, focusing your attention keenly on the task at hand. After you buy a stock, for example, you fixate on the possibility that it will keep going up. That thrill is limited only by your imagination. But the outcome itself—let's say the stock does go up after you buy it—is less exciting, especially when you expected it all along. By the time you pocket the money, the thrill of greed has faded into something that resembles a neurological yawn—even though you got the gains you wanted. Making money feels good, all right; it just doesn't feel as good as expecting to make money. In a cruel irony that has enormous implications for financial behavior, your investing brain comes equipped with a biological mechanism that is more aroused when you anticipate a profit than when you actually get one.

Twain's Gains

Long before neuroeconomists could track the interior activity of the human brain, Mark Twain knew that anticipating a jackpot feels even better than hitting it. In his early memoir, *Roughing It,* Twain recalls what happened when he and a partner struck a vein of silver ore in Nevada in 1862. Twain stayed up all night "just as if an electric battery had been applied to me"—fantasizing so vividly about the two-acre estate he would build in the heart of San Francisco, and the three-year tour of Europe he would take, that "my visions of the future whirled me bodily over in bed." For ten glorious days, Twain was a multimillionaire on paper. Then the partners' claim on the silver vein was suddenly invalidated by a legal technicality. "Sick, grieved, broken-hearted," Twain never forgot the elation of thinking that he had struck it rich.

Late in his life, in the brilliant parable "The $30,000 Bequest," Twain returned to the same theme, poking fun at the castles in the air that Saladin and Electra Foster build when they learn that they might inherit the grand sum of $30,000 (roughly $600,000 in today's money). Electra invests the inheritance they do not yet have, repeatedly "giving her imaginary broker an imaginary order" until the Fosters "roll in eventual wealth" and their fantasy portfolio climbs to $2.4 billion (roughly the same size as Bill Gates's fortune in today's dollars). In a "delirium of bliss," the humble storekeeper and his wife daydream about living "in a sumptuous vast palace" and "dawdling around in their private yacht." But it turns out that the inheritance was a cruel practical joke and no windfall will come their way. Stunned into silence, the Fosters die of grief.

The ultimate irony: Even though he could mock the extremes of greed in himself and others, Twain repeatedly sank his own money into risky ventures that offered the hope, but never the reality, of high returns. Over the years, Twain pumped cash and dreams into a dizzying variety of daffy speculations, including a chalk-based printing process, a machine that could reproduce photographs on silk, a powdered nutritional supplement, an intricate mechanical typesetter, spiral hatpins, and an improved design for grape scissors.

Why did this profoundly intelligent, skeptical, and already rich man so often succumb to ridiculous get-rich-quick schemes? Twain probably couldn't help it. Much the way Laurie Zink plays the lottery to recapture her thrill at winning the reality-show jackpot, Twain must have been driven to relive the visceral excitement he had felt in 1862 when he hit that giant vein of silver in Virginia City. That memory kept his anticipation circuitry in overdrive whenever he thought about money. The result was a lifelong, compulsive craving for the big score that sent Twain on wild swings from wealth to debt to bankruptcy and back again.

The Wi-Fi Network of the Brain

I lived through the rush of greed in an experiment at Brian Knutson's neuroscience lab at Stanford University. Knutson is a compact, bouncy man with a quick, infectious grin. A former student of comparative religion, he now researches how emotions are generated in the brain. Knutson put me into a functional magnetic resonance imaging (fMRI) scanner to trace my brain activity while I played an investing videogame that he has designed. By combining an enormous magnet and a radio signal, the fMRI scanner pinpoints momentary changes in the level of oxygen as blood ebbs and flows within the brain, enabling researchers to map the neural regions engaged by a particular task.

Being in an fMRI machine is incredibly loud: If you imagine lying inside an inverted cast-iron bathtub while poltergeists bang on it with iron rods, assault it with dental drills, and pour bucketsful of ball bearings onto it, you'll get the idea. But I've found that you become used to all the buzzing and clanging surprisingly quickly. After a few minutes, I was ready for Knutson's task, and shapes offering different levels of gain or loss began flashing on the display monitor.

In Knutson's experiment, a display inside the fMRI machine showed me a combination of symbols. A circle indicated that I could win money; a square meant that I could lose. Within each circle or square, the position of a vertical line showed how much money was at stake (left, $0; middle, $1; right, $5), and the location of a horizontal line (bottom, middle, or top) indicated how easy or hard it would be to capture the gain or avoid the loss. Thus, a circle containing a vertical line at the right and a horizontal line at the top meant that I might win $5, but my chances were slim. A square with a vertical line down the middle and a horizontal line near the bottom showed that I might lose $1, but it should be fairly easy for me to avoid the loss.

After each shape came up, between two and two-and-a-half seconds would pass—that's the anticipation phase, when I was on tenterhooks waiting for my chance to win or lose—and then a white square would appear for just a split second. To win the amount I had been shown—or to avoid it, if a loss was at stake—I needed to click a button with my finger at the precise moment when the cue appeared. If I clicked an instant too early or too late, I would miss out on the gain (or lock in the loss). At the highest of the three levels of difficulty, I had less than one-fifth of a second to hit the button, giving me only about a 20% chance of success. After each try, the screen indicated how much I had just won or lost and updated my cumulative score.

Not much seemed to happen when a shape signaling a small reward or penalty appeared; I clicked placidly and either won or lost. But if a circle marked with the symbols of a big easy payout came up, then, no matter how calm and deliberate I tried to be, I could feel a wave of expectation sweep through me. (The tingle of greed was even greater if, on my last few tries, I had failed to win anything or was on a losing streak.) Like sportscaster Marv Albert heralding a deep jump shot swishing through a basketball net, an intense voice of hope whispered "Yesss!!!" inside my head. *"Here it comes,"* I thought. *"Here's my chance."* At that moment, the fMRI scan showed, the neurons in a reflexive part of my brain called the nucleus accumbens fired like wild. When Knutson later measured the activity tracked by the fMRI scan, he found that the possibility of winning the $5.00 reward had set off roughly twice as strong a signal in my brain as the chance at gaining $1.00 did.

The nucleus accumbens lies deep behind your eyes, in back of the hindmost part of the frontal lobe of the brain (see Figure 3.1), where it curls back toward the center of your head. (Not surprising for a part of the brain that helps anticipate reward, the nucleus accumbens is also involved in experiencing sexual pleasure.) Many other areas are part of the anticipation circuitry, which is widely distributed throughout your reflexive brain much the way networks of Wi-Fi hotspots are sprinkled throughout the heart of major cities. (At least one region of the reflective brain, the orbitofrontal cortex, also appears to be linked to this anticipatory system.) But the nucleus accumbens is one of the central switches in the reward network.

Learning the outcome of my actions, by contrast, was no big deal.

Whenever it turned out that I had clicked at the right moment and captured the reward, I felt only a lukewarm wash of satisfaction that was much milder than the hot rush of anticipation I'd felt before I knew the outcome. In fact, Knutson's scanner found, the neurons in my nucleus accumbens fired much less intensely when I received a reward than they did when I was hoping to get it. (See Figures 3.2 a and b.) Based on the dozens of people Knutson has studied, it's highly unlikely that your brain would respond much differently.

"Reward is experienced in two fundamental ways," says neuroscientist Hans Breiter of Harvard Medical School. "A good example of this is sex: There's a long process of arousal, and an end point of satiation. Or when you're really hungry and you're preparing a meal, the preparation builds up into arousal—and then when you finally eat it, the satiation may not be followed by much euphoria at all. The arousal piece is actually the main component of euphoria, and it's expectation—not satiation—that causes most of that arousal."

Testing the difference between sexual anticipation and satiation inside an fMRI tube would be difficult, not to mention uncomfortable. But the difference between expecting and actually tasting a food flavor has been tested. Once people learned that they would receive a sip of sugar water after seeing a certain shape, their nucleus accumbens fired much more powerfully when they saw that shape than when they got the glucose—confirming Breiter's observation that imagining a good meal can be at least as exciting as eating it. Money works the same way. To paraphrase the old saying, 'tis better to hope than to receive.

The Rats That Can't Wait

Why does the reflexive part of the investing brain make a bigger deal out of what we might get than what we do get? That function is part of what Knutson's mentor, Jaak Panksepp of Bowling Green State University in Ohio, calls "the seeking system." Over millions of years of evolution, it was the thrill of anticipation that put our senses in a state of high awareness, bracing us to capture uncertain rewards. The anticipation circuitry in our brains, says Paul Slovic of the University of Oregon, acts as "a beacon of incentive" that enables us to pursue longer-term rewards that can be earned only with patience and commitment. If we derived no pleasure from imagining riches down the road, we could never motivate ourselves to hold out long enough to earn them. Instead, we would grab only at those gains that loom immediately in front of us.

As the French essayist Michel de Montaigne wrote, "If we were placed between the bottle and the ham with an equal appetite for drinking and for eating, there would doubtless be no solution but to die of thirst and of hunger." In John Barth's 1958 novel *The End of the Road,* Jacob Horner is incapable of imagining future pleasure and thus is paralyzed whenever he confronts a choice. Horner's therapist, Dr. Dockey, gives him the simple rules of "Sinistrality, Antecedence, and Alphabetical Priority"—choose whichever alternative is on the left, presents itself first, or whose first letter is closest to the beginning of the alphabet. Absurd as they are, those rules at least enable Horner to begin functioning. If our own seeking systems didn't work, we would all be like Jacob Horner, frozen with indecision whenever we face more than one alternative. As Dr. Dockey tells Horner, "Choosing is existence: to the extent that you don't choose, you don't exist."

We can learn more about anticipation by studying how it works in other animals. A research team led by Taketoshi Ono of the Toyama Medical and Pharmaceutical University in Japan studies how rats anticipate rewards. Ono and his colleagues have shown that the prospect of a reward like water, sucrose, or electrical stimulation activates a part of the rat's brain called the sensory thalamus. Like a double-toggling light switch, this circuitry snaps on in two distinct stages: First comes a bolt of neural lightning that flashes to life in as little as a hundredth of a second, signaling to the rest of the brain that a reward may be on the way. (In effect, when presented with a cue that predicts a gain, the rat's brain screams, "Incoming!") Then comes a sustained, escalating response as neurons keep firing until the reward is delivered; the more desirable the potential prize, the more fiercely the neurons fire and the faster the rat pounces on the reward when it ultimately appears. During this phase, the rats appear to be figuring out just which form the reward is most likely to take.

The first stage of anticipation, then, seems to be a form of looking backward: The rats know, from previous trial runs, that a given sound or light is associated with the delivery of rewards, so the cue sets off an almost instantaneous alert. The second stage is a form of looking forward: Between the time of the cue and the delivery of the reward, the rats work on recognizing which particular kind of reward is coming. The better they expect it to be, the more their brains put the rats in a state of high readiness. (Ono's research team found that rats would lick up a sip of plain water in about a quarter of a second—but could slurp up sugar water in less than one-twentieth of a second.) As Ono puts it, the two stages of anticipation seem to be a way of comparing "learned experience" against "future outcome." Emily Dickinson expressed this idea perfectly when she wrote: "Retrospection is prospect's half, sometimes almost more."

Lately, psychologists have been exploring what happens when the anticipation circuitry is damaged. In a laboratory at the University of Cambridge in England, rats are placed into chambers where they can press either of two levers with their paws. With a push on one lever, the rats receive a reward of a single sugar pellet immediately; a nudge on the other lever yields a yummier prize of four sugar pellets, but only after a delay of between 10 and 60 seconds. Rats with fully intact brains are highly impatient, but they will still choose the larger, later rewards up to half the time.

However, rats with damage to the nucleus accumbens in their brain suffer from a novel form of ADD—not attention deficit disorder, but anticipation deficit disorder. Without a properly functioning nucleus accumbens, these rats become almost completely incapable of delaying gratification, and they choose the earlier but smaller prize more than 80% of the time. The lost ability to anticipate future reward forces these rats into what psychologist Rudolf Cardinal of Cambridge calls "impulsive choice." As he puts it, the anticipation circuits may enable normal rats to "focus their cognitive resources" on what lies just over the horizon. But for rats without a functioning nucleus accumbens, the here and now is all that matters; the future is never.

So the seeking system in our own brains functions partly as a blessing and partly as a curse. Our anticipation circuitry forces us to pay close attention to the possibility of coming rewards, but it also leads us to expect that the future will feel better than it actually does once it arrives. That's why it's so hard for most of us to learn that the old saying is true: Money doesn't buy happiness. After all, it forever feels as if it *should.*

Why Good News Can Be So Bad

One of Wall Street's oldest proverbs is "Buy on the rumor, sell on the news." The theory behind this cliché is that stocks go up as whispers spread among the "smart money" that something big is about to happen. Then, as soon as the general public learns the good news, the sophisticates sell out at the top and the stock goes down the toilet.

There's something to this, but it probably has more to do with the anticipation circuits in everyone's brain than it does with the supposedly superior brainpower of a few big investors. A vivid example of a stock that soared on hope and foundered on reality is Celera Genomics Group. On September 8, 1999, Celera began sequencing the human genome. By identifying each of the 3 billion molecular pairings that make up human DNA, the company could make one of the biggest leaps in the history of biotechnology. As Celera's dazzling enterprise began to attract attention, investors went wild with anticipation. In December 1999, Eric Schmidt, a biotechnology stock analyst at SG Cowen Securities, summarized the market's state of mind: "There is tons of investor excitement now for this sector. They want to own stories today that will drive the economy tomorrow." Celera's stock went flying up from $17.41 at the start of the sequencing project to a peak of $244 in early 2000.

And then, on June 26, 2000, in a grandiose press conference held at the White House with President George W. Bush and British Prime Minister Tony Blair in attendance, Celera's chief scientist J. Craig Venter announced what he called "an historic point in the 100,000-year record of humanity." And how did Celera's stock react to the official word that the company had completed cracking the human genetic code? It tanked, dropping 10.2% that day and another 12.7% the next day.

Nothing had happened to change the company's fortunes for the worse. Quite the contrary: Celera had achieved nothing short of a scientific miracle. So why did the stock crash? The likeliest explanation is simply that the fires of anticipation are so easily quenched by the cold water of reality. Once the good news that investors have awaited for so long is out, the thrill is gone. The resulting emotional vacuum almost instantly fills up with a painful awareness that the future will not be nearly as exciting as the past. (As Yogi Berra famously said, "The future ain't what it used to be.") Getting exactly what they wished for leaves investors with nothing to look forward to, so they get out and the stock crashes.

By late 2006, shares in Celera Genomics—one of two classes of stock under the Applera Corp. name—traded around $14, more than 90% below their all-time high. That shows the dangers of buying into a company whose single greatest asset is the greed of the people who trade its stock.

Memories Are Made of Money

In a remarkable recent experiment, researchers in Germany tested whether anticipating a financial gain can improve memory. A team of neurologists scanned people's brains with an MRI machine while showing them pictures of objects like a hammer, a car, or a cluster of grapes. Some images were paired with the chance to win half a euro (about 65 cents in U.S. money), while others led to no reward at all. The participants soon learned which pictures were reliably associated with the prospect of making money, and the MRI scan showed that their anticipation circuits fired furiously when those images appeared.

Immediately afterward, the researchers showed the participants a larger set of pictures, including some that had not been displayed inside the MRI scanner. People were highly accurate at distinguishing the pictures they had seen during the experiment. And they were equally adept at recognizing which of those pictures had predicted financial gain and which had not been associated with any payoff.

Three weeks later, the participants came back into the lab, where they were shown the pictures again. This time, however, something amazing happened: People now could even more readily distinguish the pictures that had signaled a financial gain from those that had not—although they hadn't laid eyes on them in twenty-one days! Astounded by this discovery, the researchers went back and reexamined the fMRI scans from three weeks earlier. It turned out that the potentially rewarding pictures had set off more intense activation not only in the anticipation circuits but also in the hippocampus, a part of the brain where long-term memories live.

The original fire of expectation, it seems, somehow sears the memory of potential rewards more deeply into the brain. "The anticipation of reward," says neurologist Emrah Düzel, "is more important for memory formation than is the receipt of reward." Once you learn that a given gamble could make money, you will remember those circumstances—and the thrill of expecting the jackpot—far better and much longer than you will recall the bets that never paid off. Like the spray that artists apply onto pastels to make the colors stick, anticipation acts as a fixative to keep your memory of how to earn rewards from fading.

"For some people," says Peter Shizgal, a neuroscientist at Concordia University in Montreal, "those memories of that good feeling can crowd out all sorts of more financially significant information." Shizgal relates this story: "A psychologist I know had a patient with a compulsive gambling problem, and one weekend the patient won something like $100,000. So he asks the patient, 'Well, what's your net gain or loss overall?' And the patient says, 'Oh, minus $1.9 million. I was down $2 million, but now I won $100,000!'" Explains Shizgal: "The first part of his answer had no emotion associated with it at all. It was as if the information was in there but had no impact. It was only the big killing that was really memorable and that was going to continue to control his behavior."

No wonder so many of us look back on our past investments and see a performance that could rival Warren Buffett's—when, in fact, our real track record is strewn with error and loss. Because expecting a gain helps us remember our gains, hindsight can often turn our blurry 20/100 vision into what feels like perfect 20/20 acuity.

Thus, as Taketoshi Ono's experiments with rats suggested, anticipation appears to be a two-stage process. The first stage looks backward with memory, the second looks forward with hope. That would explain why Laurie Zink, who never bought a lottery ticket before she won on a reality show, now loves playing lotto—and why Mark Twain, despite all his wealth, kept trying to strike it rich anyway.

The Anticipation of Anticipation

Experiments in rats have shown that neurons in the nucleus accumbens shoot out their signals in as little as one-tenth of a second after the animals spot a symbol that predicts reward. Over the next five to fifteen seconds, those signals goad the rats into pursuing whatever reward was predicted by the cue. This mental leap between a reward and the cue that predicts it helps explain why the sight of a syringe makes a heroin addict feel an irresistible craving for a hit. It's also why the clanging glitter of a casino floor makes a compulsive gambler whip out his wallet. As Dostoyevsky wrote in his novella *The Gambler,* "Even on my way to the gambling hall, as soon as I hear, two rooms away, the clink of the scattered money I almost go into convulsions." Because the mere sight of a predictive cue can trigger an impulsive rush, convenience stores put their lotto machines right next to the cash register—and brokerage firms put a whizzing electronic stock ticker just inside the front door, or place a television in the waiting room tuned permanently to CNBC.

When rewards are near, the brain hates to wait. Neurons in the caudate nucleus, a region in the center of the primate brain, become active even before the predictive cue is presented. Monkeys learned that they would earn a drink of water when they moved their eyes toward a particular shape. They also learned roughly when the next cue was likely to come along. Amazingly, the monkeys' caudate neurons began firing up to 1.5 seconds before the cue appeared. In other words, once we become aware that rewards may be in the offing, our attention is riveted not only by a gain or the signal that a gain may be coming—but even by a hint that the *signal* may be coming! Hiroyuki Nakahara of the RIKEN Brain Science Institute in Wako, Japan, calls this early-warning flare "the anticipation of anticipation of reward." It's as if Pavlov's dogs salivated not when the bell rang, but as soon as they saw Pavlov start walking toward the bell.

That helps explain why, in the late 1990s, day traders got a buzz simply from sitting down in front of their computers if the previous day's trades had made money. And when Cisco Systems had beaten Wall Street's earnings forecasts by exactly one penny per share for ten calendar quarters in a row, just the approach of its next earnings announcement made investors feel euphoric. In the five days leading up to its earnings release in February 2000, Cisco's stock jumped 10.5% on trading volume more than one-third heavier than normal. This Pavlovian price rise was driven by investors salivating over the earnings announcement they knew was coming. (In the end, Cisco did beat the forecast by one penny, just as expected—and managed to do so for three more quarters. Then it imploded, losing roughly $400 billion in stock market value by 2002, the biggest decline by a single stock in financial history.)

Twilight of the Odds

Anticipation has another unusual neural wrinkle. Brian Knutson has found that while your reflexive brain is highly responsive to variations in the amount of reward at stake, it is much less sensitive to changes in the probability of receiving a reward. In effect, your investing brain is better at asking "How big is it?" than "How likely is it?" Thus, the bigger the potential gain, the greedier you will feel—regardless of how poor the odds of earning that gain might be.

If the jackpot in a lottery were $100 million and the posted odds of winning fell from 1-in-10-million to 1-in-100-million, would you be ten times less likely to buy a ticket? If you're like most people, you probably would shrug, "A long shot's a long shot," and be just as happy buying a ticket as before. That's because, as economist George Loewenstein of Carnegie Mellon University explains, the "mental image" of $10 million sets off a burst of anticipation in the reflexive regions of your brain. Only later will the reflective areas calculate that you're about as likely to win as Ozzy Osbourne is to be elected the next pope.

"Money is a basic form of reward that is rapidly processed in the reflexive system," explains Loewenstein. "You're very likely to have some vivid imagery of piles of money and fantasies about how you'll spend it. But your brain isn't designed to form mental imagery of a probability. The pleasure you anticipate feeling if you win the money would change enormously if we multiplied or divided the amount of the jackpot by 10 or 100 or 1000, but your reaction to similar changes in probability would trigger very little emotion." Because anticipation is processed *reflexively* while probability is processed *reflectively,* the mental image of winning $100 million crowds out the calculation of just how unlikely that jackpot really is. In short, when possibility is in the room, probability goes out the window.

In the movie *Dumb and Dumber,* Jim Carrey's character, Lloyd, asks the love of his life what his chances are of making her love him, too. "Not good," replies Mary Swanson. "Not good like one out of a hundred?" asks Lloyd haltingly. Mary answers, "I'd say more like one out of a million." Exclaims Lloyd: "So you're telling me there's a chance? *Yeah!*"

It's no different when you buy a stock or a mutual fund: Your expectation of scoring a big gain will typically elbow aside your ability to evaluate how likely you are to earn it. That means your brain will tend to get you into trouble whenever you're confronted with an opportunity to buy an investment with a hot—but probably unsustainable—return.

The Risks Not Taken

There's another thing you should realize about your anticipation circuitry: It doesn't evaluate potential gains in isolation. In theory, we should all prefer winning more money to winning less. But in practice, it often doesn't work out that way. Psychologist Barbara Mellers of the University of California at Berkeley has shown that people can derive more "relative pleasure" from a gamble that offers the chance either to win or lose money than they do from a gamble that offers only upside. "We're so attuned to change," says Mellers, "that we assess our potential outcomes not just against what did happen but also against what might have happened." The chance that we could have lost money makes earning money even sweeter.

Working with psychologist Daniel Kahneman, a team of neuroscientists led by Hans Breiter of Harvard Medical School tested how the chance of losses can affect the intensity with which we anticipate gains. The researchers created wheels of fortune that were each segmented into three possible outcomes with equal odds: A spin of the first wheel offered the chance to win $10, $2.50, or nothing; the second offered a $2.50 gain, nothing, or a $1.50 loss; the third offered the chance to win nothing, lose $1.50, or lose $6.00. Sometimes the "good" wheel would come up, sometimes the wheel with the middling payouts, and sometimes the "bad" one, all in random order.

In this experiment, your brain doesn't look at the potential gains in isolation, but compares them to the other possible outcomes. The good wheel could get you $10—but with no risk of loss. Meanwhile, the middling wheel could earn you just $2.50 at most—but that potential gain carries the risk of losing $1.50. Thus, even though the middling wheel offers a smaller upside, it engages your investing brain with equal intensity, since it's coupled with a risk of loss. Breiter and his colleagues showed that in several brain regions, including the nucleus accumbens, neurons are just as active anticipating the outcome of a spin on the middling wheel as they are while waiting for the good wheel to pay off. (The bad wheel, with no chance to make money at all, activates one of the brain's fear centers, the amygdala.)

So the possibility of loss makes the hope of gain even more tantalizing. If you think about it, this makes perfect sense. Evolution has naturally designed us to pay closer attention to rewards when they come surrounded by risks—just as we all know we need to be more careful when picking a rose than when picking a daisy.

The Expectation Game

Marketers and the hordes of people who want to separate you from your money understand full well how the anticipation circuits work in your brain. In almost every casino, the slot machines are located just inside the main entrance so that the first sound you hear when you walk in will be those bells ringing or that brassy clatter of coins pouring out—filling you with the thrill of possible gains. Meanwhile, get-rich-quick scam artists have long played on the emotional power of what they call "flash the cash and dash"—hyping a huge potential profit, usually by waving actual money around, then skipping town before the bilked victims catch on.

Among stock promoters, promising a big score down the road is one of the oldest tricks in the book. In 1720, the South Sea Co. pumped up its stock price by announcing a future dividend increase—which, in the words of one contemporary observer, "perhaps contributed more to intoxicate the minds of the people than anything done besides." In the late 1990s, Wall Street's cynical investment bankers systematically underpriced initial public offerings of stock, or IPOs, enabling the shares to soar by as much as 697% on their first day of trading. That, in turn, made investors desperate to get in on the ground floor of the next IPO. It's no coincidence that the official disclosure document of an IPO is called a "prospectus," from the Latin term for "looking forward."

Furthermore, even when the market as a whole is going down, plenty of stocks are still going up. In 2000, as the Wilshire 5000 index (the broadest measure of U.S. stock returns) lost 10.7%, 185 stocks at least tripled and 23 rose at least tenfold. The next year, as the average fell another 11%, 231 stocks more than tripled and 16 rose more than tenfold. In 2002, even as the market sank a miserable 20.8%, 58 stocks at least tripled and 3 rose at least tenfold. Since there is always some stock making somebody rich, it seems like a cinch to be able to find one of them yourself. But your seeking system cares more about what might happen than about what does happen, so it's hard to remember that many of the stocks that go way up one year go down disastrously the next year. That's why, for all too many people, investing yields heartburn and heartbreak—years spent chasing one "hot" stock or fund after another, only to have them go cold just as riches seem within reach.

It's not just individual investors who get carried away by greed. Hedge funds, those giant pools of money available only to millionaires and institutions like pension plans and endowments, offer the exclusive privilege of investing in "proprietary" or confidential strategies. Big investors, who would indignantly refuse to pay more than 1% in fees to a conventional mutual fund, are happy to pay at least 2% of assets and 20% of profits to a hedge fund that provides virtually no information about what it does with their money. (At many hedge funds, the investing approach is so obscure that it's called a "black box.") In fact, it's precisely by keeping their strategies secret that many hedge funds can get away with charging such high fees. For the clients, having no idea what's inside makes the prospect of hitting the jackpot more irresistible. Your birthday wouldn't be much fun if all your presents came wrapped in clear plastic.

Getting a Grip on Your Greed

So how can you keep your seeking system from landing you in financial trouble? The first thing to realize is that your anticipation circuitry *will* get carried away. That's its job. So, if the rest of your brain doesn't impose checks and balances, you'll end up chasing every hot return that flares up in front of you—and capturing nothing, in the long run, but risk and loss. Here's how to do better.

➢ **THERE'S ONLY ONE SURE THING ON WALL STREET.** And that's that there are no sure things. Remember that your seeking system is especially turned on by the prospect of a big score—and that this sense of arousal will hinder your ability to calculate realistic odds. Be on your guard against anyone who tries to lure you with jackpot jargon like "double your money," "the sky's the limit," "this baby's really gonna take off." The higher the return that an investment supposedly offers, the more questions you should ask. Start with this one: "Why is the person who knows about this great investment willing to let anyone else in on the secret?" Then try this one: "How come I'm being offered the rare privilege to share in this great opportunity?" Furthermore, never—repeat, *never*—make an investment based on an unsolicited phone call from a broker you haven't met. Just say no, and hang up. Never—repeat, *never*—respond to an unsolicited e-mail encouraging you to invest in anything. Just delete, without opening.

➢ **LIGHTNING SELDOM STRIKES TWICE.** If you've ever had the taste of a big gain, you're likely to be tempted to spend the rest of your life trying to get that feeling back. Although it's easy to spot the stocks that have been going up, it's a lot harder to spot the ones that will keep going up. Be especially wary of investing in stocks that remind you of that one you made a killing on long ago; chances are, any similarities to another stock, living or dead, are purely coincidental. You should sink a lot of money into a single stock only if you have studied the underlying business carefully and would be happy to own it if the stock market shut down for a five-year vacation.

➢ **LOCK UP YOUR "MAD MONEY" AND THROW AWAY THE KEY.** If you can't stop yourself from taking a gamble in the market, then at least limit the amount you put at risk. Like a casino gambler who caps his potential losses by locking his wallet in the hotel safe and bringing only $200 down to the gaming floor, you should put a cap on how much you will risk on speculative trading. Put at least 90% of your stock money into a low-cost, diversified index fund that owns everything in the market. Put 10%, tops, at risk on speculative trades. Be sure this "mad money" resides in an entirely separate account from your long-term investments; never mingle them. No matter how much it goes up or down, never add more money to the speculative account. (It's especially important to resist the temptation to put more money in when your trades have been doing well.) If you get wiped out, close it out.

➢ **CONTROL YOUR CUES.** Just as the dogs in Pavlov's laboratory salivated whenever the bell signaled that food was on its way, just as the sound of beer gushing into a glass across the room can make an alcoholic crave a drink, the stock market constantly generates signals that can goad you into trading. Psychologist Howard Rachlin of the State University of New York at Stony Brook has shown that one of the best first steps toward quitting tobacco is to try smoking the same number of cigarettes each day—and that offers a hint for us. Fewer opportunities for greed, plus less variability in the amount of satisfaction you can look forward to, will equal more self-control. Suggests Brian Knutson: "Ask yourself, 'How can I clean up my environment?' (Think of a smoker who's trying to quit and hides all the ashtrays.) 'How can I expose myself to fewer cues and less variability in the cues?'" Try watching CNBC with the sound turned off, so none of the hullabaloo about what the market is doing this second can distract you from your long-term financial goals. Or, if you find yourself walking past the local brokerage firm every day so you can peek through the window at the electronic ticker, then start walking a different route. If you catch yourself obsessively checking a stock price online, use the "history" window on your Web browser to count how many times you've updated the price each day. The number may shock you, and knowing how often you do it is the first step toward doing it less often.

Another simple and powerful way to control your cues is to write a checklist of standards that every investment must meet before you buy or sell it. Each year's annual report for Berkshire Hathaway Inc. (available at www.berkshirehathaway.com) includes a list of six "acquisition criteria" showing the standards that chairman Warren Buffett and vice chairman Charles Munger apply to any business they consider buying. Make sure your own checklist includes some things you *don't* want to take into consideration, so that you can quickly rule out many of the bad ideas that might otherwise tempt you. For a checklist of investing dos and don'ts, see Appendix 2.

➢ **THINK TWICE.** At least when it comes to investing, Malcolm Gladwell's advocacy of "thinking without thinking" is a recipe for disaster. Instead, you need to think twice. "It's important to realize," says Stanford's Knutson, "that the magnitude of a long-shot reward is going to drive your behavior far more than the probabilities, which are minuscule. If you can recognize that, then you should be able to say to yourself, 'I should walk away and play with my kids for an hour and then think about it.'" Making a financial decision while you're inflamed by the prospects of a big gain is a terrible idea. Calm yourself down—if you don't have kids to distract you, take a walk around the block or go to the gym—and reconsider when the heat of the moment has passed and your anticipation circuits have cooled off. So don't just blink; *think*.

CHAPTER FOUR

Prediction

> Pecuniary motives either do not act at all—or are of that class of stimulants which act only as Narcotics.
>
> —*Samuel Taylor Coleridge*

From Babel to Bubble

In the Mesopotamian galleries of the British Museum in London sits one of the most startling relics of the ancient world. A life-size clay model of a sheep's liver, it served as a training tool for a specialized Babylonian priest known as a *baru,* who made predictions about the future by studying the guts of a freshly slaughtered sheep. The model is a catalog of the blemishes, colors, and differences in size or shape that a real sheep's liver might display. The *baru* and his followers believed that each of these variables could help foretell what was about to happen, so the clay model is painstakingly subdivided into sixty-three areas, each marked with cuneiform writing and other symbols describing its predictive powers.

What makes this artifact so astounding is that it is as contemporary as today's coverage of the financial news. More than 3,700 years after this clay model was first baked in Mesopotamia, the liver-reading Babylonian *barus* are still with us—except now they are called market strategists, financial analysts, and investment experts. The latest unemployment report is "a clear sign" that interest rates will rise. This month's news about inflation means it's "a sure thing" that the stock market will go down. This new product or that new boss is "a good omen" for a company's stock.

Just like an ancient *baru* massaging the meanings out of a bloody liver, today's market forecasters sometimes get the future right—if only by luck alone. But when the "experts" are wrong, as they are about as often as a flipped coin comes up tails, their forecasts read like a roster of folly:

- Every December, *BusinessWeek* surveys Wall Street's leading strategies, asking where stocks are headed in the year to come. Over the past decade, the consensus of these "expert" forecasts has been off by an average of 16%.

- On Friday the 13th in August 1982, the *Wall Street Journal* and the *New York Times* quoted one analyst and trader after another, all spewing gloom and doom: "A selling climax will be required to end the bear market," "investors are on the horns of a dilemma," the market is gripped by "outright capitulation and panic selling." That very day, the greatest bull market in a generation began—and most "experts" remained stubbornly bearish until the rebound was long under way.

- On April 14, 2000, the NASDAQ stock market fell 9.7% to close at 3321.29. "This is the greatest opportunity for individual investors in a long time," declared Robert Froelich of Kemper Funds, while Thomas Galvin of Donaldson, Lufkin & Jenrette insisted "there's only 200 or 300 points of downside for the NASDAQ and 2000 on the upside." It turned out there were no points on the upside and more than 2,200 on the downside, as NASDAQ shriveled all the way to 1114.11 in October 2002.

- In January 1980, with gold at a record $850 per ounce, U.S. Treasury Secretary G. William Miller declared: "At the moment, it doesn't seem an appropriate time to sell our gold." The next day, the price of gold fell 17%. Over the coming five years gold lost two-thirds of its value.

- Even the Wall Street analysts who carefully study a handful of stocks might as well be playing "eeny meeny miny moe." According to money manager David Dreman, over the past thirty years the analysts' estimate of what companies would earn in the next quarter has been wrong by an average of 41%. Imagine that the TV weatherman said it would be 60 degrees yesterday, and it turned out to be 35 degrees instead—also a 41% error (on the Fahrenheit scale). Now imagine that's about as accurate as he ever gets. Would you keep listening to his forecasts?

All these predictions fall prey to the same two problems: First, they assume that whatever has been happening is the only thing that could have happened. Second, they rely too heavily on the short-term past to forecast the long-term future, a mistake that the investment sage Peter Bernstein calls "postcasting." In short, the "experts" couldn't hit the broad side of a barn with a shotgun—even if they stood inside the barn.

As a matter of fact, whichever economic variable you look at—interest rates, inflation, economic growth, oil prices, unemployment, the Federal budget deficit, the value of the U.S. dollar or other currencies—you can be sure of three things. First, someone gets paid lots of money to make predictions about it. Second, he will not tell you, and may not even know, how accurate his forecasts have been over time. Third, if you invest on the basis of those forecasts you are likely to be sorry, since they are no better a guide to the future than the mutterings of a Babylonian *baru.*

The futility of financial prediction is especially frustrating because it seems so clear that analysis *should* work. After all, we all know that studying beforehand is a good way to improve our (or our children's) test scores. And the more you practice your golf or basketball or tennis shot, the better player you will become. Why should investing be any different? There are three main reasons why investors who do the most homework do not necessarily earn the highest grades:

1. **The market is usually right.** The collective intelligence of tens of millions of investors has already set a price for whatever you're trading. That doesn't mean that the market price is always right, but it's right more often than it's wrong. And when the market is massively wrong—as it was about Internet stocks in the late 1990s—then betting against it can be like trying to swim into a tidal wave.
2. **It takes money to move money.** The brokerage costs of buying and selling a stock can easily exceed 2% of the amount you stake. And the tax man can take up to 35% of your gains if you trade too frequently. Together, those expenses wear away profitable ideas like sandpaper.
3. **Randomness rules.** No matter how carefully you research an investment, it can go down for reasons you never anticipated: a new product fails, the CEO departs, interest rates rise, government regulations

change, war or terrorism bursts out of the blue. No one can predict the unpredictable.

So why, despite all the evidence that their efforts are futile, do today's financial *barus* keep on predicting? Why do investors keep listening to them? Most important of all, if no one can accurately foresee the financial future, then what practical rules can you use to make better investing decisions? That's what Chapter Four is all about.

What Are the Odds?

It took two psychologists, Daniel Kahneman and Amos Tversky, to deal a death blow to the traditional view that people are always "rational." In economic theory, we process all the relevant information in a logical way to figure out which choice offers the best tradeoff between risk and return. In reality, Kahneman and Tversky showed, people tend to base their predictions of long-term trends on surprisingly short-term samples of data—or on factors that are not even relevant. Consider these examples:

1. Two bowls, hidden from view, each contain a mix of balls, of which two-thirds must be one color and one-third must be another. One person has taken 5 balls out of Bowl A; 4 were white, 1 was red. A second person drew twenty balls out of Bowl B; twelve were red, 8 were white. Now it's your turn to be blindfolded, but you can take out only one ball. If you guess the right color in advance, you will win $5. Should you bet that you will draw a white ball from Bowl A, or a red ball from Bowl B?

   Many people bet on getting a white ball, since the first person's draw from Bowl A was 80% white, while the second person drew only 60% red from Bowl B. But the sample from Bowl B was four times larger. That bigger drawing means that Bowl B is more likely to be mostly red than Bowl A is to be mostly white. Most of us know that large samples of data are more reliable, but we get distracted by small samples nevertheless. Why?

2. A nationwide survey obtains brief personality descriptions of 100 young women, of whom 90 are professional athletes and 10 are librarians. Here are two personality profiles drawn from this group of 100:

Lisa is outgoing and lively, with long hair and a tan. She is sometimes undisciplined and messy, but she has an active social life. She is married but has no children.

Mildred is quiet, with eyeglasses and short hair. She smiles often but seldom laughs. She is a hard worker, extremely orderly, and has only a few close friends. She is single.

- What are the odds that Lisa is a librarian?
- What are the odds that Mildred is a professional athlete?

Most people think Lisa must be an athlete, and Mildred must be a librarian. While it seems obvious from the descriptions that Lisa is more likely than Mildred to be a jock, Mildred is probably a professional athlete, too. After all, we've already been told that 90% of these women are. Often, when we are asked to judge how likely things are, we instead judge how alike they are. Why?

3. Imagine that you and I are flipping a coin. (Let's flip six times and track the outcomes by recording heads as an H and tails as a T.) You go first and flip H T T H T H: a 50/50 result that looks exactly like what you should get by random chance. Then I toss and get H H H H H H: a perfect streak of heads that makes us both gasp and makes me feel like a coin-flipping genius.

   But the truth is more mundane: In six coin flips, the odds of getting H H H H H H are identical to the odds of getting H T T H T H. Both sequences have a one-in-64, or 1.6%, chance of occurring. Yet we think nothing of it if one of us flips H T T H T H, while we both are astounded when H H H H H H comes up. Why?

Pigeons, Rats, and Randomness

The answers to these riddles about randomness lie deep in our brains and far back in the history of our species. Humans have a phenomenal ability to detect and interpret simple patterns. That's what helped our ancestors survive the hazardous primeval world, enabling them to evade predators, find food and shelter, and eventually to plant crops in the right place at the right time of year. Today, our skill at seeking and completing patterns helps us navigate many of the basic challenges of daily life.

("Here comes the train I have to catch." "The baby's hungry." "My boss is always a butthead on Mondays.")

But when it comes to investing, our incorrigible search for patterns leads us to assume that order exists where it often doesn't. It's not just the *barus* of Wall Street who think they know where the stock market is going. Almost everyone has an opinion about whether the Dow will go up or down from here, or whether a particular stock will continue to rise. And everyone *wants* to believe that the financial future can be foretold.

The pursuit of patterns in random data is a fundamental function in our brains—so basic to human nature that our species should not be known only as *Homo sapiens,* or "man the wise." We might better be named *Homo formapetens,* or "man the pattern-seeker." Although most animals have the ability to identify patterns, humans are uniquely obsessive about it. Our knack for perceiving order even where there isn't any is what the astronomer Carl Sagan called the "characteristic conceit of our species," and what others have called *pareidolia,* from the Greek for incorrect or distorted imagery. Some people see an image of the Virgin Mary in the scorch marks on a ten-year-old grilled-cheese sandwich—and one was even willing to pay $28,000 for it on eBay. Others sift through mountains of stock market data to find "predictable patterns" that might enable them to beat the market:

- It became a common belief, based on historical numbers, that U.S. stocks tend to go up on Fridays and down on Mondays—but, in the 1990s, they did the exact opposite.

- October (the month of the 1987 market crash) is widely supposed to be the worst month to own stocks—but, over the long sweep of history, it has actually averaged the fifth-best returns of any month.

- Millions of investors believe in technical analysis, which supposedly predicts future prices on the basis of past prices, and in market timing, which purports to enable you to get out of stocks before they go down and back in before they go up. There is little, if any, objective evidence that either tactic works in the long run.

- Every year, many Wall Streeters root for National Football Conference teams to win the Super Bowl, based on the widely held—and wildly inaccurate—belief that when teams originating in the old NFL take the championship, the stock market goes up the next year.

What drives this behavior? For decades, psychologists have demonstrated that if rats or pigeons knew what a stock market is, they might be better investors than most humans are. That's because rodents and birds seem to stick within the limits of their abilities to identify patterns, giving them what amounts to a kind of natural humility in the face of random events. People, however, are a different story.

In a typical experiment of this kind, researchers flash two lights, one green and one red, onto a screen. Four out of five times, it's the green light that flashes; the other 20% of the time, the red light comes on. But the exact sequence is kept random. (One run of 20 flashes might look like this: RGRGGGGGRGGGGRGGGGGG. Another might be: GGGGRGGGGGGGRRGGGGGR. You can view a simplified version of this task at www.jasonzweig.com/uploads/matchvmax.ppt.) In guessing which light will flash next, the best strategy is simply to predict green every time, since you stand an 80% chance of being right. And that's what rats or pigeons generally do when the experiments reward them with a crumb of food for correctly guessing what color the next flash of light will be.

Humans, however, tend to flunk this kind of experiment. Instead of just picking green all the time and locking in an 80% chance of being right, people will typically pick green four out of five times, quickly getting caught up in the game of trying to call when the next red flash will come up. On average, this misguided confidence leads people to pick the next flash accurately on only 68% of their tries. Stranger still, humans will persist in this behavior even when the researchers tell them explicitly—as you cannot do with a rat or pigeon—that the flashing of the lights is random. And, while rodents and birds usually learn quite quickly how to maximize their score, people often perform worse the longer they try to figure it out. The more time they spend working at it, the more convinced many people become that they have finally discovered the trick to predicting the "pattern" of these purely random flashes.

Unlike other animals, humans believe we're smart enough to forecast the future even when we have been explicitly told that it is unpredictable. In a profound evolutionary paradox, it's precisely our higher intelligence that leads us to score lower on this kind of task than rats and pigeons do. (Remember *that* the next time you're tempted to call somebody a "birdbrain.")

A team of researchers at Dartmouth College, led by psychology professor George Wolford, has studied why we think we can spot patterns where there are none. Wolford's group ran light-flashing experiments on "split-brain patients," people in whom the nerve connections between the hemispheres of the brain have been surgically severed as a treatment for severe epilepsy. When the epileptics viewed a series of flashes that they could process only with the right side of their brains, they gradually learned to guess the most frequent option all the time, just as rats and pigeons do. But when the signals were flashed to the left side of their brains, the epileptics kept trying to forecast the exact sequence of flashes—sharply lowering the overall accuracy of their predictions.

"There appears to be a module in the left hemisphere of the brain that drives humans to search for patterns and to see causal relationships," says Wolford, "even when none exist." His research partner, Michael Gazzaniga, has nicknamed this part of the brain "the interpreter." Wolford explains: "The interpreter drives us to believe that 'I can figure this out.' That may well be a good thing when there is a pattern to the data and the pattern isn't overly complicated." However, he warns, "a constant search for explanations and patterns in random or complex data is not a good thing."

That's the investment understatement of the century. The financial markets are almost—though not quite—as random as those flashing lights, and they vary in incredibly complex ways. Although no one has yet identified exactly where in the brain the interpreter is located, its existence helps explain why the "experts" keep trying to predict the unpredictable. Facing a constant, chaotic storm of data, these pundits refuse to admit that they can't understand it. Instead, their interpreters drive them to believe they've identified patterns from which they can project the future.

Meanwhile, the rest of us take these seers more seriously than their track records warrant, with results that are often tragic. As Berkeley economist Matthew Rabin points out, just a couple of accurate predictions on CNBC can make an analyst seem like a genius, because viewers have no practical way to sample the analyst's entire (and probably mediocre) forecasting record. In the absence of a full sample, a small streak of random luck looks to us like part of a longer pattern of reliable foresight. But listening to an "expert" who made a couple of lucky calls is one of the surest ways for an investor to get unlucky in a hurry.

It's vital to recognize the basic realities of pattern recognition in your investing brain:

- **It leaps to conclusions.** Two in a row of almost anything—rising or falling stock prices, high or low mutual fund returns—will make you expect a third.
- **It is unconscious.** Even if you think you are fully engaged in some kind of sophisticated analysis, your pattern-seeking machinery may well guide you to a much more instinctive solution.
- **It is automatic.** Whenever you are confronted with anything random, you *will* search for patterns within it. It's how your brain is built.
- **It is uncontrollable.** You can't turn this kind of processing off or make it go away. (Fortunately, as we'll see, you can take steps to counteract it.)

How We Got Our Brains

Why are we cursed with this blessing—or blessed with this curse—of compulsively seeking patterns in random data? "It's a really weird thing," exclaims Paul Glimcher, a neurobiologist at New York University's Center for Neural Science. "I hang out with my economist friends, and they analyze financial decision-making as if it were a Platonic problem in reasoning. They don't have a clue that it's a biological problem. We've got millions of years of primate evolution behind us. We are biological organisms. Of course there's something biological going on! Evolution *must* drive the decisions we make when we face the kinds of situations we evolved to encounter."

For nearly our entire history as a species, humans were hunter-gatherers, living in small nomadic bands, seeking mates, finding shelter, pursuing prey and avoiding predators, foraging for edible fruits, seeds, and roots. For our earliest ancestors, decisions were fewer and less complex: Avoid the places where leopards lurk. Learn the hints of coming rainfall, the clues of antelope just over the horizon, the signs of fresh water nearby. Understand who is trustworthy, figure out how to collaborate with them, learn how to outsmart those who are not. Those are the kinds of tasks our brains evolved to perform.

"The main difference between us and apes," explains anthropologist Todd Preuss of Emory University, "seems to be less a matter of adding new areas [in the brain], and more a matter of enlarging existing areas and modifying their internal machinery to do new and different things. The 'what if' questions, the 'what will happen when' questions, the short-term and long-term consequences of doing X or Y—we have *lots* more of the brain where that kind of processing goes on." Humans are not the only animals that make tools, show insight, or plan for the future. But no other species can match our phenomenal ability to forecast and extrapolate, to observe correlations, to infer cause from effect.

Our own advanced species, *Homo sapiens sapiens,* is less than 200,000 years old. And the human brain has barely grown since then; in 1997, paleoanthropologists discovered a 154,000-year-old *Homo sapiens* skull in Ethiopia. The brain it once held would have been about 1,450 cubic centimeters in volume. That is at least three times the volume of a gorilla or chimpanzee brain—but no smaller than the brain of the average person living today. Our brains are deeply rooted in the primeval environments in which our earlier ancestors evolved, long before *Homo sapiens* arose. Evolution has not stopped, but most of the "modern" areas of the human brain, like the prefrontal cortex, developed largely during the Stone Age.

It's easy to visualize the ancient East African plain: a highly variable and camouflaged environment, with alternating dapples of sun and shade, patches of dense foliage, and rolling open ground broken by sharply banked streambeds. In that landscape, extrapolation—figuring out the next link that would complete a simple pattern of repeating visual cues—became a vital adaptation for survival. Once a sample of information yielded the correct answer (ample food, safe shelter), it would never have occurred to the early hominids to look for more proof that they had made the right decision. So our ancestors learned to make the most of small samples of data, and our investing brains today still specialize in this kind of "I get it" behavior: perceiving patterns everywhere, leaping to conclusions from fragmentary evidence, overrelying on the short run when we plan for the long-term future.

We like to imagine that a long history of technological advancement stands behind us, but domesticated food crops and the first cities date back only about 11,000 years. The earliest known financial markets—in which products like barley, wheat, millet, chickpeas, and silver were sporadically traded—sprang up in Mesopotamia around 2500 B.C. And formal markets with regular trading of stocks and bonds date back only about four centuries. It took our ancestors more than 6 million years to progress to that point; if you imagine all of hominid history inscribed on a scroll one mile long, the first stock exchange would not show up until four inches from the end.

No wonder our ancient brains find the modern challenges of investing so hard to manage. The human mind is a high-performance machine—"a Maserati," says Baylor College of Medicine neuroscientist P. Read Montague—when it comes to solving prehistoric problems like recognizing simple patterns or generating emotional responses with lightning speed. But it's not so good at discerning long-term trends, recognizing when outcomes are truly random, or focusing on a multitude of factors at once—challenges that our early ancestors rarely faced but that your investing brain confronts every time you log on to a financial website, watch CNBC, talk to a financial advisor, or open the *Wall Street Journal.*

Why Do You Think They Call It Dopamine?

Wolfram Schultz, a neurophysiologist at the University of Cambridge in England, has closely cropped grey hair and a neatly trimmed silver mustache. He is so fastidious that he turns his office teacups upside down on a towel when he's not using them, lest they get dusty. The day I visited him, the only notable decoration in his office was a poster of the Rosetta Stone, a reminder of how enormous a task neuroscientists face as they try to drill down to the biological bedrock of how we make decisions. A German who spent years teaching in Switzerland, Schultz seems tailor-made to explore the microstructure of the brain by monitoring the electrochemical activity of one neuron at a time.

Schultz specializes in studying dopamine, a chemical in the brain that helps animals, including humans, figure out how to take actions that will result in rewards at the right time. Dopamine signals originate deep in the underbelly of the brain, where your cerebral machinery connects to your spinal cord. Of the brain's roughly 100 billion neurons, well under one-thousandth of one percent produce dopamine. But this minuscule neural minority wields enormous power over your investing decisions.

"Dopamine spreads its fingers all over the brain," as neuroscientist Antoine Bechara of the University of Southern California describes it. When the dopamine neurons light up, they don't focus their signals as if they were flashlights aiming at isolated targets; instead, these neural connections shoot forth their bursts like fireworks, sending vast sprays of energy throughout the parts of the brain that turn motivations into decisions and decisions into actions. It can take as little as a twentieth of a second for these electrochemical pulses to blast their way up from the base of your brain to your decision centers.

In the popular mind, dopamine is a pleasure drug that gives you a natural high, an internal Dr. Feelgood flooding your brain with a soft euphoria whenever you get something you want. There's more to it than that. Besides estimating the value of an expected reward, you also need to propel yourself into the actions that will capture it. "If you know that a reward might happen," says psychologist Kent Berridge of the University of Michigan, "then you have knowledge. If you find that you can't just sit there, but that you must do something, then that's adding power and motivational value to knowledge. We've evolved to be that way, because passively knowing about the future is not good enough."

Researchers Schultz and Read Montague, along with Peter Dayan, now at University College London, have made three profound discoveries about dopamine and reward:

1. Getting what you expected produces no dopamine kick. A reward that matches expectations leaves your dopamine neurons in a kind of steady-state hum, sending out electrochemical pulses at their resting rate of around three bursts per second. Even though rewards are meant to motivate you, getting exactly what you expected is neurally unexciting.

   That may help explain why drug addicts crave an ever-larger "fix" to get the same kick—and why investors have such a hankering for fast-rising stocks with "positive momentum" or "accelerating earnings growth." To sustain the same level of neural activity, they require a bigger hit each time.

2. An unexpected gain fires up the brain. By studying the brains of monkeys earning "income" like sips of juice or morsels of fruit, Schultz confirmed that when a reward comes as a surprise, the dopamine neurons fire longer and stronger than they do in response to a reward that was signaled ahead of time. In a flash, the neurons go from firing 3 times a second to as often as 40 times per second. The faster the neurons fire, the more urgent the signal of reward they send.

   "The dopamine system is more interested in novel stimuli than familiar ones," explains Schultz. If you earn an unlikely financial gain—let's say you make a killing on the stock of a risky new biotechnology company, or you strike it rich by "flipping" residential real estate—then your dopamine neurons will bombard the rest of your brain with a jolt of motivation. "This kind of positive reinforcement creates a special kind of attention dedicated to rewards," says Schultz. "Rewards are what keep you coming back for more."

   The release of dopamine after an unexpected reward makes us willing to take risks in the first place. After all, taking chances is scary; if winning big on long shots didn't feel good, we would never be willing to gamble on anything but the safest (and least rewarding) bets. Without the rush of dopamine, explains Montague, our early ancestors might have starved to death cowering in caves, and we modern investors would keep all our money under the mattress.
3. If a reward you expected fails to materialize, then dopamine dries up. When you spot the signal that a reward may be coming, your dopamine neurons will activate—but if you then miss out on the gain, they will instantly cease firing. And that will deprive your brain of its expected shot of dopamine. Instead of a fundamental "I-got-it" response, your brain will experience a wrenching swing into a motivational vacuum. It's as if someone yanked the needle away from an addict just as he was about to give himself his regular fix.

The Prediction Addiction

Just as nature abhors a vacuum, people hate randomness. The human compulsion to make predictions about the unpredictable originates in the dopamine centers of the reflexive brain. I call this human tendency "the prediction addiction."

That's more than a metaphor. When patients with Parkinson's disease are treated with drugs that make the brain more receptive to dopamine, some are swept up by an insatiable urge to gamble; after the patients drop the dopamine-enhancing medication, this compulsive gambling recedes almost immediately, "like a light switch being turned off." Alcohol, nicotine, marijuana, cocaine, and morphine all hook their users by affecting, in a variety of ways, the trigger zones for dopamine in the brain. A hit of cocaine, for instance, jolts the brain into releasing dopamine roughly fifteen times faster, suggesting that dopamine may somehow help convey the euphoric kick of cocaine.

If laboratory rats are wired up to receive tiny pulses of electrical stimulation in the dopamine centers of the brain when they press a lever, they often begin tapping it nonstop—to the exclusion of all other activities, including eating and drinking. They would rather starve to death than live without that dopamine surge inside their brains. Humans who have gotten electrical or magnetic stimulation to the equivalent areas of the brain report intense, even ecstatic pleasure, which some research suggests is related to the release of dopamine.

Neuroscientists do not yet know for sure how pleasure is transmitted in the brain, or exactly why rewards feel so, well, *rewarding.* What we do know for certain is that animals (including humans) without functioning dopamine circuits are incapable of taking the actions necessary to reap rewards.

At Harvard Medical School, neuroscientist Hans Breiter has compared activity in the brains of cocaine addicts who are expecting to get a fix and people who are expecting to make a profitable financial gamble. The similarity isn't just striking; it's chilling. Lay an MRI brain scan of a cocaine addict next to one of somebody who thinks he's about to make money, and the patterns of neurons firing in the two images are "virtually right on top of each other," says Breiter. "You can't get a better bull's-eye hit than those two." (See Figure 4.1.) "If there's an addictive process around chemicals and currency can be used to buy chemicals," asks Breiter, "is it possible that the same process applies to money? It's a very, very good question that has not been answered, but there's a developing mass of anecdotal data that suggests that perhaps it is." In other words, once you score big on a few investments in a row, you may be the functional equivalent of an addict—except the substance you're hooked on isn't alcohol or cocaine, it's money.

What is the underlying force that drives the prediction addiction in our brains? The dopamine researchers led by Montague and Schultz have made a discovery that may remind you of Pavlov's dogs. Once you've learned what kind of cue can signal a reward, your dopamine neurons no longer fire up in response to the gain itself; instead, they will be triggered by the appearance of the cue.

And if the reward is big enough, dopamine appears to carry a kind of long-lasting "memory" for the cue. After rats have learned that a specific sound predicts the impending delivery of a reward, neurons in their nucleus accumbens will fire the next time the rats hear the cue, even if four weeks have gone by since they last heard it or earned the reward. (Four weeks to a rat is 80 to 100 weeks in "people years.")

It's impossible to overstate the importance of these discoveries. After you learn a pattern or set of circumstances that made you money, the release of dopamine in your brain will be triggered by that stimulus—not by actually earning the money itself. Anyone using a stock-picking "system" like technical analysis has fallen prey to this problem. As soon as a stock seems to conform to a pattern that has made money before, an "I-got-it" effect kicks in, making investors feel sure they know what's coming next—regardless of whether there's any objective reason to believe that they do.

This effect ramps up with repetition. The more experience you have, the more likely it becomes that the rush you get from making money will be shifted back from the moment you actually earn it to the instant when you think you can first see it coming.

"Learning Without Awareness"

I learned how automatically my own brain makes predictions when Read Montague and his frequent research partner, neuropsychiatrist Gregory Berns of Emory University, ran me through an astounding experiment in Berns's lab at Emory's medical center. Montague and Berns make a kind of Odd Couple of brain research. Montague is an outgoing, excitable Georgia native with chiseled cheeks, a lantern jaw, and muscles that nearly burst the seams of his clothing; he seems to have stepped right off the football field. Berns, who grew up in Southern California, is small, pale, and quiet, with a soothing air that must have come in handy in his former job providing psychiatric counseling to heroin addicts. Together, the two have what feels like a four-digit IQ.

On that sweltering Atlanta day, Berns strapped me onto a gurney and rolled me into an MRI scanner. Although my head was immobilized, my index fingers were free to press a touchpad on either side of my body. On a display area above my face, the scientists projected a simple experiment: I could choose a red square, on the left, by pressing my left index finger, or a blue square, on the right, with my right index finger. A "slider bar" between the squares rose and fell to show whether my choice was a winner or a loser. My goal was to drive the slider up, earning the maximum of forty dollars. Meanwhile, the magnetic field of the MRI machine would trace the ebb and flow of oxygen in the blood inside my brain, creating a map of my neural hot spots as I thought my way through the experiment.

As I lay in the MRI tube, I sucked on a baby pacifier. No, I didn't miss my mommy; it was part of the procedure. As I picked either the left or right square, trying to figure out which sequence of them would earn me the most money, little squirts of liquid trickled into my mouth through tubes hooked up to the pacifier. One tube squirted Kool-Aid (flavor: Tropical Fruit Burst). The other delivered plain water, but the squirts did not come every time I pressed the buttons.

The boxes were as irksomely random as the squirts were. Picking the left square sometimes gave me a gain, sometimes a loss; same with the right square; same with two in a row for either square, or three in a row, or any other combination I could string together. It was frustrating, almost maddening: No matter how hard I tried, I couldn't figure out which square to pick next, and the slider bar on my overhead display just kept jitterbugging up and down around the midpoint, instead of rising toward the jackpot.

And then, suddenly, as I stared in wonder, my slider bar shot straight up. With a shock of recognition, I realized that my left index finger had been clicking madly while my right hand lay still. And then I realized that my mouth was full of Kool-Aid.

What on earth was going on? Quite simply, the reflexive part of my brain had figured out the pattern intuitively while the reflective part was still struggling to analyze it. Although the squirts seemed random, I got water only when I picked the right box and Kool-Aid only when I picked the left. (The pattern seemed chaotic because, on average, the liquid was delivered only one-third of the time.) While the reflective parts of my brain were trying to figure out which sequence of choices would work best, the nucleus accumbens in my reflexive brain suddenly recognized that sweet liquid arrived only if I picked the left-hand box. (Figure 4.2 shows my nucleus accumbens firing at this exact moment.) So, without the slightest idea I was doing it, I had begun picking the left square every time—the only response, it turns out, that would move the slider up to hit the jackpot.

"You're probably 99.9% unaware of dopamine release," says Montague, "but you're probably 99.9% driven by the information it conveys to other parts of your brain. I think it's important that you're not aware of most of it. If you had to wait around for things to consciously feel good, then you would never behave right and you would die." Fueled by dopamine signals flooding up from the base of the brain, the nucleus accumbens specializes in this subliminal, lightning-fast recognition of patterns, what Berns calls "learning without awareness." You have no idea you're doing it, but this biological imperative forces your investing brain into forecasting.

Three's Company

It's remarkable how little it can take to kick your prediction circuits into gear. Scott Huettel, a neuroeconomist at Duke University, has shown that the brain begins to anticipate another repetition after a stimulus occurs only twice in a row. Huettel and his colleagues showed people sequences of circles and squares, and explicitly told them that one shape or another would appear randomly. (After ten tries, for example, the outcome might look like this: ●■●●●■■●■■, or this: ■■■●■●●■●■.) The researchers made a simple but stunning discovery. When people saw either a single ■ or a single ●, they didn't know what to expect next. But after ■■, they automatically expected a third ■; after ●●, they subconsciously anticipated another ●.

In short, there's neuroeconomic reality in sayings like "three's a trend" or "third time's a charm"—even when, in the world around us, there's no trend at all. A run of two in a row flips a switch in your investing brain, compelling you to expect that three in a row is coming. That's why slot machines are designed so that a pull of the handle or push of the button often results in a pair of jackpot shapes on the first two spinning reels, leaving players waiting breathlessly to see if the third reel will land on a matching shape. Many "instant rub-off" lottery tickets are preprinted with two matching symbols already exposed, leading players to feel that they are bound to get a third—and winning—match when they scratch the coating off the last symbol. And the great investment analyst Benjamin Graham criticized the way traders pounce on any stock that goes up twice in a row: "The speculative public is incorrigible. In financial terms it cannot count beyond 3." Now we can finally understand why.

"The brain forms expectations about patterns," explains Huettel, "because events in nature often do follow regular patterns: When lightning flashes, thunder follows. By rapidly identifying these regularities, the brain makes efficient use of its limited resources. The brain can expect a reward even before it is delivered." But, he adds, "the downside to this process is that in our modern world, many events don't follow the natural physical laws that our brains evolved to interpret. The patterns our modern brains identify are often illusory, as when a gambler bets on 'hot' dice or an investor bets on a 'hot stock.'"

Huettel's findings shed new light on old habits of investing behavior. By definition, in any period, half of all stock pickers will do better than average and half will do worse. Thus, in the phrase popularized in Burton Malkiel's book *A Random Walk Down Wall Street,* "a blindfolded chimpanzee throwing darts at the *Wall Street Journal*" has a 50% chance of beating the market in any given year. Over three years, a chimp flinging darts in the dark has a 12.5% chance of outperforming the market average. Yet investors think a fund manager who beats the market for three years in a row must be a stock-picking genius. All too often, the public flings money at a "genius" only to find out that he's a chimp:

- From 1991 through 1993, the American Heritage Fund, run by a charismatic German named Heiko Thieme, was the best-performing mutual fund in the U.S., earning a market-stomping 48.9% average annual return. Investors poured roughly $100 million into Thieme's fund. But then American Heritage lost 35% in 1994, another 31% in 1995 and, after decent years in 1996 and 1997, went on to lose between 12% and 60% of its value every year from 1998 through 2002.

- In 2000, the manager of the Grand Prix Fund, Robert Zuccaro, bragged that "You're going to do very well in five to 10 years with Grand Prix." After all, his fund had racked up astonishing annual returns of 112% in 1998 and 148% in 1999, and it had already gained 33% in just the first three months of 2000. So investors flung roughly $400 million into the fund. Instead of the three-in-a-row effect kicking in, investors got their teeth kicked in, as every $1,000 invested in Grand Prix at the beginning of 2000 shriveled to $180 by the end of 2004.

- Stock traders often make the same mistake. In 2003, shares of Taser International, which makes a kind of "stun gun" used by police, electrified the market with a 1,937% gain. Then, in 2004, Taser's stock rose an almost equally shocking 361%. By early 2005, more than 10 million shares a day were changing hands as new investors piled into the stock, convinced it would go up for the third straight year. Instead, Taser knocked its newest investors to their knees, losing nearly two-thirds of its value in the first six months of 2005.

The biggest investors on earth fall just as hard for the "three's a trend" fallacy. A recent study of the hiring and firing of money managers by pension funds, endowments, and foundations found that these "sophisticated investors" consistently hire firms that are on a three-year hot streak. They also fire money managers that are on a three-year cold streak. Ironically, the firms they hire go on to underperform the market, while the firms they fire end up outperforming. These so-called experts who run the world's largest investment pools would earn much better returns if, rather than going by the "three's a trend" fallacy, they froze their portfolios in place and did absolutely nothing.

What Have You Won for Me Lately?

There's more to investing than just thinking you've spotted a stock or a fund on a hot streak. In the real world, investments rarely go up in a smooth, uninterrupted line; more often, they bounce up and down on a jagged path. How does the brain make sense of these short-term movements that seem to have no clear direction? In a brilliant experiment done with rhesus macaque monkeys, scientists have shown that dopamine signals predict the future by calculating a kind of moving average of the past.

As wiry as a sprite, neurobiologist Paul Glimcher of New York University sparkles with mental energy as he describes the discovery. Imagine, he says, that you're seeking a reward; let's say it's a ride in a taxicab, so you don't have to stand in a stifling subway car. As Glimcher explains, "It starts with a prediction: I believe that if I go out to Broadway right now, I will have to wait five minutes for a cab. This is followed by an observation: I actually had to wait seven. There's a mismatch between the prediction and the observation when your prediction isn't right. We call that mismatch your prediction error." And taxicabs suffer from the same problem that plagues the financial markets: unpredictability. The next time you need a cab, it might take five minutes, seven, or twenty—or you might not be able to find one at all. So your brain needs a way to figure out the average time you should expect to wait for a cab, without being overly influenced either by rare results like an extremely long or short wait, or by results from the distant past that are less likely to be relevant today. And you need to update the average quickly, so you can adapt to shifting circumstances.

How do the dopamine neurons do that? Instead of responding only to the most recent prediction error—that mismatch between the last value you expected and the last value you got—they take a running average of all your past predictions and rewards. This calculation, says Glimcher, "has bundled into it the influence of every cab you ever took." But the further back in time the predictions and rewards are, the smaller the response they evoke in the dopamine neurons. On the other hand, these neurons fire faster if you've gotten more positive surprises lately, producing a surge of dopamine that signals the rest of the brain to expect more of the same.

Because your most recent experience carries more weight, these neurons evaluate the likelihood of a gain based mainly on the average result of your last five to eight attempts at making money—with almost all the influence coming from your latest three or four tries. As Glimcher puts it: "When Fred is sitting in his living room watching the Ameritrade ticker and deciding whether to buy or sell, when someone is using the recent history of what's happened to a stock to decide what to do next—that's exactly what's happening."

This is the first biological explanation of what psychologists have called "recency," or the human tendency to estimate probabilities not on the basis of long-term experience but rather on a handful of the latest outcomes. As Glimcher's findings show, whatever has happened most recently will largely determine what you think is most likely to happen next—even if, in reality, there's no logical reason to assume that the recent past will have any impact on the future. (Only when a reward was extraordinarily large, as we saw in "Twain's Gains" in Chapter Three, will the distant past carry much weight.)

You might wonder whether humans think like rhesus monkeys, but at least when it comes to this kind of decision there's not much doubt that we do. Terrance Odean, a finance professor at the University of California at Berkeley, has studied more than 3 million stock trades by over 75,000 American households. "The average person," says Odean, "buys more aggressively in response to recent price rises, but not just to yesterday's big boom. What makes people buy is a combination of very recent rises and any longer-term 'trend' of rising prices that they might perceive." (Odean pronounces the word "trend" with weary irony because he knows that most of what seem to be patterns in stock prices are just random variations.) A survey of forecasts by hundreds of individual investors found that their expectations of stock returns over the next six months were more than twice as dependent on what the stock market did last week than on what it did over the previous few months.

Back in December 1999, after five straight years in which the U.S. stock market had gained at least 20%, investors expected to earn 18.4% on their stocks over the next twelve months. But by March 2003, after annual losses in 2000, 2001, and 2002, investors expected their stocks to go up 6.3% in the coming year. Their reliance on the recent past led investors to get the future exactly backwards: In 2000, rather than going up 18.4%, stocks dropped a shocking 9%; in the twelve months following March 2003, the U.S. market rose not a tepid 6.3% but a whopping 35.1%.

Similarly, investors chase whichever mutual funds have been hottest lately. By doing so, they ignore a basic law of financial physics: What goes up must go down, and what goes up the most usually goes down the most. Firsthand Technology Value Fund, for instance, gained 61% in 1996, 6% in 1997, 24% in 1998, and then an astounding 190% in 1999. At that point, in early 2000, people didn't just invest in the fund; they attacked it. In just three months, $2.1 billion of new money poured into a portfolio that, less than a year earlier, had totaled less than $250 million. But technology stocks promptly crashed, and investors in the Firsthand fund lost billions of dollars over the next three years.

Likewise, the investors who poured more than $2 billion into energy funds in just the first five months of 2005—based mainly on the hot returns of 2003 and 2004—are almost certain to end up getting burned. Sadly, when funds go on a streak of high returns, investors tend to get in right before the peak; then, when the hot streak goes cold, too many shareholders bail out at the bottom.

Are experts any better at resisting "recency"? Many investment newsletter editors turn bullish after as little as four weeks of hot returns in the stock market. Among mutual fund managers, the clearest sign of bullishness is how much cash they keep in reserve. Decades of data show that it takes only a few weeks of high stock returns for fund managers to reduce their cash on hand; as stock prices go up, the managers buy more, rather than the other way around.

The bottom line: Instead of buying low and selling high, all too many investors, amateurs and professionals alike, buy high and sell low, driven as powerfully by recent results as the monkeys in Paul Glimcher's neuroeconomics lab at NYU. The mental wiring that helped our ancestors thrive can blow a fuse in our investing lives today.

How to Fix Your Forecasts

What we've learned about prediction might make you wonder whether investing is a fool's game or we're all doomed to destroy our wealth with our own idiocy. Neither is true. By using the latest neuroeconomic insights, you can assure yourself of far better investing results than ever before.

The first step is to realize how much you are governed by intuition and automatic behavior. While the reflective regions like the prefrontal cortex are also vital to the process, you make predictions about future rewards largely with the more emotional, reflexive parts of your brain.

"We like to think that we're 'thinking' when we estimate probabilities," says Eric Johnson, a psychology professor at Columbia Business School, "but a surprisingly large portion of the process appears to occur automatically, below the level of consciousness." That's why it's so vital to put sound practices in place before your investing decisions can be whipsawed by the whims of the moment. Here are some proven ways to make your brain work for you instead of against you.

➢ **CONTROL THE CONTROLLABLE.** Rather than devoting your time and energy to a doomed attempt at finding the next Google or figuring out which fund manager is the next Peter Lynch, you should instead focus on what I call "controlling the controllable." You have no control over whether the stocks or funds you pick will earn higher than average returns, but you *can* control:

- **your expectations,** by setting realistic goals for future performance based on past evidence. If, for example, you think you can earn more than an average of 10% a year from U.S. stocks, you're kidding yourself—and setting yourself up for almost guaranteed disappointment.
- **your risk,** by remembering to ask not only how much you might make if you are right but also how much you will lose if you are wrong. Even the greatest investors say they are wrong nearly half the time. So you need to consider, in advance, what you will do if your analysis turns out to be mistaken.
- **your readiness,** by making sure you have used an investing checklist like the "THINK TWICE" guide in Appendix 1. Thinking twice, instead of "blinking" with intuition, is the best way to keep yourself from getting carried away by the prediction addiction.
- **your expenses,** by refusing to buy mutual funds with high management fees. Hot returns come and go, but expenses never die. So skip hedge funds entirely and rule out any mutual fund with annual expenses higher than these thresholds:
  - Government bond funds: 0.75%
  - U.S. stock funds: 1.00%
  - Small-stock or high-yield bond funds: 1.25%
  - International-stock funds: 1.50%.

- **your commissions,** by shopping for a low-cost stockbroker or financial planner and by declining to trade stocks more than a few times a year. It can easily cost 2% of your money to buy a stock, and another 2% to sell it. So a stock has to go up by more than 4% just for you to break even after paying your broker. The less you pay the broker, the more you can keep.
- **your taxes,** by holding your investments for at least one year at a time, which will minimize your liability for capital gains tax. If you hold stocks or funds for less than one year, they are taxable in the U.S. at ordinary income rates of up to 35%; hang on for the long run and your tax rate could be as low as 10%.
- **your own behavior,** by handcuffing yourself before you fall prey to the prediction addiction.

➢ **STOP PREDICTING AND START RESTRICTING.** Presented with almost any data, your investing brain will feel it knows what's coming—and it will usually be wrong. So your best bet is to prevent yourself from making too many bets. An ideal way to handcuff yourself is "dollar-cost averaging," a strategy offered by many mutual funds, which enables you to invest a fixed amount every month with a safe, automatic electronic transfer from your bank account. That way, you can't stay out of the market on a hunch it's going down, or fling all your money at the market on a guess that prices are about to soar. Buying a little bit every month keeps you from acting on the whims of your reflexive brain, and it commits you to building wealth over the long run. That puts your money to work on autopilot, so the heat of the moment can't melt your resolve.

➢ **ASK FOR THE EVIDENCE.** The ancient Scythians discouraged frivolous prophecies by burning to death any soothsayer whose predictions failed to come true. And the Bible explicitly banned "divination," calling it "an abomination unto the Lord." Investors might be better off if modern forms of divination like market forecasts and earnings projections were held to biblical standards of justice.

J. Scott Armstrong, a marketing professor at the Wharton School of Business, likes to cite what he calls "the seer-sucker theory: for every seer there is a sucker." The obvious corollary to Armstrong's theory is "neither a seer nor a sucker be." Whenever some analyst brags on TV about making a good call, remember that pigs will fly before he will broadcast a full list of his past predictions, including the bloopers. Without that complete record of his market calls, there's no way for you to tell whether he knows what he's talking about—so you should assume he doesn't.

Bob Billett, a mechanical engineer in Southern California, has learned how to stop being a sucker for the seers. In the mid-1990s, while Billett was "looking for a quick fix" for his portfolio, a stockbroker with a small local firm called and bragged about how good the firm's past stock picks had been. Billett took the bait, buying five stocks—and lost his shirt. Shady brokers still call Billett, but now, he says, "I keep a little log. I keep track of who called when from where. Then when they say, 'I called you three months ago and recommended ABC right before it doubled,' I look up my notes. And I always find that either the conversation had never happened or that they had recommended something else." Billett hasn't bought a shady stock ever since, but he has gotten a bit of free entertainment.

Sherwood Vine, a retired physician in New Jersey, was skeptical when his financial advisor told him to sell two of his mutual funds and "buy two better ones"—and that "it wouldn't cost a cent." So Vine asked for two things: a calculation of how much the new funds would have to outperform in order to overcome the capital gains taxes he would incur by selling the old ones, and a complete list of all the funds that the advisor had recommended as "better" one, two, three, and five years earlier. Vine's broker said he would get back to him, but Vine never did get an answer. He still owns his old funds.

"A foolish faith in authority," warned Albert Einstein, "is the worst enemy of truth." As Billett and Vine have learned, investors can shed their "foolish faith" by vigilantly keeping records and thinking for themselves.

➢ **PRACTICE, PRACTICE, PRACTICE.** Since your brain is programmed to detect patterns even in random data, it's important to learn whether your hunches and predictions are valid before you risk real money on them. "Find cheap situations in which you can test your biases," recommends Richard Zeckhauser, an economist at Harvard University. "Keep track in a hypothetical world of cheap experiments."

In one of the last interviews he gave before he died, the great investment analyst Benjamin Graham suggested that every investor should spend a year running a paper portfolio—devising strategies, making stock picks, and testing the results—before venturing any real money. That's a lot easier to do nowadays than it was in 1976: At websites like Yahoo! Finance and morningstar.com, you can use "portfolio tracker" software to set up an interactive list of all your imaginary investments. By monitoring all your buys and sells, and then comparing them to an objective benchmark like the S&P 500 index of blue chips, you can see how good your decisions are before you have to put any actual money behind them. Better yet, an online portfolio tracker won't let your memory play tricks on you or selectively "bury" your mistakes, so it will provide a complete and accurate record of your decisions. Investing experimentally before you try it for real is like learning how to pilot an airplane in a flight simulator—nearly as informative as the real thing, and a lot safer. Just be careful not to form a habit of checking the portfolio's value too often.

➢ **FACE UP TO BASE RATES.** One of the best ways to improve your predictions is by training your brain to ask, "What's the base rate?" The base rate is a technical term for the outcome you would logically expect to see in a very large sample of long-run results. But anything unusual or vivid can distract us from thinking about base rates. Psychologist Howard Rachlin explains the base-rate problem this way: Imagine that you are on a beach, guessing the profession of the people you see. If a man emerged from the water wearing flippers, a mask, and a wetsuit, you might assume he is a professional diver or a Navy frogman because they are more likely to be "dressed" that way than are people in other professions. But you would stand a better chance of winning the bet if you guessed that he is a lawyer, since there are far more lawyers than professional divers in America. We're just not used to thinking of lawyers in snorkels and rubber suits.

And if you witness someone who has just managed to flip heads 31 times in a row, you may no longer be able to focus on the obvious fact that, in the long run, the base rate for flipping coins has to be 50% heads. If your favorite baseball team is a perennial loser that suddenly dominates a five-game series against the team with the best record, you'll probably expect a continued comeback—but, based on long-term averages, the worst team in baseball will win a five-game series against the best roughly 15% of the time by luck alone.

In much the same way, a short but vivid streak of success in the stock market can make you overlook how unlikely it is to last. The stock of Google, for example, tripled in the twelve months after its initial public offering, or IPO, when outside investors could first own tradable shares in the company. That huge gain set tens of thousands of investors off on a frantic search for "the next Google" among other high-tech IPOs. But these people forgot to ask, "What's the base rate?" When you buy an IPO, you're not buying the next Google; you're just buying the next IPO, and your results are far more liable to resemble the long-term gain of the average IPO than they are to resemble the performance of Google. Jay Ritter, a finance professor at the University of Florida, periodically updates the long-term average performance of IPOs at his website, http://bear.cba.ufl.edu/ritter/ipodata.htm (look for "Long-Run Returns on IPOs"). Ritter's authoritative data show that over five-year periods since 1970, IPOs have underperformed the shares of older companies by an average of at least 2.2 percentage points annually. That base rate should tell you that by buying IPOs, you are less likely to beat the market than to be beaten by it.

If a financial advisor or website is hawking a hedge fund, mutual fund, or other investment vehicle that "seeks to beat the market," ask this simple question: Over the long run, what percentage of money managers outperform the market average? The answer: Over ten-year periods, only about one in three money managers will beat the market. If that base rate doesn't sound appealing to you, do what I do: Invest in an index fund, which merely aims to match the performance of the overall market at rock-bottom cost.

Finally, as the old Wall Street saying goes, never confuse brains with a bull market. If somebody boasts about how good his stock picks are, remember to check whether the segment of the market he invests in performed even better. (A technology-stock picker might brag that he earned 48% in 2003, but the Goldman Sachs tech-stock index rose 53% that year.) Remember, anybody can look like a genius when most investments are going up. As psychologist Daniel Kahneman quips, "In a rising market, enough of your bad ideas will pay off so that you'll never learn that you should have fewer ideas."

➢ **CORRELATION IS NOT CAUSATION.** One of the oldest tricks in Wall Street's book is to produce a graph of stock prices and then overlay a second graph on top of it. Then the person wielding them points out the uncanny way one graph predicts the other. In the 1990s, for instance, market pundit Harry Dent estimated how many 46½-year-olds there were in the U.S. each year. Then he took the Dow Jones Industrial Average, adjusted it for inflation, and *voila!* The price of the Dow since 1953 was predicted almost perfectly by the number of 46½-year-olds (because, contended Dent, that's the age at which American consumers spend the most). Based on the expected number of 46½-year-olds in the future U.S. population, Dent projected that the Dow would hit 41,000 by the year 2008. He even launched a mutual fund that would pick stocks based on his theory.

But there are thousands of stocks, dozens of market indexes, and an almost unlimited number of time periods to choose from, so there's nothing remarkable at all about a historical chart on which one line of data appears to predict another. The real miracle would be if someone *couldn't* find a statistical variable that seemed to predict the financial future. After all, if 1953 didn't work as a starting point, you could just switch to 1954, 1981, 1812, or any year that did work. If the Dow wouldn't give the result you wanted, you could use the S&P 500 or any other index.

In 1997, money manager David Leinweber wondered which statistics would have best predicted the performance of the U.S. stock market from 1981 through 1993. He sifted through thousands of publicly available numbers until he found one that had forecast U.S. stock returns with 75% accuracy: the total volume of butter produced each year in Bangladesh. Leinweber was able to improve the accuracy of his forecasting "model" by adding a couple of other variables, including the number of sheep in the United States. Abracadabra! He could now predict past stock returns with 99% accuracy. Leinweber meant his exercise as satire, but his point was serious: Financial marketers have such an immense volume of data to slice and dice that they can "prove" anything. And they never tell you about all the theories and data they tested and rejected—since that might give you a notion of how dangerous and dumb most of their ideas really are.

Here are some questions to ask whenever someone (including yourself!) tries to convince you that he's found the Holy Grail of market forecasting:

- How would these results change if the starting and ending dates were moved forward or backward?
- How would these results change under only slightly different assumptions? (Has consumer spending always peaked at age 46½? Might not people's peak spending years shift in the future?)
- Is the factor that supposedly predicts future returns something that you would reasonably expect to drive the market? (Why would consumer spending necessarily be more important than, say, health care spending or corporate expenditures?)

These steps will help you remember that correlation is not causation, and that most market prophecies are based on coincidental patterns. That was the problem in the late 1990s at the Motley Fool website. Its Foolish Four portfolios were based on research claiming that factors like the ratio of a company's dividend yield to the square root of its stock price could predict future outperformance. In the long run, however, a company's stock can rise only if its underlying business earns more money. Can you imagine being more eager to buy a company's products or use its services because you liked the number you got when you divided the square root of its stock price into its dividend yield? No customer has ever thought like this in the history of capitalism, and none ever will.

Since the Motley Fool's goofy ratio couldn't possibly be *causing* stock prices to rise, the only sensible conclusion was that its predictive power was an illusion. The Foolish Four portfolio made investors feel like idiots when it lost 14% in the year 2000 alone. Meanwhile, after six years of underperforming the market by nearly two percentage points annually, the Harry Dent–inspired mutual fund shut down in mid-2005 with the Dow mired about 31,000 points below his forecast.

➢ **TAKE A BREAK.** In his experiments on pattern-seeking (see "Pigeons, Rats, and Randomness," page 57), psychologist George Wolford found that people were better at estimating probabilities when they were distracted with a "secondary task" like trying to recall a series of numbers they'd recently seen. Interruptions may improve their performance by keeping the brain too busy to find spurious patterns in the data.

One of our oddest mental quirks is what's known as "the gambler's fallacy"—exemplified by the belief that if a coin has come up heads several times in a row, then it's "due" to come up tails. (In truth, of course, the odds that a fair coin will turn up tails are always 50%, no matter how many consecutive times it's come up heads.) When a process seems obviously random—like, say, coin flips or spins of a roulette wheel—then the gambler's fallacy takes hold of our minds. That leads us to believe that a streak of luck is likely to reverse. (It's when human skill appears to play a major role, as in sports, for example, that we tend to believe a hot streak will persist.) Sometimes the gambler's fallacy has tragic results: In Italy, more than two years went by in which the number 53 never came up a winner in the Venice lottery. After that long spell, 53 was finally drawn in early 2005—but not before a woman drowned herself and a man shot his wife, his son, and himself, after spending all their money betting in vain on number 53.

Since most professional investors concede that the stock market is partly random, belief in the gambler's fallacy is as common on Wall Street as dustballs under a couch: Some pundits will say that Stock X is sure to rebound because it's been doing badly for years, while others proclaim that Stock Y is doomed to crash because it has risen so much lately.

There's a simple way to free yourself from the grip of the gambler's fallacy. Twenty years ago, researchers at Carnegie Mellon University showed that people will intuitively tend to bet on tails if a coin comes up heads several times in a row—unless you let the coin "rest" for a while before flipping it again. Then people will bet on heads, as if the passage of time somehow makes them feel that the chance of getting heads again has reverted to its true ratio of 50/50. This experiment, along with Wolford's findings, suggests that one of the best ways to prevent your investing brain from fooling you into perceiving patterns that aren't there is simply to take a break from studying a stock or the market. Diverting yourself with another activity for twenty minutes or so should do the trick.

DON'T OBSESS. Gone are the days when an individual investor could track stock prices only by telephoning his broker, visiting a brokerage office that had a stock ticker, or waiting to read the stock market listings in the next morning's newspaper—which would sum up an entire day's trading activity with a single, static line of agate type that might look like this:

"40.43 +.15 47.63 30.00 0.6 23.5 18547."

Now, thanks to the wonders of electronic technology, stocks have become dynamic visual displays, almost living organisms. Every trade shows up as another blip on a rising or falling stream of prices; every few ticks appear to validate or reverse a "trend." Visual information—especially imagery that conveys change—fires up your reflexive system and crowds out reflective thought. A streak of electric green climbing up your computer monitor, or a lurid red line tearing down across your screen like a scar, will set off your brain's emotional circuits with a force that no dry row of type in a newspaper ever could.

By using technology to turn investing into a kind of Wall Street Game Boy—price paths snaking up and down a glowing screen, arrows pulsating in vivid hues of red and green—online brokerages have tapped into fundamental forces at work in the human brain. The similarity to Nintendo or PlayStation has a chilling implication: Researchers have found that when players do well at a video game, the amount of dopamine released in their brains roughly doubles, and that this surge can linger for at least a half-hour afterward.

So the more "price points" you can see, the more your brain will fool itself into thinking it has detected a predictable pattern in the numbers—and the more powerfully your dopamine system will kick in. As we've seen, it can take as few as three price changes to make you think you've spotted a trend; in years past, when investors got their stock prices out of the newspaper, it could take three days to gather that much data, while today a market website will get you there in less than sixty seconds. No wonder, by the late 1990s, the typical "investor" in popular tech stocks like Qualcomm, VeriSign, and Puma Technology owned them for an average of less than eight days at a time.

"If owning stocks is a long-term project for you," warns psychologist Daniel Kahneman, "following their changes constantly is a very, very bad idea. It's the worst possible thing you can do, because people are so sensitive to short-term losses. If you count your money every day, you'll be miserable." By obsessively monitoring the prices of your holdings, you raise the odds that you will see an interim loss or an apparent trend worth trading on—when, in fact, there's probably nothing in the data but a swarm of random squiggles. Considering what we've learned about how the dopamine system works, these stimuli hit your brain like a can of kerosene dumped onto a campfire. Several experiments by Kahneman and other researchers have found that the more often people watch an investment heave up and down, the more likely they are to trade in and out over the short term—and the less likely they are to earn a high return over the long term.

An episode of *Seinfeld* captures this sort of misery when Jerry buys a junky stock called Sendrax, then compulsively keeps tracking the price into the evening. After Jerry grabs a newspaper, his date tells him, "The stock is the same as when you checked it earlier. There are no changes after the market closes. The stock is still down." Replies Jerry: "I know. But this is a different paper. I thought maybe they have, uh, different . . . sources." If we laugh at Seinfeld, it's the nervous laugh of self-recognition: A recent survey by *Money* magazine found that 22% of investors say they look up the prices of their investments every day, and 49% check on them at least once a week.

So, instead of driving yourself crazy by constantly monitoring your stocks or funds, you should cut back gradually until you look at their value only four times a year—either at the end of each calendar quarter or on four easily memorable and roughly equidistant dates.

After all, time is money—but money is also time. If you're compulsively checking up on the prices of your investments, you're not only hurting your financial returns, you're unnecessarily taking precious time away from the rest of your life.

CHAPTER FIVE

Confidence

> Before I was Pope, I believed in papal infallibility. Now that I am Pope, I can *feel* it.
>
> —*joke told of Pope Pius IX*

What, Me Worry?

IN 1965, TWO PSYCHIATRISTS AT THE UNIVERSITY OF WASHington, Caroline Preston and Stanley Harris, published a study in which they asked fifty drivers in the Seattle area to rate their own "skill, ability, and alertness" the last time they were behind the wheel. Just under two-thirds of the drivers said they were at least as competent as usual. Many described their most recent drive with terms like "extra good" or "100%." But there was something remarkably strange about these results. Preston and Harris conducted all their research interviews in the hospital—because each of the drivers had begun his most recent trip in his own car, but ended it in an ambulance.

According to the Seattle police department, 68% of these drivers were directly responsible for their crashes, 58% had at least two past traffic violations, 56% totaled their vehicles, and 44% would ultimately face criminal charges. (Only 5 of the 50 drivers admitted to Preston and Harris that they were even partially responsible for the crash.) They had a ghastly gamut of injuries ranging from concussions, facial trauma, a crushed pelvis and other broken bones to severe spinal damage—and three of their passengers had died. It's hard to imagine a sorrier bunch of reckless, careless, clueless people—and it seems almost incredible that they could insist that they were driving so well at the very moment they wreaked such havoc.

Were these drivers crazy? Not at all. They were nothing more or less than normal human beings. One of the most fundamental characteristics of human nature is to think we're better than we really are. While the pathetic folks interviewed by Preston and Harris might seem uniquely delusional about their abilities, a later survey of people with a clean driving record found that 93% believed themselves to be above-average drivers.

Just ask yourself: Am I better looking than the average person?

You didn't say no, did you?

Not many folks do. If you ask a group of 100 people, "Compared with the other 99 people here, who's above average at *x*?" roughly 75 will stick up their hands, regardless of whether *x* is driving a car, playing basketball, telling jokes, or scoring well on an intelligence test—and despite the fact that, by definition, half the people must be below average within the group. When I give speeches to groups of investors, I sometimes hand out a slip of paper asking people to jot down how much money they think they will save by the time they retire and how much money they think the average person in the room will save. Invariably, people think they will save at least 1.8 times as much as the average person.

It's as if Homer Simpson, when he looks in the mirror, sees Brad Pitt gazing out at him—while Marge sees the reflection of Nicole Kidman shining back at her. Psychologists call this view of life "overconfidence," and it can get investors into a heap of trouble.

Of course, being overconfident isn't all bad. Princeton University psychologist Daniel Kahneman is fond of saying that if we were always realistic about our chances of success, we would never take any risks; we'd be too depressed. A survey of nearly 3,000 entrepreneurs who had recently launched a new business showed how true that is. When asked to rate the chances of success for "any business like yours," only 39% said the odds were at least 7 out of 10. But when asked to estimate the odds that their own venture would succeed, 81% of the entrepreneurs said they stood at least a 7 out of 10 chance, and an amazing 33% of them said there was zero possibility that they would fail. (On average, roughly 50% of new businesses fail within their first five years.)

There's no doubt that most of these entrepreneurs were kidding themselves. But how else could any of them muster the courage to start a new business? Without an extra dose of confidence, it would be much harder to make decisive choices in an uncertain world and to overcome the setbacks that barricade the path to success. As Kahneman puts it, "The combination of optimism and overconfidence is one of the main forces that keep capitalism alive."

Positive thinking can be useful, but extreme optimism is dangerous. For investors, there are several ways in which overconfidence leads to underperformance:

- We can make a level-headed estimate of the typical person's odds of succeeding at something, but we usually have our heads in the clouds when we size up our own chances at success. That leads us to take risks we regret down the road.

- We put too much trust in whatever is familiar. This "home bias" leads us to invest too little money outside the industry we work in, the region we live in, and the rest of the world beyond our national borders. It also prompts people to invest too much money in the stock of the firm they work at.

- We overstate how much power we wield over our own circumstances. This "illusion of control" causes us to become complacent, to put too little effort into planning ahead, and to be overcome by surprise when our investments tank.

- We tell ourselves that we foresaw what was going to happen—even if, in the past, we had no idea what the future would hold. By misleading us into thinking we did see what *was* coming, this "hindsight bias" makes us believe we can see what *is* coming. Worse, it keeps us from learning from our mistakes.

- Above all, we have a terrible time admitting when we don't know something. As nature abhors a vacuum, the human mind despises the words "I don't know." The more we know, the more we think we know even more than we do. We're even overconfident about our ability to overcome our own overconfidence!

The single biggest step you can take to improve your investing results is to stare long and honestly into a mirror to see whether you really are the investor you think you are. Are you truly above average? Are your own decisions the main force that shapes your returns? Is "buying what you know" the best investment approach? Did you really predict where the market was going? Do you know as much as you think you do?

The bad news is that, for most of us, the answer to most of these questions is No. The good news is that neuroeconomics can help you bring your confidence into synch with reality, making you a better investor than you may ever have imagined.

I'm the Greatest

Every month in the late 1990s, the Gallup poll contacted nearly 1,000 investors across the U.S. to ask how much they thought the stock market, and their own portfolios, would go up over the coming twelve months. In June 1998, investors thought the market would rise 13.4%—but their own accounts would go up 15.2%. By the peak of the bull market in February 2000, they expected the market as a whole to jump by 15.2%—but they thought the stocks they had picked would soar 16.7%. Even in the dark days of September 2001, when people expected the market to rise just 6.3%, they still expected their own stock picks to earn 7.9%. No matter how well the market might do, investors expected that their personal portfolios would earn about 1.5 percentage points better.

Of course, the stock market is not the only contest in which every person seems to think that he is a winner in a world of losers. Unrealistic optimism is everywhere:

- A nationwide survey of 750 investors found that 74% expected that their mutual funds would "consistently beat the S&P 500 every year"—even though most funds fail to outperform the Standard & Poor's 500-stock index in the long run, and many funds fail to beat it in any year.

- Only 37% of corporate managers believe that mergers create value for the buyers, and a paltry 21% think that mergers meet the strategic goals set by acquirers. However, when it comes to their *own* mergers and acquisitions, 58% of experienced managers say they have created value, and 51% believe they have met their strategic targets.

- College students were asked how likely various events were to occur in their own lives—and the lives of other undergraduates. The typical student said she was 50% more likely than her peers to be happy in her first job. She also thought she was 21% more apt to earn a good salary, had a 13% better chance of seeing her home double in value within five years, and stood 6% better odds of having a gifted child. What's more, she thought she was 58% less likely to become an alcoholic, 49% less likely to get divorced, 38% less likely to suffer a heart attack, and even 10% less likely to buy a car "that turns out to be a lemon."

- Finally, in what may be the ultimate form of unrealistic optimism, 64% of Americans believe they will go to heaven after they die; only one-half of one percent expect to go to hell.

In short, to evaluate ourselves is to lie to ourselves, especially when the evaluation requires us to compare ourselves to the average person. Inside each of us, there lurks a con artist who is forever cajoling us into an inflated sense of our own powers. The less skilled or experienced you are at something, the harder your inner con man works at convincing you that you are brilliant at it.

And that's good, up to a point. By fibbing to ourselves, we can give a needed boost to our self-esteem. After all, none of us is perfect, and daily life brings us into constant collision with our own incompetence and inadequacies. If we did not ignore most of that negative feedback—and counteract it by creating what psychologists call "positive illusions"—our self-esteem would go through the floor. How else could we ever get up the nerve to ask somebody out on a date, go on a job interview, or compete in a sport?

There is only one major group whose members do not consistently believe they are above average: people who are clinically depressed. These chronically sad people rate their own abilities very accurately—which, of course, may be one of the main reasons they stay so sad. Their depression has robbed them of the ability to fool themselves. As psychologists Shelley Taylor and Jonathon Brown put it: "The mentally healthy person appears to have the enviable capacity to distort reality in a direction that enhances self-esteem . . . and promotes an optimistic view of the future."

Fibbing to yourself is one thing, but telling outrageous whoppers is another thing altogether. Imagining that you're a good basketball player if you're merely competent probably won't get you into much trouble.

But if you think you're LeBron James even though you couldn't score a basket from a ladder propped up against the backboard, then a game of hoops against tough competition is sure to leave you with either a crushed ego or a snapped Achilles tendon.

Investing works the same way. A pinch of confidence encourages you to take sensible risks and keeps you from storing all your money in the concrete bunker of cash. But if you think you're Warren Buffett or Peter Lynch, your inner con man is not telling little fibs; he's a big fat liar. You will never make the most of your investment potential if you think you have far more potential than you actually do. The only way to achieve everything you're capable of is to accept what you are *not* capable of.

That's surprisingly hard for most investors to do. Two studies have recently tracked what happens when the typical investor listens to his inner con man. In both cases, like fishermen spreading their arms and proclaiming that their catch was "this big," investors wildly overstated how well they had done.

In a survey conducted by *Money* magazine in late 1999, more than 500 investors reported whether their stocks or stock funds had beaten the market (as measured by the Dow Jones Industrial Average) over the previous twelve months. All told, 131 investors, or 28%, said their portfolio had beaten the Dow. They were then asked to estimate their rate of return. Roughly a tenth reported that their portfolio had gone up by 12% or less, around a third claimed to have earned between 13% and 20%, another third or so said they had gained between 21% and 28%, and a quarter thought they had made at least 29%. Finally, 4% of the investors admitted they had no idea how much their portfolios had gone up—but they were sure they had beaten the market anyway! And yet, over those twelve months, the Dow was up 46.1%—vastly outperforming at least three-quarters of the investors who claimed they had beaten it.

In a second study, 80 investors received clear updates on how their mutual funds were faring against the S&P 500. Then, at the end of the experiment, the investors were asked how well they had done. Nearly a third claimed that their funds had beaten the market by at least 5%, and 1 of every 6 said they had outperformed by more than 10%. But when the research team checked the portfolios of the people who claimed to have beaten the market, it turned out that 88% had exaggerated their returns. More than a third of those who thought they had beaten the market actually lagged it by at least 5%, and a fourth of all the self-described market beaters finished at least 15% behind the S&P 500.

"This shows a shocking lack of learning," says psychologist Max Bazerman of Harvard Business School. "You have the right to try to beat the market, but you need to learn the odds—which suggest that you're taking a sucker's bet. And it's easy to have illusions about the future if you don't even have a grip on your own recent past."

Together, these studies suggest that most people are kidding themselves when they claim to beat the market. "Everybody wants to believe that they're special and better than average," says psychologist Don Moore of Carnegie Mellon University. "They think they can beat the market with their own special something. And it's remarkable how this illusion persists even in the face of evidence to the contrary."

It's also perfectly understandable. Most people would much rather listen to the blarney of their inner con man than measure their financial performance accurately. After all, their inner con man never has anything bad to say.

That brings us to a much larger lesson. When you watch financial TV, visit market websites, or read the financial press, you will be told that "down here in the trenches," investing is a contest, a fight, a duel, a battle, a war, a struggle for survival in a hostile wilderness. But investing is not you versus "Them." It's you versus *you.* As Jack Nicholson hollered at Tom Cruise in the movie *A Few Good Men,* "You can't handle the truth!" In fact, the single greatest challenge you face as an investor is handling the truth about yourself.

There's No Place Like Home

In March 2002, three months after Enron Corp. went bankrupt, I gave a speech to a group of individual investors in Boston. I reminded the group that when Enron went bust, its employees didn't just lose their jobs; their retirement funds were also wiped out. Enron's employees had put 60% of their retirement savings into the company's stock. When the shares crashed, Enron's 20,000 workers lost at least $2 billion. "You already work at your company," I told the audience. "The last thing you should do is take on a double layer of risk by putting your retirement money there, too." After all, I warned, "The company where any of us work might turn out to be the next Enron. And the only way to protect against that is to diversify—own the whole stock market, and make sure you don't put more than 10% of your retirement money into your own company's stock."

I was unprepared for what happened next. A man jumped to his feet and jabbed his index finger toward me. "I can't believe what you just said," he called out. "I completely agree that any company could be the next Enron. But that's why your advice makes no sense. Why should I move my money from the one company I know *everything* about to hundreds of stocks I don't know *anything* about? Diversifying doesn't protect me from the next Enron, it exposes me to every next Enron—and the stock market is full of them! I want my money where I know it's safe—in the company I work for and the company I understand better than any other one around. *That's* how I control risk."

As gently but firmly as I could, I responded that most Enron employees were sure, until the eleventh hour, that it was one of the world's greatest companies and that nothing could go wrong. Enron was ranked No. 7 on the Fortune 500 list of America's biggest companies, its stock had vastly outperformed the market averages over the long run, and nearly every Enron employee believed the firm's official slogan, "The World's Leading Company." They never dreamed that Enron could implode. At a meeting for hundreds of employees in December 1999, Enron's head of human relations had been asked, "Should we invest all of our 401(k) in Enron stock?" Her answer: "Absolutely!"

My questioner, like the employees at Enron, suffered from "home bias," the tendency to think that the most familiar investments are the best. Around the world, amateur and "expert" investors alike are chronic homebodies:

- After AT&T broke up into eight regional telephone companies in 1984, investors chose to keep three times more shares in their local phone company than in all the other "Baby Bells" combined.

- Mutual fund managers prefer to invest in firms that are based nearby. The typical fund owns stocks of companies that are based ninety-nine miles closer to the fund's headquarters than the average U.S. company is.

- Mutual fund investors in France have 55% of their money in French stocks, even though the Paris bourse accounts for only 4% of the world's total stock market value. New Zealanders keep 75% at home, even though Kiwi stocks are less than 1% of the world's total. Greeks, whose stock market also makes up less than 1% of the global total, keep 93% of their investments at home. (It could be worse: Fifteen years ago, Japanese investors had 98% of their portfolios in domestic companies, right before Japanese stocks sank into one of the worst bear markets the world had ever seen.)

- Investors in 401(k) plans allocate only about 5% of their stock portfolios to companies based outside the U.S.—even though other markets make up half the value of the world's stocks.

- A study run jointly in the U.S. and Germany found that German investors expected their stock market to outperform the U.S. by two to four percentage points per year. Meanwhile, U.S. investors expected the Dow to beat the German market by almost exactly the same margin.

- Only 16.4% of 401(k) investors said they believe that the stock of the company they work for is riskier than the stock market as a whole.

Why do we feel so safe at home? The frontier between the familiar and the alien, between ourselves and the rest of the world, is incredibly narrow and near. The saliva in your mouth, for example, is so much a part of you that you take it for granted. This beneficial fluid aids digestion, relieves thirst, and helps keep your mouth clean. But what if someone asked you to spit into a clean cup, count to five, and then drink your saliva back up? Suddenly—now that it was "out there," instead of inside your body—even a tiny sip of your own saliva would seem disgusting. A few moments and a few inches outside your mouth is all that it takes to turn an integral part of you into something alien and revolting. That's how little time and distance there is between the comfort zone of the familiar and the danger zone of the outside world.

It makes sense, of course. If our early ancestors had not learned to steer clear of the germs, predators, and other dangers lurking outside their own bodies and beyond their immediate home ground, they would not have survived. Too much curiosity could kill the cave dweller. Over the course of countless generations, a preference for the familiar and a wariness toward the unknown were ingrained into the human instinct for survival. Familiarity became synonymous with safety.

The Eerie Power of Mere Exposure

Almost forty years ago, a psychologist named Robert Zajonc (pronounced "zye-ontz") began a series of extraordinary experiments. Zajonc started by having Americans listen to words like *afworbu, kadirga,* and *dilikli.* Then he asked the listeners to guess whether each word meant something good or bad in Turkish. The more often a word was repeated, the more likely a listener was to feel that it stood for something positive. (In fact, most of these words were nonsense syllables with no meaning in either Turkish or English.) Next Zajonc displayed Chinese ideographs to people who had no familiarity with the Asian alphabet—and found that whether they thought each ideograph represented something good or bad depended merely on how often they had been exposed to it.

Zajonc dubbed his discovery the "mere-exposure effect." He flashed twenty irregular, octagonal shapes onto a dimly lit screen in a blazingly fast exposure of only one millisecond apiece—about one three-hundredth the duration of the blink of an eye. At that speed, no one can make out the shape, and most people are unsure whether they saw anything at all. Then he showed pairs of octagons—one new, and one from the set he had flashed before—for one full second and in better light. When Zajonc asked which shape they liked better, people overwhelmingly preferred the one they had already been exposed to, even though they had no idea that they had ever seen it before.

Zajonc kept flashing. He showed one group of people a set of Chinese ideographs, five times each, in random order. He exposed a second group to another set of Chinese characters, each shown only once. All the ideographs were flashed for just four to five milliseconds—again, such a short exposure that most people can register it only subliminally. Then Zajonc showed the original ideographs again, randomly jumbled up with similar but new ideographs and a set of totally unrelated shapes. He let people look at each image for a full second—long enough to be conscious of seeing it—then asked how much they liked each one. It turned out that if people had first gotten five subliminal exposures to an ideograph, they now liked it much better than if it had been flashed only once before.

Then the story got stranger. Compared to those who had seen each Chinese character a single time, the people who had undergone repeated exposures didn't just like the old ideographs better. They also liked the new ones—and even the unrelated shapes—more. And the participants who had been exposed to repeated subliminal flashes ended up in a measurably happier mood than those who had been exposed to each image only once.

Being in the presence of familiar things (even when we are unaware of them) simply makes us feel better. "The repetition of an experience is intrinsically pleasurable," says Zajonc. "It augments your mood, and that pleasure spills over anything which is in the vicinity." Aesop had it wrong when he said "familiarity breeds contempt." In truth, familiarity breeds contentment.

You might think that your likes and dislikes are conscious choices, that your preferences are based on inferences you make from studying the evidence. Instead, Zajonc's findings show, our preferences come out of our experiences—regardless of whether they ever were conscious. Whatever we experience most often, we are most likely to end up liking. (One of the few exceptions is abstract art: No matter how many subliminal exposures to it they get, people still don't like it any better.) That helps explain why investors tend to overpay for the stocks of companies with famous brand names. It also shows why Peter Lynch's advice to "buy what you know" rang so true to people—even though many investors have lost their shirts doing just that.

Mere exposure might remind some people of the Orwellian idea of "subliminal seduction" popularized by Wilson Bryan Key in his book of the same name, published in 1973. Key argued that advertisers, by exposing us to endless repetition of subliminal images, dictate our spending habits and our very lives. Key claimed (falsely) that the word *sex* was covertly imprinted on Ritz crackers and that images of naked women lurked within the ice cubes in liquor ads. Others have claimed that moviegoers will gorge on popcorn if the words "BUY MORE POPCORN" are flashed onto the screen for a split-second.

But Zajonc points out that sentences cannot be understood subliminally. And while repeated exposure to the word *sex* certainly could improve your mood, it's unlikely to make you lust after salty crackers. The popular concept of subliminal seduction is nonsense.

The mere-exposure effect, on the other hand, is real. Just as the gravitational fields of the moon invisibly drive the tides, you cannot sense mere exposure governing your behavior—but it does. "Evolutionary theory suggests that you should pay the strongest attention to something which is novel or unknown to you," says Zajonc. "And if you keep encountering that novel stimulus over and over again and it does not bite you, then you are safe in the presence of this object. That makes your attitude toward it more positive, although you are probably not aware of that fact."

What's going on inside your brain during the flash of familiarity? Neuroscientists have scanned people's brains during one of Zajonc's Chinese-ideograph experiments. The scans show that even though you are unaware that you have already seen a particular Chinese character, your brain's memory centers are automatically activated by mere exposure. In effect, the repetition of images appears to submerge them deep into your brain: You have no idea these sunken memories exist, but they are there, waiting for a longer repetition in the outside world to float them up from the ocean floor of your consciousness.

Another study found that people who say they like Coke better than Pepsi cannot reliably tell which one they prefer in a blind taste test—and that the brains of Coke fans and Pepsi fans light up in essentially the same way if they drink either brand without knowing which one it is. When people see a labeled Pepsi can before they sip, the memory centers in the hippocampus region and the emotional circuitry in the reflexive brain are only mildly activated. But when people get a glimpse of that flaming red logo on a Coke can, their memory centers and emotional circuitry kick into high gear. You might think you like Coke better for the taste, when in fact you like it better mainly because it's more familiar. Likewise, investors plunk money into brand-name stocks—precisely because the brand name makes them feel good.

The "Halle Berry" Neuron

There are tantalizing hints that familiarity may be built one brain cell at a time. The hippocampus (from the Latin for "seahorse") is a lumpy crescent of cells buried toward the middle of your brain about one inch inward from your ears. A key part of the reflexive brain, the hippocampus is a hotbed of emotional memories. It is also packed with neurons that fire individually when you move through, see, or even imagine a specific location. These neurons are called "place cells" for their uncanny ability to tell each feature of your environment apart from the next. They constitute your inner map of the outer world—complete with dazzling detail, accessible with no effort. Place cells enable you to reach out in the dark and find a light switch without any conscious thought. Many experiments in rats have shown that the firing of place cells helps keep the brain focused on attaining a goal.

Once anything in your environment becomes relevant and important, a place cell in your hippocampus will activate whenever you encounter it again. The level of precision is nothing short of miraculous: Your brain appears to dedicate a single cell uniquely to every tangible element in your surroundings. That environmental cue might be a particular face, a given name, a specific building, the exact color and typeface of a company logo. It's as if one single neuron is dedicated, like a microscopic sentry, to recognizing each of these objects—and little or nothing else. Astonishing experiments at UCLA and Tel Aviv University have shown that some people have a single neuron dedicated to recognizing the distinctive schooner-shaped profile of the Sydney Opera House. Others have a place cell that fires exclusively when an image of movie star Halle Berry appears. The same cell will respond to her photograph, a drawing of her, or even the letters of her name—but not to an image of, say, Pamela Anderson. The names and faces of other celebrities, like Jennifer Aniston, Julia Roberts, and Kobe Bryant, also trigger a separate and unique response—each centered on a single neuron in or near the hippocampus.

London taxi drivers have an unusually large posterior hippocampus; their constant need to recognize city landmarks and memorize travel routes apparently causes place cells to proliferate in this area of their brains. In other people's memory centers, one cell is almost completely dormant around images of Ariel Sharon and Saddam Hussein, but fires up to nine times per second in the presence of pictures of Mother Teresa.

If you have a "Halle Berry neuron" in your brain, it stands to reason that you may have a whole host of "the company where I work" neurons.

Each separate feature of your company may trigger its own distinct signal of familiarity within your brain, causing your place cells to fire in a rapid fusillade as you move among the landmarks in your workplace. The notion that place cells may fire in multipart harmony when you are in your "home base" helps explain the eerie power of mere exposure.

Brain scans in Peter Kenning's neuroeconomics lab at the University of Münster in Germany show that when investors consider putting money in foreign markets, the amygdala—one of the brain's fear centers—kicks in. These findings suggest that keeping our money close to home generates an automatic feeling of comfort, while investing in unfamiliar stocks is inherently frightening. Those responses originate in the biological bedrock of the reflexive brain. (No wonder that man in Boston got so mad when I told him to diversify.)

Once you understand the mere-exposure effect, it's easy to see why 401(k) investors consistently put too much money into the stock of the company they work for. Every day, employees are bombarded by their company's name and logo and products and services: on ID passes, computer screens, pens and pencils, memo pads, coffee mugs, key chains, baseball caps; in the parking lot, the cafeteria, the reception desk, the mailroom, the bathroom.

Just as a person standing outside in a thunderstorm could not possibly count the number of raindrops that hit him, you could never consciously track all the different ways you are exposed to your company during a typical day. But they all can activate your place cells, saturating you with a sense of familiarity toward the firm you work at.

In a recent survey, 55% of employees at roughly 100 different companies insisted that "owning my employer's stock does not affect my attitudes and feelings." Yet 4 in 10 felt that their employer's stock had "about the same level of risk" as a diversified fund—even though the shares of these companies had, on average, lost nearly twice as much money as the overall market over the previous five years! The best way to explain this blindness is that the constant reinforcement of the mere-exposure effect turns company stock into a "feel-good" investment. As employees become saturated with exposures to images of their company, owning its stock generates a subconscious pleasure that crowds out the question of whether the stock is actually worth owning.

Now it's clear why so many workers overinvest in their own

company's shares. Roughly 5 million U.S. investors have more than 60% of their retirement funds in their employer's stock. Nearly one-tenth of 401(k) investors who can invest in their company's own stock have at least 90% of their retirement savings riding on it. And, while Merrill Lynch's brokers are supposed to prevent clients from putting too much money into the shares of the companies they work for, 27% of the money in Merrill Lynch's 401(k) is invested in—you guessed it—shares of Merrill Lynch stock.

The mere-exposure effect underlies other kinds of investment mind games, too. Professional financial analysts following the Tel Aviv Stock Exchange rated "more familiar" stocks as much less risky than "less familiar" ones—even though the pros should know that the stocks you think you know a lot about can lose at least as much money as the ones you've never heard of. In the stock market, the shares that are most popular and familiar to investors change hands more often, and that, in turn, attracts still more attention to these stocks as they turn up in the day's list of "most active shares." Because of all that extra exposure, the stocks with the highest trading volume have higher returns in the short run—but, in the long term, they tend to underperform by two to five percentage points per year. In the stock market, which is a game of inches, that's a country mile.

Mere exposure helps explain why so many investors are comfortable "buying what they know," investing in the stocks of companies whose products or services they use. Robert Zajonc, the psychologist, lives in Palo Alto, often drives past Google's headquarters in nearby Mountain View, and is a shareholder in the company. "I think I probably bought the stock," he admits a little sheepishly, "because Google is based right here and I use it a lot." In other words, the man who discovered the mere-exposure effect realized only after the fact that his own portfolio may have been shaped by it. It's easy to see why any investor would love Google stock after clicking on the website countless times every day; each additional exposure to Google's services makes its users more familiar with it and more inclined to like it. The warm glow people get from the website casts a kind of halo over the stock price.

Unfortunately, history shows that good companies do not always make good investments. Whether a stock is a bargain depends not only on how great the company's potential is, but also on how many other people already know about that great potential. If you've become familiar with the high quality and growing popularity of a company's products or services, then plenty of other folks probably already are, too. And once hordes of people like a company, it becomes what finance professor David Hirshleifer calls a "celebrity stock." At that point, just like Kathie Lee Gifford or Mr. T, it is almost sure to end up overpriced, overexposed, and overdue for a collapse in popularity. No matter how great a business may be, its stock can't be an enduring moneymaker once an investor stampede drives the price up too high. Thus, over the long run, familiarity breeds failure.

I'm in Charge Here

Juanita Edwards is a bright, well-educated corporate designer who goes on vacation late every summer. On her last day at work, she takes her 401(k) money out of stocks and bonds and moves it all into cash. Two weeks later, when she returns to the office, she moves the whole thing back. "That way I know I'm in charge while I'm on vacation," she says, "and I can relax instead of worrying that I might lose money." Juanita suffers from what's known as the illusion of control. Like a bowler who thinks he can roll a strike by waving toward the middle of the lane after he lets the ball go, Juanita thinks she can control the outcome with her actions. But, while she's away on vacation, the market could just as well go up as down—meaning that Juanita is at least as likely to miss a gain as she is to avoid a loss. She's not really in control; she has simply created the self-delusion that she is.

The illusion of control is that uncanny feeling that we can exert some authority over random chance with our physical actions. There's no better example than watching someone rolling the dice in a board game or at a gambling table in a casino. When people want to roll a high number, they shake the dice for a long time and throw them hard. When they want a low number, they give the dice a quick shake and toss them more softly.

Long ago, the psychologist B. F. Skinner wondered what would happen if his lab equipment automatically swung a food dispenser in front of a hungry pigeon at fixed intervals. Before the first "meal" arrived without warning, the pigeons did what we all do when we have empty stomachs and nothing to eat: They fidgeted. In the final moment before the food appeared, one bird happened to be turning to the left, another was tossing its head up and down, a third was hopping from its right to its left foot. After that first peck at the food, the pigeons began repeating whatever they had been doing when it arrived. Acting as if their physical action had caused the food to come, they kept it up even when Skinner cut off the food. One pigeon hopped from one foot to another in the same way more than 10,000 times before it finally learned that hopping had no effect on whether it got fed. After all, food really had been delivered right after that earlier flap or hop or head toss, so it's easy to see why the birds "believed" that their prior actions had caused the rewards to arrive. But there was no causation, only what Skinner called "accidental correlations."

Humans make the same mistake whenever we confuse correlation with causation: We buy a stock because some website recommends it, then it goes up, and we conclude that the website is a good place to get profitable stock tips. But the stock didn't go up because the website recommended it. Unless we can see an independently audited track record of all the site's stock picks, there's no way to know whether this one call was the result of luck or skill. We might think we have found a way to exert control over the stock market, but it may be sheer coincidence.

To see how sensitive most of us are to the illusion of control, consider the following two bets:

A. I will pick a name at random from the stock tables in the *Wall Street Journal.* You will guess whether this stock will go up or down tomorrow. If you're right, you win $10; if you're wrong, you lose $10.

B. I will pick a name at random from the stock tables in the *Wall Street Journal.* You will guess whether it went up or down yesterday. (You are not allowed to look up the price.) If you're right, you win $10; if you're wrong, you lose $10.

Which bet do you like better?

In experiments at Stanford University, two-thirds of the participants took the first one. Most of these people knew that a stock chosen at random cannot be much more likely to go up tomorrow than it was to rise yesterday. But the first bet seems more comfortable, because it does not make you feel as if the result is out of your hands. An earlier study found that people would wager more money, and even be willing to accept worse odds, if they bet before, rather than after, a pair of dice was thrown.

Psychologist Ellen Langer illustrated the illusion of control in classic experiments she conducted thirty years ago. Testing her theories on workers at two companies, Langer gave them each the chance to buy a lottery ticket for $1. The first group of workers got to pick their own tickets; those in the second group had their tickets chosen for them. Before the winning ticket was drawn, people were asked if they would sell their ticket. Those who had gotten to pick their own demanded a selling price four times higher, on average, than those who had not chosen the tickets themselves. The simple magic of making their own choice made them feel that they could somehow beat the odds—even though everyone had been told that the winning ticket would be drawn randomly from a cardboard box.

Thanks to the "I'm in charge here" feeling, we believe our own decisions are inherently better than the choices other people make for us. In an experiment at a Spanish university, students who won money by tossing dice themselves were much more confident that they would keep on winning than were those who won when someone else threw the dice for them. A study of retirement-plan investors in the U.S. who could either choose their own mutual funds, or permit someone else to decide for them, found that both groups exaggerated their actual returns over the previous year. The people who hadn't picked the funds themselves overstated their gains by 2.4 percentage points—but those who had made their own choices inflated their actual returns by a whopping 8.6 percentage points! The illusion of control explains why one of the worst nightmares for a financial advisor can be a client who picks his own stocks alongside the portfolio the advisor puts together: Even if the client's stock picks and those of the advisor go up by the same amount, the client will instinctively feel that "my" return is higher than "yours."

Study after study has asked people how confident they are of winning both before and after they place a bet. The mere act of putting money down makes people more certain they will win—and it takes only a few seconds for that certainty to kick in. Wagering as little as 25 cents can make bettors up to three times more confident than people who have not yet risked their own money.

Commitment raises our confidence, even when the odds of winning don't budge. It's as if taking the plunge somehow warms up the water; choosing makes us like our choice. No wonder so many investors say about a stock they already own, "I might not buy more at this price, but I wouldn't sell it at this price, either."

The illusion of control is stronger when an activity:

- appears at least partly random,
- offers multiple choices,
- involves competition against other people,
- can be practiced over time,
- requires effort, and
- feels familiar.

More than almost any other activity except sports or gambling, investing meets all these tests. Many investors suffer so badly from the illusion of control that they end up resembling Colonel Klink, the hapless commander on the old TV sitcom *Hogan's Heroes,* who thought he had an iron grip on every detail but was often oblivious to the chaos that constantly swirled around him.

From small-timers with just a few thousand dollars in a 401(k) to some of the biggest money managers in the world, investors everywhere fall prey to the Colonel Klink effect:

▪ "I never, ever, ever write with a red pen," institutional equity trader James Park of Brean Murray & Co. told a reporter in 2003; "red signifies losses." He added, "I also keep my desk completely organized. I feel the more organized I am, the better my stocks trade."

▪ According to research by Duke University anthropologist Mack O'Barr, many managers of pension funds seem to "have a neatness fetish," as if tidying up their offices could magically keep billions of dollars from getting messy.

▪ "There's lots of superstition," an institutional trader in London remarked about his colleagues. "If they had a bad day they'll not wear the same suit or tie again, or not drive to work by a particular route. If I had a new suit on a bad trading day, I wouldn't wear it again, even if it was brand new."

- In 2001, Blaine Rollins, manager of the $31 billion Janus Fund, announced that "I'm never going to take a vacation again until I improve the performance at my fund"—as if the portfolio's 14.9% loss the year before had somehow been caused by his personal holiday.

- Online stock traders often check their portfolios ten or twenty times an hour—as if, by not letting their stocks out of their sight for more than a few minutes, they could somehow keep their prices from dropping.

- Many 401(k) investors load up on their own company's stock in the apparent belief that they can levitate it all by themselves. "When you own a significant number of shares, you want to work hard and make the company prosper so you can prosper," said an employee of Global Crossing, the fiber-optic phone network, in 2001. (His hard work, unfortunately, did not keep Global Crossing from going bust and wiping out his retirement savings.)

Is It Safe?

Neuroeconomists are now exploring the forces that drive the Colonel Klink effect. The caudate area—twin curlicues of tissue, roughly the size and shape of your pinkie finger, deep in the center of the brain—may monitor the relationship between cause and effect. The caudate serves as your brain's coincidence detector. In this part of the reflexive, emotional brain, we match our actions against the outcomes in the world around us—whether they are actually connected or not. "It's not just receiving money that fires up these areas," says neuroscientist Caroline Zink of the National Institutes of Health, "but how you receive it. There seems to be a separate pleasure or arousal from feeling you did something to get it." (Perhaps it's no coincidence that the caudate is one of the brain regions most intensely activated when you learn to trust a stranger—and when you fall passionately in love.)

In one recent experiment, people tried pressing one of several buttons to earn a reward. Sometimes, they would win $1.50 if "they guessed the correct button." At other times, they could win $1.50 regardless of which button they pressed. Whenever a button press was followed by $1.50, the caudate flared into action for up to four seconds, but it was quiet whenever people perceived no connection between the button press and the reward. (See Figure 5.1.) Explains psychologist Mauricio Delgado: "Being in control—or at least believing that we are—makes us that much more invested in our actions and their consequences."

Researchers at the University of Wisconsin have shown that imagining you are in control of a situation—even when it is entirely out of your hands—can reduce the neural activity in areas of your brain that process pain, anxiety, and conflict. By helping to dampen your brain's distress network, the illusion of control can create actual comfort.

That seems to be a basic part of how animals think. In a lab at Columbia University run by Nobel Prize–winning neurobiologist Eric Kandel, mice learn that they might sometimes get a mildly uncomfortable electrical shock to their feet when they are in a recording chamber. But Kandel also gives them the opportunity to learn something else: When they hear a series of beeps, that is a signal that they will *not* receive a shock. It takes about ten repetitions for the mice to learn what Kandel calls "safety conditioning"—much the way a human investor, after either breaking even or making money on all his latest trades, starts assuming that the market environment is safe.

Then Kandel puts the safety-conditioned mice into an unfamiliar open space, and something amazing happens. If you've ever come across a mouse scurrying through your basement or attic, you know they always hug the wall, where their instinct tells them they are safer from predators. But Kandel's safety-conditioned mice barge right out into the middle of open space when they hear those beeps that they've learned to associate with the absence of immediate danger. Boldly going where no mouse has gone before, they wander much farther afield in a bout of what Kandel calls "adventurous exploration."

What makes these mice turn lion-hearted? When safety-conditioned mice hear the series of beeps, neurons in the caudoputamen—a part of the mouse brain similar to our caudate area—go into overdrive, firing at nearly three times their normal intensity. At the same time, neurons in the amygdala—a fear center in the mouse brain, as it is in ours—quiet down. It's as if the perception of safety leads to a feeling of mastery over the environment, numbing the brain's ability to be afraid. No wonder investors take more risk when their own gains fool them into thinking the market has gotten safer. Much like Kandel's mice, you may end up getting safety-conditioned, too, whether you realize it or not. By turning off the fear response in the amygdala, a streak of profitable trades can fill you with a sense of false security. That illusion of safety can lead you straight into investing danger.

I'm on a Roll

In late 1999 and early 2000 Brad Russell, an air traffic controller in New Hampshire, invested in a hot Internet stock called CMGI Inc. At first he nibbled, buying just a few shares. The stock shot straight up; he bought more; it went up more. Now Russell bought over and over and over again, until he had jumped into the stock with at least ten separate purchases, at prices up to $150 a share. At the peak, he had 40% of all the money outside his retirement fund in CMGI stock. Then the Internet bubble burst, and CMGI came crashing down like a boulder knocked off a cliff. Russell finally sold when the stock hit $1.50, a 99% loss at its worst.

Looking back on his financial kamikaze flight, the only thing that amazes Russell more than how much money he lost by the time he hit bottom is how exhilarated he was at the top. "After the stock went higher," he recalls, "I thought I knew what I was doing, plain and simple—and the pull of the craze was just too much for me to resist."

As Russell's story shows, a hot streak can pump up your investing confidence like helium rushing into a balloon, leading you to take on even more risk until the whole thing bursts. What drives that sensation of "I'm on a roll"?

First of all, a streak of gains makes you feel that you are "playing with the house money." That's the term gamblers use when they mentally divide their bucks into different buckets: the cash they started out with (which remains their "own money") and any winnings they've made on top of that ("the house money").

Let's say you put $1,000 into a stock that triples; now that it is priced at $3,000, you've got $2,000 of "house money." So long as any of that $2,000 gain is left, you may shrug off any losses as a reduction of the house money—rather than a depletion of your own. Somehow, losing the house money hurts less than losing your "own"—even though, strictly speaking, all the dollars are the same. As Russell found, this "house-money effect" can egg you on into taking an ever-escalating series of risks until you get wiped out.

Secondly, a streak makes the future feel more predictable. Like many kinds of repeating patterns, a financial hot streak can make your brain automatically expect more of the same. In Gregory Berns's neuroeconomics lab at Emory University, people tried to guess which of four squares would next turn blue. Sometimes the sequence of color changes was random; sometimes it ran in a pattern that was fixed, but too complex for people to be consciously aware that it was predictable. When the blue square appeared randomly, the prefrontal cortex and parietal cortex were active; these centers of the analytical, reflective brain launched a conscious struggle to figure out what was going on. When people saw the fixed sequence, however, the caudate area in their brains kicked in; this region in the reflexive, emotional brain recognized a repeating pattern without triggering any conscious awareness of it. Explains Berns: "Unambiguous sequences [like a streak of financial gains] can be learned without attention." Thus a hot streak is easier for your investing brain to process than a random, complex jumble of gains and losses. Once you're on a roll, structures like the caudate put your expectations on autopilot: *More of the same is on the way!*

Next, a hot streak makes you feel as if luck, not just random chance, is on your side. Years ago, psychologists tossed a coin thirty times and asked college students to guess whether it would come up heads or tails each time. Some were told they got most of their early guesses right; others were informed that most of their first calls were wrong. Those with the early hot streaks ended up believing that, if they got the chance to call another 100 coin flips, they could get 54 of them right. What's more, an amazing 50% of the guessers with early hot streaks thought they could improve their score with "practice." The thrill of being on a roll had made them forget the obvious: It's impossible to get better at guessing whether a coin will come up or heads or tails.

The upshot: An early run of success makes people feel they suddenly have power over a purely random process. Instead of attributing the results to an abstract force like "chance," they now believe in "luck," a personal force that watches over them (at least temporarily) like a guardian angel. So long as luck seems to be lingering in the air, people feel compelled to make the most of it—and that can lead investors to take reckless amounts of risk.

It's not just small investors like Brad Russell who end up spiraling out of control once they think they are "on a roll." Professional stock analysts who forecast a company's earnings accurately just four times in a row then go on to make progressively riskier predictions that end up 10% worse than average. Research on more than four thousand corporate takeovers in Britain shows that when a firm's first acquisition makes money, future deals are more likely to destroy value. In the U.S., the typical company that earns a hot return on its first acquisition is much more likely to take over at least one more firm within the next five years—even though, on average, that will knock 2% off its stock price.

Even Jack Welch, the former leader of General Electric who is widely regarded as one of the best CEOs ever, admits that he got carried away on his own hot streak of deal-making. When he bought the Wall Street brokerage Kidder Peabody, Welch later admitted, "I didn't know diddly about it. I was on a roll." And that sensation, quipped the short, balding Welch, made him feel as if "I was 6-foot-4 with hair." GE lost more than $1 billion on its investment in Kidder Peabody after an arcane trading scheme went bad.

Nobel Prize–winning economist Vernon Smith has shown that business executives and professional traders who earn big gains as stocks become overvalued—and who then get wiped out when prices come crashing down—will come right back and do it all over again. Smith's research shows that in this case the old adage "Once burned, twice shy" is wrong. Because being on a roll is so thrilling, it generally takes at least two scaldings before even so-called experts can begin to learn not to touch a "market bubble."

People may also be in thrall to the "prediction addiction." In a lab at the California Institute of Technology, researchers recently tested what happens when inexperienced gamblers get hot results early on. In the game, players chose between two piles of cards: One deck provided big wins mixed with occasional small losses, while the other offered smaller gains that were sometimes coupled with larger losses. Let's call the first deck the hot one and the second the cold one. Halfway through, however, the experiments secretly reversed the payoffs, so the deck that had originally been hot now was cold, and vice versa. Strikingly, the players that had tasted a hot streak could not recognize that the rules of the game had changed. "They don't get it," says Cal-Tech neuroscientist John Allman. "The people who've had the biggest gain have the hardest time reversing their choice. It's as if their ability to correct errors has been anesthetized, and they have become addicted to the favorable outcome."

People who have suffered injuries to parts of the prefrontal cortex are particularly bad at learning when a hot streak has gone cold. That suggests that when the reflective brain is impaired, then the reflexive, emotional regions take over. Triggered by an earlier hot streak, reflexive structures like the caudate, nucleus accumbens, and the hippocampus stay stuck in high gear, incapable of quickly recognizing that the payoff pattern has changed.

Once you understand that monetary gains have this narcotic power, it no longer seems surprising that people can get so carried away financially. Gamblers know this, of course. Colonel Tom Parker, the legendary manager of Elvis Presley, was such a slot-machine junkie that late in his life, after his shoulder was crushed in an elevator accident, he ordered his personal assistant to come with him to casinos and pull the handles on the one-armed bandits for him. More recent, some casino players have chained themselves to slot machines, while others have even strapped on adult diapers so they would not lose their seat at a "lucky" machine by having to get up and walk to the bathroom.

What does a financial hot streak look like inside your brain? When you're on a roll, winning money time after time, three regions of the reflexive brain light up like a Christmas tree: the thalamus, globus pallidus, and subgenual cingulate. The thalamus, a gizzard-shaped lump toward the center of your skull, acts as a switching station, relaying sensory impressions of the outside world to the rest of your brain. The globus pallidus, a small, pale blob near the thalamus, helps track reward and punishment. And the subgenual cingulate, on the inner edge of your prefrontal cortex where it curves back behind your forehead, is the most interesting of all.

The subgenual cingulate helps regulate sleep, and it tends to be smaller and less active in severely depressed people. On the other hand, it appears to kick into overdrive when people suffer the manic phase of bipolar disorder. Typically, a person in a manic state is impulsive, turbocharged with euphoria, often unable to sleep, and endowed with a grandiose ability to "perceive" the underlying significance of everything around him. People with severe mania can become so rash that they hurt themselves, or so brash that they are impossible for others to live with. Mania appears to result when too much dopamine squirts into the circuits of the subgenual cingulate.

Finding out that this mental illness can be traced to the same area of the brain that is activated by a string of financial gains isn't just an intriguing way to explain the behavior of market lunatics like James J. Cramer of CNBC. It's also alarming. For it suggests that when investors think they're "on a roll," when they're filled with that sensation that they can see into the future and nothing can stop them, they can end up making the same kinds of mistakes that chronically manic patients make. With your subgenual cingulate inflamed by a financial hot streak, it's hard not to turn euphoric, restless, and carefree about risk. That manic feeling can make it almost impossible for you to let go of a stock that strikes you as hot. And like most extreme mood swings, a burst of mania is bound to end badly. Brad Russell, who got carried away by his early gains on CMGI, has never quite gotten over the aftermath. "Man," he says today, "those four letters [CMGI] still make me cringe!" It's only natural that when a bull market turns into a stampede, we call it a "mania."

I Knew It All Along

In the waning days of the Soviet empire, when history textbooks were constantly being rewritten to cover one disgrace or another, dissidents in Eastern Europe used to joke that the past was as difficult to predict as the future. That's true in the stock market, too. In 2002, more than 800 investors responded to a survey about the bull market of 1999–2000. Looking back, nearly half of these investors now said that the rise in technology and telecommunications stocks had "definitely" been a "bubble." Almost a third more believed that it had "probably" been a "bubble." So had they all sat prudently on the sidelines throughout that market mania? Not at all: During the very time when they now insisted that stocks had been overvalued, every single one of these people had eagerly invested in one of the most insanely priced telecommunications stocks in the U.S. market.

Despite the common cliché, hindsight is not 20/20. Without corrective lenses, hindsight is close to legally blind. Once we learn what did happen, we look back and believe that we knew it was going to happen all along—even if we were utterly in the dark at the time. That's what psychologists call "hindsight bias."

This quirk of human behavior was first clearly diagnosed in 1972, when Richard M. Nixon was about to travel to Beijing, the first visit by a U.S. president since mainland China became communist. No one knew what might happen: Would Mao Zedong meet with Nixon or snub him? Might Taiwan, Japan, or the Soviet Union raise a ruckus? Could the visit worsen the Vietnam War?

Few expected what actually happened: The visit went so well that the U.S. and China signed a joint communiqué in which they pledged to work toward normalizing diplomatic relations. Right before Nixon's trip, dozens of Israeli university students were asked to predict the probability that his visit would be a success. Then, at two intervals after his trip, they were asked to recall their earlier forecasts of what would happen. Less than two weeks after Nixon's visit, 71% remembered predicting a higher probability of success than they actually had. Four months after the visit, 81% claimed to have been more certain it would succeed than they really were at the time.

Psychologist Baruch Fischhoff, now at Carnegie Mellon University, cowrote the original study. "When you hear something," he explains, "you immediately incorporate it into what you already know. That seems like a more efficient and sensible thing to do than trying to put new information into some kind of intellectual limbo, waiting until it proves itself before you can use it. But it's not particularly helpful if you want to go back and figure out what was the extent of your knowledge and what is your ability to predict things."

"Hindsight bias makes surprises vanish," says psychologist Daniel Kahneman. "People distort and misremember what they formerly believed. Our sense of how uncertain the world really is never fully develops, because after something happens, we greatly increase our judgments of how likely it was to happen."

Hindsight bias is another cruel trick your inner con man plays on you. By making you believe that the past was more predictable than it really was, hindsight bias fools you into thinking that the future is more predictable than it can ever be. That keeps you from feeling like an idiot as you look back—but it can make you act like an idiot as you go forward.

How does hindsight bias play out when you invest?

Like this: In the fall of 2001, after the terrorist attacks of September 11, you tell yourself, "Nothing will ever be the same again. The U.S. isn't safe anymore. Who knows what they'll do next? Even if stocks are cheap, nobody will have the guts to invest." Then the market goes on to gain 15% by the end of 2003, and what do you say? "I *knew* stocks were cheap after September 11th!" And suddenly it seems you're more omniscient than Alan Greenspan.

Or like this: Google sells its stock to the public for the first time in August 2004. You say to yourself, "Hmmm, great website . . . maybe I should try to buy the stock? But what about all the money I lost on all those other Internet stocks over the past few years? . . . Naah, I better pass." Then the stock goes from its initial price of $85 to $460 by the end of 2006, and what do you say? "I *knew* I should have bought Google!" This "I-told-you-so" rant from your inner con man makes it hard to remember that he never told you any such thing. And conning yourself about Google may well make you more eager to take the plunge the next time you have a chance to get in on the ground floor of a risky high-tech start-up. Of course, "the next Google" may turn out to be the next Enron instead.

Pundits have the same problem. "I knew that it was going to crash, I really did," insisted investing guru George Gilder as he looked back, in 2002, at the bursting of the technology-stock bubble in 2000. But he had to admit that in his newsletter he had never once warned of the bloodbath to come.

"We all knew it was a mania," intoned *Money* magazine in June 2000. "From August 1999 to March 2000 . . . tech stocks like Ariba and VerticalNet surged by more than 800%, despite the fact that neither of them had ever made a profit. You never know when this kind of frenzy will end, but it wasn't hard to see that we were headed for a reckoning." And yet, in its December 1999 cover story, the magazine had urged investors to "gain from the Web's blazing growth" and had touted Ariba as one of its top "stocks to buy now."

Hindsight bias also skews the way we view money managers. Imagine that 2006 has just ended and your broker calls with exciting news: He can get you into the Numbers Growth Fund, which has just beaten the market (as measured by the S&P 500 index) for the tenth year in a row. Your broker declares that the fund manager, Randy Numbers, is a genius—and it's hard to imagine otherwise, since in the typical year well over half of all funds fail to beat the market. To beat it every year for a decade sounds like a miracle.

How should you decide whether Randy Numbers really is a genius? Since a fund manager must either beat the market or be beaten by it, the odds of outperforming the average in any given year are 50/50. (That's before expenses and taxes, of course.) The easiest way to picture how this unfolds over time is by flipping a coin. The odds of getting ten heads in a row are one in 1,024—making Randy Numbers sound like more of a genius than ever. But this kind of hindsight is so vivid that it can blind you in two ways. It keeps you from seeing that past returns do not foreshadow the future, and that picking a winning mutual fund is vastly harder in real time than it seems when you look backward.

To see why, think back to the end of 1996, when Numbers's winning streak began. There were 1,325 U.S. stock funds at that point. Imagine that one manager from each fund flipped a coin each year. The odds that at least one of those 1,325 managers would get heads every single year, by random chance alone, are 72.6%. So is Randy Numbers's track record the result of genius—or dumb luck? If he is just lucky, his future returns will fade as soon as his luck goes away. Even if he is skillful, that still doesn't mean that you could have spotted him without the trickery of hindsight. Back in 1996, no one had ever heard of him; he was just one in a crowd of 1,325. Trying to pick him out as the future winner in that throng would have been as hard as trying to spot the eventual champion of a marathon by peering down from a helicopter into the horde of bodies milling around behind the starting line before the gun goes off.

For another angle on hindsight, visit http://viscog.beckman.uiuc.edu/grafs/demos/15.html. This online movie features two teams of college students—some in white shirts, some in black—passing a basketball. Pick either team and carefully count how many times the players pass the ball; to start the movie, click on the green arrow. Did you notice anything else? If not, play the video again. Only moments after you were bewildered by this visual puzzle, the solution will now seem obvious to you. All that will bewilder you now is how anybody could have found it hard to figure out in the first place. (If this experiment failed on you, try it on a friend.) That's the power of hindsight bias—yet another way in which your inner con man works at trying to make you feel smarter than you really are. By preventing you from learning the truth about the investing past, hindsight bias keeps you from getting a firm grip on the financial future.

I Know, I Know

Here are three quick trivia questions designed not just to test your knowledge, but also to test how much you know about your knowledge.

1. Driving due south from Detroit, what is the first country you will come to when you leave the U.S.?
   a. Cuba
   b. Canada
   c. Mexico
   d. Guatemala

Now think for a moment about how certain you are that your answer is right. Are you 100% sure? 95%? 90%? Or less?

2. Which country derives more than 75% of its energy from nuclear power?
   a. United States
   b. France
   c. Japan
   d. none of the above

Again, take a moment to think about how certain you are that your answer is right. Are you 100% sure? 95%? 90%? Or less?

3. Approximately how many microbes live in the typical person's gastrointestinal tract?
   a. 100 trillion
   b. 100 billion
   c. 100 million
   d. 100,000

Once more, think about how certain you are that your answer is right. Are you 100% sure? 95%? 90%? Or less?

Many people who are not from the Midwestern U.S. say they are at least 90% certain that the answer to the first question is Mexico; they are often 100% positive. In fact, the correct answer is Canada; the city of Windsor, Ontario, lies directly south across the Detroit River from Motown. (If you don't believe it, visit http://maps.google.com and enter "Detroit" in the search window.)

With the second question, most folks tend to be about 70% sure that the correct answer is "none of the above." It's actually France, which relies more heavily on nuclear energy than any other nation.

As to the third question, most people seem to be around 50% certain that they've got about 100 million microorganisms living inside their guts. The simple but disgusting truth is that your digestive system harbors a teeming universe of roughly 100 trillion microbes.

Years ago, college students in Oregon were asked: Was Adonis the god of love or vegetation? Do most of the world's cacao beans come from Africa or South America? A quarter of the students were at least 98% sure that Adonis was the god of love; over a third were at least 98% certain that most chocolate originates in South America. (Try it yourself: How sure of your answers are you?) Even when they were coached about how inaccurate most people's judgments are, many students were so sure of their answers that they were willing to bet $1 that they were right. But only 31% correctly identified Adonis as the god of vegetation; just 4.8% rightly pinpointed Africa as the leading source of cacao.

The same ignorance of our own ignorance haunts our financial judgments. Among American workers who say they are "very confident" that they will have enough money to live comfortably in retirement, 22% are currently saving nothing for that goal, and 39% have saved less than $50,000. Another 37% have never even estimated how much money they will need to retire comfortably.

It's bad enough to be "very confident" of a cozy retirement when you are not saving for it now. It's even worse when you have no idea how much money you will need to retire comfortably—but assume you will have enough anyway. This kind of overconfidence can lead to drastic undersaving and a threadbare retirement riven with regret.

That's why the old proverb, "It ain't what we don't know that gets us into trouble, it's what we know that ain't so," ain't exactly true. What really gets us into trouble is not even knowing what we don't know.

That's as true for professional money managers as it is for supposed amateurs like individual investors. In some ways, the "experts" may be even more ignorant of their own ignorance. There's powerful evidence that the more you know, the more you will think you know even more than you actually do. Furthermore, people are often wrong even when they are virtually certain that they're right:

- A study in Stockholm found in 1972 that experts (like bankers, stock market analysts, and financial researchers) were no better, and in some cases slightly worse, at predicting stock prices than university students were.

- At the University of Michigan, undergraduates turned out to be more accurate—or, rather, less inaccurate—at forecasting future stock prices and corporate earnings than graduate finance students (including some who had worked as financial analysts).

- Two groups of people in Sweden tried to pick which of two stocks would be a better investment—and to estimate their own odds of getting it right. The first group was made up of fund managers, analysts, and brokers with an average of twelve years of investing experience. The second group was a bunch of college kids majoring in psychology. On average, the amateurs put their odds of success at 59%—and picked the correct stock 52% of the time. The professionals, puffed up with a sense of their own expertise, thought they would pick the better stock 67% of the time—and got it right on a paltry 40% of their tries!

- Because they so often think they know what's coming, overconfident investors are forever buying or selling something. And yet they know less than they think they do. The portfolios of those who trade the most underperform the holdings of those who trade the least by an amazing 7.1 percentage points per year.

- A German study looked at forecasts of future stock returns by investing professionals. These experts were supposed to space their estimates widely enough so that they should be right 90% of the time. Fully 62% of them failed to be right even half the time. The more experience they had, the more overconfident they were.

- In 1993, the mutual funds editor at *Forbes* magazine was contributing 5% of his salary, just half the allowable maximum, to his 401(k). When a friend asked why he didn't put more into his 401(k), the editor shot back,

"Because I can do better with my money than they can, that's why." Looking back more than a decade later, the former editor calculated how much this decision had cost him. I know the answer, because I was that editor. The cost of my overconfidence is more than a quarter of a million dollars—so far.

Our confidence has another quirk. The harder a task is and the closer the odds of success come to 50/50, the more inclined we are to be overconfident about our chances. One experiment found that only 53% of people could reliably tell whether a drawing had been done by a European or Asian child, but the average person was 68% sure that he could. Likewise, just under 50% of college students can accurately say which U.S. states have higher high school graduation rates, but on average they are 66% certain that they can do it.

That's why frequent traders tend to be more overconfident than buy-and-hold investors. If you hang on to investments for only a few hours or days or weeks, your results may be fabulous for a while, but in the long run a frequent trader has to be extremely lucky to make money even 50% of the time. Along the way, however, your short streaks of luck will give you an unwarranted sense of confidence.

Experiments with monkeys have shown that when the odds of earning a reward are around 50/50, neurons deep in their brains emit a surge of dopamine that steadily escalates for nearly two seconds. Rewards that are highly certain trigger a much shorter, flatter response. The extra excitement that an "even gamble" generates in the dopamine system may be nature's way of getting us off the fence. Otherwise, we'd never be able to choose between actions that have roughly equal odds of success. The extra rush of dopamine helps tip the balance. But it also pushes our confidence far in front of the evidence.

The human refusal to say "I don't know" has huge consequences. It explains everything from the design of O-rings that cracked on the space shuttle *Challenger* to the delusion of Time Warner CEO Gerald Levin that merging with AOL was a brilliant idea—and the Pentagon fantasy that Iraqis would strew flowers at the feet of American troops. It leads people to stop asking questions because they assume they already know the answers—or are afraid of what will happen if they admit they don't.

Some psychologists contend that overconfidence is a minor problem that can easily be cured by presenting data differently. In truth, the road to investing hell is paved with overconfidence. Back in 1993, county treasurer Robert L. Citron had borrowed roughly $13 billion to leverage the $7 billion investment portfolio of Orange County, California, with complex securities that could earn high returns if interest rates fell or stayed flat. When a banker asked what would happen to the county's portfolio if interest rates rose instead, Citron retorted that rates would not go up. The banker asked how he could be so sure. "I am one of the largest investors in America," snapped Citron. "I know these things." Nine months later, interest rates shot up—and Citron's portfolio lost $2 billion, driving Orange County into the biggest municipal bankruptcy in American history.

In 2005 polls taken among institutional investors who oversee $700 billion found that 56% of them were sure that they could invest prudently in hedge funds. Yet, in the next breath, 67% conceded that they lacked the necessary tools to analyze and manage the extra risks of these arcane holdings. In September 2006, the Amaranth Advisors hedge fund lost 50% of its value in one week when a risky natural-gas trading strategy blew up—foisting more than $5 billion in losses onto supposedly sophisticated investors like units of Morgan Stanley and Goldman Sachs, along with the $7.7 billion San Diego County Employees' Retirement Association. Only months earlier, the San Diego pension fund had bragged about its ability to bypass outside expertise to invest in hedge funds directly.

Probably the main reason we insist we know more than we do is that admitting our ignorance undermines our self-esteem. A little knowledge really is a dangerous thing: Learning even a tiny bit about something fills us with a sense of power, and that feeling would be intensely threatened if we admitted how much more we don't know. That's why it takes so much confidence to admit that you aren't confident; three of the hardest words in the world to say are "I don't know." Asked whether light takes less than one minute to travel to the earth from the sun, 13% of college students in Scotland said they were "almost certain" that it does, and 21% were "completely certain." Only 17% would admit that they didn't know the answer (which is 8 minutes, 20 seconds, on average).

Warren Buffett was supremely right when he wrote: "What counts for most people in investing is not how much they know, but rather how realistically they define what they don't know. An investor needs to do very few things right as long as he or she avoids big mistakes."

Right-Sizing Your Confidence

Many investors act as if they were imitating the Harvard philosopher Willard Van Orman Quine. Early in his career, Quine removed the question mark key from his typewriter. When someone asked, late in Quine's long life, how he had managed to write for seventy years without ever typing a question mark, he responded, "Well, you see, I deal in certainties." But there are very few certainties in the financial markets, and your investing brain is designed to exaggerate your abilities, favor the familiar, and imagine far more mastery over the past and future than you may ever have. That's why, instead of taking the question mark off our keyboards, we would probably be better off removing most of the other keys and replacing *them* with question marks.

But you don't want to downsize your confidence to nothing. An investor who has no confidence will never invest at all, since investing requires taking a stand on at least some of the uncertainties that the future always holds. So your goal is to be as sure as possible that you don't think you know more than you really do. How much you know is less important than how clearly you understand where the borders of your ignorance begin. It's not even a problem to know next to nothing, as long as you *know* you know next to nothing. Here's how you can right-size your confidence:

➢ **"I DON'T KNOW, AND I DON'T CARE."** There's no shame in saying "I don't know." To succeed as an investor, you don't have to outsmart Wall Street at its own guessing game. You don't need to know what eBay's earnings will be to the nearest penny, whether energy and gold stocks will keep going up, what next month's unemployment report will say, or which way interest rates or inflation are headed. A total stock market index fund (TSM) is a basket that holds a piece of virtually every stock worth owning, at rock-bottom cost. If you buy and hold two TSM funds—one for U.S. and one for international stocks—then you no longer need to have an opinion about which stocks or industries or sectors will do well or poorly. This way, you can win the prediction game by simply refusing to play it. A

TSM fund enables you to say seven magic words: "I don't know, and I don't care." Will the next great growth sector turn out to be Internet companies—or coal mines? I don't know, and I don't care: My index fund owns both. Will small stocks do better than big ones? I don't know, and I don't care: My index fund owns both. Who will dominate the future of computing: Microsoft, Google, or some upstart as yet unknown? I don't know, and I don't care: My index fund will own them all. What will be the world's best stock market over the decades to come? I don't know, and I don't care: My index funds own them all.

➢ **CREATE A "TOO HARD" PILE.** Many people think Warren Buffett became the world's most successful investor by knowing more than anyone else. But Buffett himself believes that the key to his success has been knowing what he doesn't know. "We have a ton of doubt on all kinds of things," says Buffett, "and we just forget about those." Buffett adds, "The essential principle is what [Buffett's business partner] Charlie [Munger] calls the 'too hard' pile, where we put things we don't know how to evaluate. We get to that pile with about 99% of the ideas that come our way." So your investing workspace should have a small "In" box with a handful of ideas to consider, a little "Out" box holding another few ideas you've already approved or rejected—and an enormous "Too Hard" pile for everything else. (If you're really serious about investing, tape the words "TOO HARD" to the side of a wastebasket and toss your pile in there. If you do everything on your computer, create a "Too Hard" folder on your desktop or a special area in your recycle bin.) Nothing should go into the "In" box until you've first asked whether it belongs in the "Too Hard" pile. (If you're not sure, then that's a sure sign that it's too hard!)

➢ **MEASURE TWICE, CUT ONCE.** If you've ever watched a master carpenter at work, you've noticed how little scrap or waste wood he creates—because he measures very carefully before he cuts. So, after you take a first cut at your estimates of what a stock is worth, crop them again. Behavioral finance writer Gary Belsky and psychologist Thomas Gilovich suggest using an automatic "overconfidence discount" of 25%. Apply it to both the high and low end of the range, making the upside smaller and the downside bigger. If, for instance, you think a stock is worth between $40 and $60, then crop each number by 25%, yielding a new range of value between $30 and $45. This conservative cropping will help keep you from being carried away by your own overconfidence.

➢ **WRITE IT DOWN, RIGHT AWAY.** Psychologist Baruch Fischhoff, who has studied overconfidence for more than thirty years, suggests using an investing diary. "Keep a record of what was on your mind when you made predictions," he says, "and try to make those predictions as explicit as possible." Think in probabilities, and include a price range and date; for example, "I think there's a 70% chance that this stock will be at $20 to $24 within a year from now." (Don't forget to apply the 25% overconfidence discount to your price range.) Finally, spell out your investing theory by filling in the blank in the following sentence: "I think this investment will go up because ______________________________."

It's important to document your reasons for investing before you buy. Memory researcher Elizabeth Loftus has shown that your recall of how you felt earlier can be easily "contaminated" by what happens later. If you make your diary entry after the fact, your memory of your original motivation may be affected by any later changes in the price ("I bought it at $14 because I knew it would instantly go to $15").

Tuck your diary entries away for a year. Then go back and see how accurate your predictions were, whether you tended to underestimate or overestimate, and how good your theories were. You're not looking to see whether or not your investments went up; you already know that. Instead, you're looking to see whether they went up *for the reasons you expected.* That will help you learn whether you were right, or just lucky. And that, in turn, will help keep your inner con man in check. Furthermore, with the benefit of true hindsight, you now will be able to ask what kinds of additional information would have raised your probability of being right or resulted in a more accurate price range.

➢ **LEARN WHAT WORKS BY TRACKING WHAT DOESN'T.** We all learn best when we get prompt and unambiguous feedback on our actions. That's why teachers mark homework immediately, with specific grades and suggestions for improvement. (Imagine what school would have been like if you never knew when you would get your homework back and if, when you did, it was marked "Not bad, so far.")

Unfortunately, the financial markets are full of messy feedback. Let's say you buy a stock at $10 and it immediately goes to $11; you pat yourself on the back for being smart. But then, before your hand can even make it back to your hip, the stock drops to $9, and now you feel stupid. Whether it's a good or bad decision depends partly on when you evaluate it; in the short run, the evidence is a constant flux of up and down, right and wrong, smart and stupid. To get a reliable picture of your stock-picking ability, you need to measure it over a long time and many choices. You also need to look at the roads not taken. Psychologist Robin Hogarth of Pompeu Fabra University in Barcelona suggests that you monitor the performance of three groups of stocks: those you own, those you have already sold, and those you thought about buying but ultimately didn't. Monitor all three baskets with an online portfolio tracker like http://money.cnn.com/services/portfolio/ or http://portfolio.morningstar.com.

This exercise is particularly useful for professional investors, financial planners, or any frequent trader. It goes beyond telling you whether the stocks you buy go up. It also shows whether the stocks you sell keep going up after you no longer own them, and whether the stocks you almost bought performed better than the ones you did buy—information that you *must* have in order to know for sure how good you are at buying and selling.

➢ **HANDCUFF YOUR INNER CON MAN.** You can help avoid lying to yourself about your own abilities by asking three questions:

1. How much better than average do I think I am?
2. What rate of performance do I think I can achieve?
3. How well have other people performed on average over time?

Let's say, for example, that you figure you're 25% better than average at picking stocks, and you think you can earn 15% a year on your portfolio. That sounds realistic enough—until you consider the third question. The long-term average annual return on the Standard & Poor's 500 index of blue-chip stocks is 10.4%. If, however, you adjust that number for the cash that people added to and subtracted from their portfolios, the average return drops to just 8.6% annually since 1926. Factor in taxes, trading costs, and inflation, and the annual return of the typical investor drops below 4%. If you really are 25% better than average, you shouldn't expect to earn much more than 5% annually after all your costs. You still might be able to earn 15% a year—if you are at least *three times better* than average. Only by asking all three questions can you tell just how crazy your inner con man is.

➢ **EMBRACE THE MISTAKE.** Christopher Davis, who oversees more than $60 billion in mutual funds at Davis Selected Advisors in New York, adorns the wall outside his office with stock certificates. The stocks that usually earn this honor are not the firm's best investments, but its worst. Nicknamed "The Mistake Wall," the area displays stock certificates from sixteen companies—so far. "One just got hung up, and one's at the framer's," says Davis wryly. "When we started it I didn't expect it to become quite the mural it's turned into." One company, Waste Management, is represented on the Mistake Wall twice, because Davis not only bought it for the wrong reasons but sold it under the wrong circumstances. To Davis, a "mistake" means being significantly wrong about the value of a business, based on faulty information or flawed analysis that could have been prevented.

When a stock turns out to be a mistake, then Davis frames and hangs the certificate, along with a brief summary of what can be learned from what went wrong. Many money managers talk about "ROI" (return on investment) or "ROE" (return on equity). Christopher Davis also talks about "ROM" (return on mistakes). Having the Waste Management certificate on the Mistake Wall kept Davis from selling Tyco International at "the point of maximum pessimism." (Tyco went on to triple in price.) Lucent's certificate, conveying the lesson "Don't be satisfied with answers you don't understand," kept the Davis funds from buying Enron before it imploded.

There's no need to frame a stock certificate. Simply write the name of your investing mistake and the lesson you learned from it on a Post-It note. By embracing your mistakes instead of burying them, you can transform them from liabilities into assets. Studying your mistakes and keeping them in plain sight will help you avoid repeating them.

➢ **DON'T JUST "BUY WHAT YOU KNOW."** Peter Lynch, the legendary manager of the Fidelity Magellan Fund, famously counseled investors to "buy what you know." You can put "the power of common knowledge" to work for you, wrote Lynch, by investing in companies whose goods and services you consume yourself. Lynch, for example, bought shares in Taco Bell because he liked the burritos, Volvo because he drove one, Dunkin' Donuts because he enjoyed the coffee, and Hanes because his wife liked its L'eggs pantyhose. What investors often forget, however, is that Lynch didn't invest in Dunkin' Donuts just because he likes jelly-squirting globs of fried dough. He also spent hours analyzing its financial statements and studying everything imaginable about the company and its business. Buying a stock solely because you like a company's products or services is like deciding to marry someone just because you like the way he or she dresses. It's fine to become interested in a company because you're familiar with what it sells—but you should never buy a stock without first consulting your investing checklist (see Appendix 2).

➢ **DON'T GET STUCK ON YOUR OWN COMPANY'S STOCK.** No matter how familiar it feels or how warm a glow you get from owning it, your company's stock is one of the riskiest investments you can possibly make. On September 30, 2004, Merck & Co. announced that it was withdrawing Vioxx, its popular arthritis drug, after research showed that Vioxx could increase the risk of heart disease. Merck's stock had its own coronary on the news, crashing 27% in a matter of moments. Because Merck's employees had kept a quarter of their 401(k) in the company's stock, over 5% of their retirement savings were wiped out in a single day.

Precisely two weeks later, New York State Attorney General Eliot Spitzer sued Marsh & McLennan, the giant brokerage, on allegations of insurance fraud. Employees held $1.2 billion of the company's stock in their retirement plan. After Spitzer's announcement, the price of Marsh's stock plunged by 48% in four days, wiping out more than a half-billion dollars in retirement savings. Less than a month later, Marsh laid off 3,000 employees; four months after that, it cut another 2,500 jobs. All these workers were left with no job—and a retirement fund that had been hacked in half.

Putting all your eggs in one basket is so risky, in fact, that finance professor Lisa Meulbroek of Claremont McKenna College estimates that a 50% allocation to company stock over a ten-year period is worth less than 60 cents on the dollar after adjusting for the extra risk it injects into your portfolio. Even if you hold "only" 25% of your assets in company stock, its risk-adjusted value is just 74 cents on the dollar.

Try thinking about it this way: Will you drop dead today? Probably not—but you still should have life insurance. Will your house burn down tomorrow? Probably not—but it's a good idea to have homeowner's insurance. Is your company the next Enron? Probably not—but it's definitely a good idea to insure your portfolio just in case. The best insurance policy is to keep no more than 10% of your money in your own company's stock (or options) and to spread the rest of your bets as broadly as possible.

➢ **DIVERSIFICATION IS THE BEST DEFENSE.** Today it seems obvious that you could have made a fortune if you'd put all your money into, say, computer stocks in the early 1980s. But hindsight bias blinds you to the truth. Back at the dawn of the personal computer age, you couldn't have bought Microsoft, which didn't go public until 1986. The superstars of technology then were Burroughs, Commodore International, Computervision, Cray Research, Digital Equipment, Prime Computer, Tandy, and Wang Laboratories. It's true that you could have bought Apple Computer after its first stock offering in December 1980—but you would probably have wanted a stake in Commodore instead, which had turned a $10,000 investment at the end of 1974 into a stupendous $1.7 million by then.

Virtually all the early stars of the computer business winked out one by one. Their innovative products lost their edge, their best talent defected, and they collapsed into bankruptcy or oblivion. Investors in nearly all of those stocks were wiped out. Looking at Microsoft and Apple, it seems clear in hindsight that anyone could have picked them as winners. But in real time, it was never clear which companies would win the race until the race was well under way.

That's why it's so important to diversify. By owning the widest possible range of stocks and bonds, at home and abroad, you can essentially eliminate the chance that a few duds like these will ruin your financial future.

➢ **PRETEND YOU'RE FOUR YEARS OLD.** As every parent knows, four-year-olds have a habit of asking "Why?" over and over again until they've exhausted Mom or Dad's store of knowledge. Asking "Why?" four or five times is a good way to test the limits of your own (or someone else's) knowledge. If a financial planner says you should put a pile of money into a mutual fund specializing in Chinese stocks because "China is the place to be," ask "Why?" If he answers, "Because it's going to be the world's fastest growing economy," ask "Why?" again. If he replies, "Because China will continue to have low manufacturing costs," ask "Why?" again. Chances are, you'll never get to a fifth "Why?" People who don't really know what they're talking about can rarely answer "Why?" more than twice. If you can't either, that's a signal that you don't yet have enough knowledge to make an informed decision. Smart investors know that it's often a good idea to act like a four-year-old.

CHAPTER SIX

Risk

> If you burn your mouth with hot milk, next time you blow on your yogurt.
>
> —*Turkish proverb*

In the Eye of the Beholder

IF EVER THERE WAS A LIKELY CANDIDATE TO LOAD UP ON risky investments, it would have to be Bobbi Bensman. After all, how much risk you take is supposed to depend on how much risk you can stomach—and Bensman has a tolerance for the kind of danger that would turn most people as pale as a fish's belly. Her favorite way to reduce stress is to inch her way a few hundred feet up the side of a cliff. Probably the premier rock climber in the U.S., Bensman won more than twenty national "bouldering" championships by the time she retired from competition in 1999. In 1992, she fell fifty feet down a rock face in Colorado, escaping catastrophic injury only because her rope pulled taut and broke her fall at the exact moment she hit the ground.

And that's not all that makes Bensman seem like a natural risk-taker. Gambling is in her blood. Her grandfather was the manager of a casino in Las Vegas; her mother grew up rubbing elbows with mobster Bugsy Siegel.

Yet Bensman shuns investment risk, keeping most of her money in what she calls "boring" mutual funds and blue-chip stocks. Does it seem odd that someone who loves clinging to cliffsides and grew up surrounded by gamblers doesn't like taking financial risk? "I'm pretty conservative, I guess," shrugs Bensman. Then again, she doesn't think it's risky to clamber hundreds of feet up a ragged wall of rock; in thirteen years of competitive climbing, she never had a major injury. "It's all about systems," she says. "If you've got the correct systems in place, it's really not dangerous at all."

As Bobbi Bensman's story shows, the conventional wisdom that every investor has a certain level of "risk tolerance" is little more than a lie. In reality, your perception of investment risk is in constant flux, depending on your memories of past experiences, whether you are alone or part of a group, how familiar and controllable the risk feels to you, how it is described, and what mood you happen to be in at the moment. The slightest change to any of these elements can turn you from a raging bull to a cowardly bear in a matter of seconds. If you unquestioningly trust your intuitive perceptions of risk, you will chronically take gambles you should avoid and back away from bets you should embrace.

Many investors who think they like big financial gambles often end up miserable once they actually lose money. Elderly widows can take huge financial risks, while some young single men invest like complete cowards. What's more, some people put their own assets into gung-ho emerging-market funds but stash their kids' college money in savings bonds. Others buy insurance *and* lottery tickets. Are these people conservative, aggressive, both, or neither? Meir Statman, a finance professor at Santa Clara University, just calls them "normal." And he's right.

Chapter Six will help you figure out how much risk you should take, how to stay calm during market storms, and how to distinguish false fears from real dangers. By mastering your perceptions of risk, you can put yourself firmly on the path toward financial peace of mind.

Risk in Real Time

For investors and their financial advisors, no question is more obvious, yet also more troublesome, than "How much risk are you comfortable taking?" To get the answer, financial planners and stockbrokers often ask investors to take a so-called risk-tolerance questionnaire. According to these people, it can take as few as a half-dozen questions to figure out "how much risk is right for you." Based on your answers, you will usually be thrown into one of three buckets: conservative (mostly cash and bonds), moderate (roughly half stocks, half cash and bonds), or aggressive (mostly stocks). Here are some questions reproduced from actual surveys:

I am willing to take a calculated risk with my long-term investments.

1. Strongly agree
2. Agree
3. Somewhat agree
4. Disagree
5. Strongly disagree

Many investments fluctuate over the short term. If a $100,000 investment that you made for ten years lost value during the first year, at what point would you sell and move to a more stable investment, rather than wait for a turnaround?

1. $95,000
2. $90,000 to $94,000
3. $80,000 to $89,000
4. less than $80,000

Which statement best describes your attitude toward investing for this goal?

1. I am extremely safety conscious and don't want the value of my investment portfolio to decline at all.
2. I realize that there are risks in investing, but I try to reduce them as much as possible.
3. I am willing to assume some investment risk to enhance the return potential of my investment portfolio.
4. I am willing to assume significant risk for a portion of my portfolio to increase my potential for high overall returns.
5. I am comfortable assuming significant risk for my overall portfolio in order to maximize the possibility of high returns.

The first problem with these questionnaires is that they assume you already know how much risk you are comfortable with. If you knew that, why would you need to take a quiz? Secondly, they are inconsistent. When 113 business students filled out risk-tolerance quizzes from six major financial companies, the average similarity among the results was only 56%. In other words, the odds that any two questionnaires would say the same person had the same risk profile were barely better than the flip of a coin. Your supposed level of risk tolerance may depend less on who you are than on whose quiz you happen to take.

But there's a more basic issue here. Does any of us really have a single level of "risk tolerance" that can be measured as precisely as our shoe size? Among the countless dumb ideas pervading the investment industry, this may be the dumbest of all.

To an astonishing degree, how much risk you can stand depends on what mood you happen to be in. Five minutes from now—perhaps only a few seconds from now—your emotions may change, and your willingness to take risk may change with it, as these examples show.

- Men viewed a series of head (and bust) shots of women downloaded from the website www.hotornot.com. The men were then offered chances to receive various amounts of money either the next day or well into the future. Men who saw pictures of "hot" women were much less willing to wait longer for more money.

- Students could choose either a safe 70% chance of winning $2 or a risky 4% chance of winning $25. In one group, which was put in a good mood by watching TV comedy skits, 60% picked the safer gamble. The people in a second group were told to sing the complete lyrics of Frank Sinatra's "My Way," solo, with no musical accompaniment, *twice.* (Most found this acutely embarrassing.) In this group, 87% chose the long shot—as if they felt the need to redeem themselves with a big financial score.

- People were made anxious by being told to imagine that they were summoned to the doctor's office to discuss an urgent medical matter. They were then asked to pick between a safe 60% chance of winning $5 or a risky 30% chance of winning $10. The anxious people preferred the safe bet far more often than people in a calm mood did. Anxiety tends to make us feel uncertain—so we shy away from extra risk.

- Students were given highlighter pens and then watched either of two videos: the death scene from the Ricky Schroeder tearjerker *The Champ,* or footage of tropical fish. Next they were asked how much they would sell their new highlighter for and what they would pay for someone else's. The folks who sat through the death scene were much more willing to pay up to buy someone else's pens. Feeling sad seems to remind us that we have lost something valuable, making us want to get a fresh start—often by taking the risk of buying something new. (If you have ever gone on a shopping spree after a romantic breakup, this may sound familiar.)

- In a disgustingly enlightening experiment, one set of people went for forty-eight hours without wearing deodorant, then "donated" their body odor, which was collected onto armpit pads while the donors watched videos that were either frightening or neutral. A second set of participants then had the armpit pads taped to their upper lip while they evaluated the emotional content of words projected onto a screen. Those who wore a pad collected from a frightened donor evaluated ambiguous words more cautiously, "as if they were motivated to avoid misses." A whiff of fear in the air may be enough to signal that you need to be careful.

- Men were asked to think of either three or eight factors that might increase their risk of heart disease. Strikingly, those who named only three factors rated their overall risk higher than did those who listed eight. Why? The men who had to think of eight separate reasons intuitively concluded something like: "Hey, how high could my risk be if it's so hard to come up with all these reasons?" But those who thought of only three factors found that shorter list much easier to bring to mind—making their own chances of getting heart disease feel higher. If it is easy to think about a risk, that alone can make the risk seem more real.

- Other researchers have shown that you may be more willing to rush into a risk if simple arguments for it are printed on red rather than blue paper. Finally, some evidence suggests that people might take risks more readily if they spend at least a half hour outside on pleasant spring days.

Add it all up, and it's clear that your mood swings can spin your "risk tolerance" around like a weathervane in a windstorm.

What We Can Learn from the Birds and the Bees

To understand why our attitudes toward risk are so easily contaminated by emotion, it helps to think about how our brains evolved.

Imagine yourself transported back eons ago to the high plains of east Africa. You glimpse a lion. A flash of neural lightning crackles through your mental alarm system, sending you scrambling up a tree to safety. If what seemed like a lion turns out to be only a patch of brown grass rippling in the wind, you have suffered no loss by scurrying up the tree. Whether the lion is real or imaginary, being afraid of it improves your chances of surviving long enough to reproduce and pass your genes on to your offspring.

Predators were not the only risk that our ancestors learned to fear. Running out of water, seeking shelter in the wrong places, betting that a supply of food would be more stable than it turned out to be—all these risks could also be a matter of life and death. They gave us an innate hatred of uncertainty. In the ancient laboratory of evolution, "sensitivity to losses was probably more [beneficial] than the appreciation of gains," psychologist Amos Tversky has said. "It would have been wonderful to be a species that was almost insensitive to pain and had an infinite capacity to appreciate pleasure. But you probably wouldn't have survived the evolutionary battle." For the early hominids, underreacting to real risks could be fatal, while overreacting to risks that turned out to be imaginary was probably harmless. Thus your brain's alarm system—centered in the thalamus, amygdala, and insula—comes with a built-in hair trigger. Over thousands of generations, a "better safe than sorry" reflex became an ingrained instinct for humans, as it is throughout the animal kingdom.

A keen response to potential danger is at the heart of the basic instinct for self-preservation that all animals share. More than two dozen species, ranging from fish and birds to rats and monkeys, have been tested for sensitivity to risk. Since other creatures don't know what money is, they don't care about losing it. But they do respond to risks like running out of food and water, or not being able to tell whether—or when—they will be able to get food or drink at all. Most animals would rather get a smaller, certain reward than the chance at a larger but uncertain one.

Ecologist Leslie Real gave bumblebees the chance to feed from two kinds of flowers. The blue flowers always contained 2 milliliters of nectar; the yellow flowers were randomly mixed so that two out of every three were empty, while one out of every three contained a triple reward of 6 milliliters of nectar. Thus a bee foraging continuously on flowers of either color would obtain the same "payoff"—an average of 2 milliliters per feeding. The only difference was that the blue flowers "paid" the same reward *every* time, while the yellow flowers paid the same average reward *over* time. The bees started out sampling both colors evenly, but soon learned to stick to blue, preferring it 84% of the time.

When Real pulled a switcheroo, reversing the setup so that every yellow flower but only one in three of the blue flowers paid off, the bees almost immediately abandoned blue—and now favored yellow an average of 77% of the time. So it's not just the amount of a gain that counts, but how consistently the gain can be counted on. Since wildflowers bearing nectar are naturally likely to be found near each other in clusters or dense patches, explains Real, "it pays for the bees to have a very strong preference for constant over variable reward."

In a lab at the University of Puerto Rico, little birds called bananaquits were offered a choice between yellow flowers, which always held 10 milliliters of nectar, and red flowers containing amounts between zero and 90 milliliters. The wider the range of reward in the red flowers, the more the birds preferred the constant payoff in the yellow flowers.

It's not just the birds and the bees that would rather have a small sure thing than a larger but less certain reward. To see whether humans think in a similar way, psychologist Elke Weber designed a simple experiment. Imagine two decks of cards in front of you. On the back of each card, a dollar amount is printed. At the end of the experiment, you will be eligible to earn the money for real by picking a card. Since you have no prior knowledge about how much money is in the cards, Weber instructs you to sample freely from either deck until you have a good sense of which one you would like to pick your final card from.

Without telling you, however, Weber has already arranged the decks so that one is a sure thing and the other is risky. In one pile, every card always pays a small amount; in the other, at least some of the cards yield nothing, while at least one offers a large payoff. For example, if every card in one deck pays $1, then nine of ten cards from the other pile might provide no gain while one pays $10.

If you are typical, you will pick about ten cards from each deck until you settle on the one that you think is better. Weber found that most people make their choice based on how widely each draw varies relative to the average draw—in other words, the deciding factor is how far off any one card is from the payout of the average card. "The experience of a loss or a gain," says Weber, "depends on what the loss or gain is relative to." People, like animals, tend to evaluate how much the outcomes vary relative to the total amount of wealth that appears to be at stake. When that difference is high, people will gravitate away from the risky deck to the certain deck. (If the cards in one deck always pay $1, while in the other deck nine cards pay nothing and one pays $10, roughly 70% of people will prefer the "sure thing" of the first deck.)

Thus, when a big jackpot does not appear to be close at hand, most of us will prefer a smaller, steadier payoff over a more variable one. In 401(k) retirement accounts, 17% of all the money sits in money-market funds, "guaranteed investment contracts," or stable-value funds. These accounts never fluctuate in value, offering the certainty of never losing money—as well as the certainty of never making very much, either.

You've Been Framed

Are there other reasons to believe our risk tolerance is not fixed?

We all know that the glass seems either half empty or half full depending on how we feel about ourselves. But it also depends on how we feel about the glass. Researchers have shown that when a four-ounce glass has two ounces of water poured out of it, 69% of people will say it is now "half empty." If the same glass starts out empty and then has two ounces of water poured into it, 88% of people will now say it is "half full." There's no difference in the size of the glass or the amount of water, but a simple twist of context changes everything.

"Equivalent ways of describing something should lead to equivalent judgments and decisions," says University of Oregon psychologist Paul Slovic. "But it's not true. People's judgments about risk are very moveable and subjective." When you face a chance to make or lose money, your decision can be pulled one way or another like a lump of Silly Putty just by a minor change in context or description—what psychologists call framing.

To see how powerful framing can be, consider these examples:

- One group of people was told that ground beef was "75% lean." Another heard that the same batch of meat was "25% fat." Everyone was asked to guess how good it would be. The group that heard about "fat" estimated the meat would be 31% lower in quality and taste 22% worse than the other group predicted. After both groups tasted burgers made from the same batch of meat, the "fat" people liked the burgers less than the "lean" people did.

- Pregnant women are more willing to agree to an amniocentesis if told they face a 20% risk of having a child with Down syndrome than if told there is an 80% chance the baby will be normal—even though those are two ways of saying the same thing.

- A study asked more than 400 doctors whether they would prefer radiation or surgery if they became cancer patients themselves. Among the physicians who were informed that 10 out of 100 patients would die from surgery, half said they would prefer to be treated with radiation. Among those who were told that 90 out of 100 patients would survive surgery, only 16% said they would choose radiation.

The classic example of framing was devised by psychologists Amos Tversky and Daniel Kahneman. They gave one group of college students the following scenario:

> Imagine that the U.S. is preparing for the outbreak of an unusual Asian disease, which is expected to kill 600 people. Two alternative programs to combat the disease have been proposed. Assume that the exact scientific estimates of the consequences of the programs are as follows:
>
> If Program A is adopted, 200 people will be saved.
>
> If Program B is adopted, there is a one-third probability that 600 people will be saved, and a two-thirds probability that nobody will be saved.
>
> Which of the two programs would you favor?

At the same time, Tversky and Kahneman gave a second group of students the same scenario with differently worded plans to combat it:

> Imagine that the U.S. is preparing for the outbreak of an unusual Asian disease, which is expected to kill 600 people. Two alternative programs to combat the disease have been proposed. Assume that the exact scientific estimates of the consequences of the programs are as follows:
>
> If Program C is adopted, 400 people will die.

If Program D is adopted, there is a one-third probability that nobody will die, and a two-thirds probability that 600 people will die.

Which of the two programs would you favor?

The results were stunning: In the first scenario, 72% of the students preferred Program A, while in the second scenario only 22% favored Program C—even though the results of both programs are identical! In either case, 200 people will live and 400 will die. But the first frame emphasizes the number of lives saved. When a choice is framed positively, as a potential gain, it's as if the glass is partly full. And that seems like an improvement over the empty glass we started with, so our instinct is to preserve what we've gained. When the glass feels partly full, the sure protection of 200 people by Program A makes the uncertainty of Program B sound like an unacceptable risk.

On the other hand, the second frame stresses the number of lives lost, making the glass feel partly empty. That makes us willing to take on extra risk to avoid losing whatever is left in the glass. Thus the certain death of 400 people under Program C makes the crapshoot of Program D sound like a justifiable gamble. Because the alternative frames play so differently on our feelings, we don't even notice that all four programs are equivalent.

Framing helps explain why so many investors can't live up to one of Wall Street's best-known sayings, "Cut your losses and let your winners ride." When you mistakenly buy a stock without doing your homework, you can limit your risk of further losses (and lock in a tax benefit to boot) by selling it. Instead, you probably hang on in a grim gamble that you will be able to sell the darn thing when it gets back to what you paid for it. That's half-empty thinking: leaving yourself exposed to extra risk in hopes of avoiding further loss.

Conversely, when a stock goes up after you buy it, there's no real reason why you should be in a rush to get rid of it, especially since your gain becomes taxable the moment you sell. But now the risk that looms large is the possibility of losing the gains you've already made. So you sell—and, all too often, watch the stock go on to double or triple after you get out. That's half-full thinking: cutting your exposure to further risk so you can hang on to what you have already gained.

Framing can lead to other freaky decisions. Imagine that you have $2,000 in the bank. I offer you a choice: Do nothing, or take a 50/50 chance of either losing $300 or winning $500. Would you stand pat or take the gamble? Think about it for a second. Now imagine, again, that you have $2,000 in the bank. Now I offer you this choice: Do nothing, or take a 50/50 chance of ending up with either $1,700 or $2,500. Would you stand pat or take the gamble?

Most people reject the first gamble but take the second one. That's because the first is framed to stress the amount you will gain or lose relative to what you started out with, while the second is framed to emphasize the total amount you end up with. The change feels bigger—and potentially scarier—in the first frame, so most people turn it down. The two gambles are economically identical. Psychologically, they are worlds apart.

In the financial world, framing is everywhere:

- Many consumers would much rather purchase something advertised as "buy one, get one free" than the same item priced at "50% off."

- In the most common form of stock split, one share is replaced by two, each valued at half the original price. (Instead of owning one share at $128, for instance, you would now own two at $64 apiece.) Although a stock split is the logical equivalent of trading a dime for two nickels, it fills many people with the false thrill of having "more" of an investment than they started with. After Yahoo! Inc. announced in 2004 that it would split two-for-one, the stock surged 16% the next day.

- If you sink 1% of your money into a single stock that goes to zero, you will probably be very upset. If your entire portfolio loses 1% of its value, you are apt to shrug it off as a routine fluctuation. And yet the effect on your total wealth is identical.

- You will be far more inclined to take a risk if you're told that the odds of success are 1 in 6 than if you're told there's a 16% chance of succeeding. If you're told there's an 84% chance of failing, you probably won't touch it.

- Most employees are happier with a 4% raise when inflation is running at 3% than they are with a 2% raise when inflation is zero. Because 4% is twice the size of 2%, it "feels" better, even though what really matters about a raise is how much of it is left after the rising cost of living.

The Frames in Our Brains

What creates the frames inside our brains? "It's the interaction between feeling and thinking," says psychologist Cleotilde Gonzales of Carnegie Mellon University. Your brain always seeks to reach decisions in the easiest possible way—with the lowest emotional cost and the least mental effort (or "cognitive cost"). Let's hark back to the "Asian disease problem," where 600 lives are at stake. In the half-full frame, which emphasizes the lives that could be gained, Program A will save 200 people; Program B offers a one-third chance of saving 600 people and a two-thirds chance of saving no one. In the half-empty frame, which stresses the lives that could be lost, Program C will result in 400 deaths; under Program D, there is a one-third chance that no one will die and a two-thirds chance that 600 people will die.

The idea of saving 200 people in Program A is literally a "no-brainer," says Gonzales. Because Program A is framed as a sure gain, "it's a simple alternative that can be evaluated at very low cognitive cost," she explains. And this frame suggests no emotional cost, since it calls your attention to the lives saved rather than the lives lost. You can see how little effort the brain takes to evaluate this choice in the top left of Figure 6.1.

On the other hand, when a risk is framed negatively—for instance, by stressing those 400 lost lives—then it incites images and ignites emotions. The thought of losing money, like losing lives, is so inherently alarming that it ends up triggering intense activation in an area of your brain called the intraparietal sulcus. This curving wrinkle of tissue, located toward the top of your head behind your ears, appears to function somewhat like a mental movie screen. It enables you to visualize and imagine the consequences of actions not yet taken. The more uncertain the consequences are, the more active the intraparietal sulcus becomes. You can see this happening in the top right image in Figure 6.1, which shows the brain of a person who is considering whether to gamble on Program B: a small chance of saving all the lives and a greater risk of saving no one. As you can see along the lower edge of the image, this vivid danger sets off fireworks in the intraparietal sulcus.

And when the frame shifts from saving to losing, your mental movie screen projects images that are painful and disturbing regardless of whether the loss is certain or merely likely. Your brain can no longer decide between the gamble and the sure thing purely on the basis of which choice arouses less emotion—since a possible and a certain loss both feel lousy. So the half-empty scenario makes the brain "work harder," says Gonzalez. As the bottom two images in Figure 6.1 show, an almost identical proportion of the brain lights up when you feel that you must choose between a certain or a likely loss of the same value. (Program C—the sure loss of most of the lives—is on the left. Program D—the small possibility that no one will die and the larger risk that everyone will die—is on the right.)

"When we make decisions," explains Gonzalez, "we balance how much we need to think about an alternative against how much we stand to lose." When your brain has to work this hard, it's the emotional stakes that tip the balance. Even a slight chance that no one will die feels better than the certainty that most people will die. That's why we pick Program D: Emotionally, it's the easy way out.

Now imagine two scenarios that are even simpler:

1. I give you $50. You now must choose between
   a. keeping $20 for sure, or
   b. taking a gamble with a 60% chance of losing $50 and a 40% chance of keeping $50.

2. I give you $50. You now must choose between
   a. losing $30 for sure, or
   b. taking a gamble with a 60% chance of losing $50 and a 40% chance of keeping $50.

You probably see that the two situations are identical. But they don't *feel* identical. The first frame focuses your attention on how much you keep; the second, on how much you will lose. Neuroscientists in London recently scanned people's brains while they faced these choices. Afterward, the participants said they had easily figured out that the alternatives were the same, and they insisted that they had always split their responses 50/50 between the sure thing and the gamble. That wasn't true. In the first frame, they had gone for the sure thing 57% of the time; in the second frame, they gambled on 62% of the trials.

When people avoided the gamble in the first frame and took it in the second, neural activity surged in the amygdala, suggesting that this fear center in the brain was steering them away from the perceived danger of loss. The amygdala apparently responds, like a very blunt instrument, only to the crude difference between "keeping" and "losing." It takes the prefrontal cortex to figure out the more subtle fact that all the choices are the same. By playing up the emotional aspects of framing, Wall Street's marketers can keep your amygdala firing—and prevent your reflective brain from intervening.

One of the cleverest forms of financial framing is called an "equity-indexed annuity" or EIA. This trendy investment—more than $27 billion were sold in the U.S. in 2005—guarantees you a minimum rate of return in the stock market while ensuring you against any losses. EIAs are often described as offering "the upside without the downside." But in exchange for putting a floor under your losses, EIAs slap a ceiling over your gains. Like Program A in the Asian disease problem ("200 people will be saved"), these annuities emphasize the certainty of avoiding losses. That makes the alternative, putting your money in the market with no downside protection, sound too risky to bother with. But this half-full thinking also makes you overlook a more subtle risk. By limiting your profits as well as eliminating your losses, EIAs keep you from capturing all the market's gains. In some EIAs, you can earn barely over half the stock market's return. If you invest $10,000 in such an EIA and the market goes up 30%, you'll earn only 16.5%. Had you not been so worried about limiting your losses, you could have earned another $1,350. Those forgone gains are a form of loss, too—but EIAs are framed in a way that blinds many investors to it.

"Who's the One?"

Besides half-full or half-empty thinking, there's another form of framing that can wreak havoc with your investing logic. There's a surprisingly big difference between how we react to odds expressed as percentages (say, 10%) and how we respond to odds expressed as frequencies ("one out of every 10").

When psychiatrists were told that "patients similar to Mr. Jones are estimated to have a 20% chance of committing an act of violence" within six months, 79% were willing to release Mr. Jones from a mental hospital. But when they heard that "20 out of every 100 patients similar to Mr.

Jones are estimated to commit an act of violence" in the same period, only 59% said they would let him out—even though the odds that he might hurt somebody were identical.

Psychologist Kimihiko Yamagishi asked people how concerned they were about various causes of death. When he informed people that cancer kills 1,286 out of every 10,000 people, they rated it as 32% riskier than they did when he told them that it kills 12.86% of the people it strikes.

Percentages are abstract and hard to think through; to get a good feel for how bad that 12.86% mortality rate is, you would need to know how many people that represents. But when you hear that the same cancer kills 1,286 out of every 10,000 people, your first thought is "Almost 1,300 people are dead!" As psychologist Paul Slovic puts it, "If you tell people there's a 1 in 10 chance of winning or losing, they think, 'Well, who's the one?' They'll actually visualize a person." More often than not, the one person you will visualize winning or losing is *you*.

So a financial advisor can goad you into taking (or avoiding) an investment risk just by changing how he describes it. If he boots up some fancy software that says you have a 78% chance of meeting your retirement goals, that sounds great. But he can reframe the same result and strike fear into your heart by declaring that "22 out of 100 people with your strategy will end up eating cat food in the dark"—and the next thing you know, he has foisted a bunch of risky stocks onto you that you never wanted.

As the study of doctors considering cancer treatment showed, experts can fall prey to framing problems just as easily as amateurs can. Whether you are a retail investor or a professional money manager, your "risk tolerance" is supposed to be an integral part of your personality. But it can be transformed by an embarrasingly basic twist in wording. That's why every investor must be eternally vigilant against the dangers of getting framed.

The Blundering Herd

Your perception of investment risk also depends on peer pressure. If I had a dollar for every time I've heard a money manager brag about being a "contrarian" or "not following the crowd" or "loving the stocks that everybody else hates," I could almost live off the interest alone. The most obvious similarity among money managers is their insistence that they are all different. In truth, they act like the multitudes in Monty Python's *Life of Brian.* When Brian exhorts the crowd, "You are all individuals," they chant, "Yes, we are all individuals!" When he tries again, telling them "You are all different," they roar in unison, "Yes, we are all different!"

Do investors really act like individuals?

- A study of thousands of buy or sell recommendations by security analysts at hundreds of brokerages found that analysts tag along with the crowd as if they were coated with Velcro. When the average recommendation is a "strong buy," for example, the next call from any given analyst is about 11% more likely to be a strong buy, too.

- When individual investors raise their average holdings in a stock by 10 percentage points, people who live within a fifty-mile radius of the buyers jack up their own stakes in the same stock by an average of two percentage points.

- In rising and falling markets alike, hedge funds—those exclusive pools of "sophisticated" money run by supposedly independent-minded investors—mimic each other's trades like teenagers trawling through a shopping mall.

- Looking at retirement-plan decisions by 12,500 employees of a major university, researchers found that people working in the same department tended to invest in funds from the same firm, even though every employee could choose freely from four different fund companies.

- In 1995, New Era Philanthropy, a scam run by John G. Bennett Jr., collapsed after conning more than $100 million out of universities, churches, and foundations. Bennett had set the trap by promising to double their money every six months. Word of his wonders had spread quickly among the tycoons who sit on nonprofit boards—and Bennett soon had bilked millions out of venture capitalist Laurance Rockefeller, former U.S. Secretary of the Treasury William Simon, and hedge fund manager Julian Robertson. Separately, these men were among the smartest investors in America; as a herd, they acted like a bunch of dolts.

- Institutional investors like insurance companies, foundations and endowments, pension funds, and mutual funds spend billions of dollars every year researching which stocks they should buy and sell. All this spadework should unearth buried treasure—rare and unusual stocks that no one else knows about or can understand. Instead, the institutions scratch the same surfaces. On average, an institutional investor is 43% more likely to increase its holdings of a stock if, over the previous three months, other big investors trading the stock were all buyers.

- When institutions own a hot stock in a trendy industry, they usually have heard of it by word of mouth rather than through original research—and they go on to talk it up with three times as many colleagues as the owners of less exciting stocks do. No wonder "everybody" often seems to be talking about the same stock.

Ideas, like yawns, are contagious. (I may well be able to make you yawn just by asking you to read this sentence containing the word *yawn.*) Let's say you need to rent a car at the airport. There are two people ahead of you in line and a sign announcing that only two kinds of cars are available: Hyundai or Fiat. Your hunch is to go with a Hyundai, but you don't know a lot about cars—and you've never driven either a Hyundai or a Fiat. So you watch the folks in front of you. The first confidently picks a Fiat—so it looks as if she knows what she's doing. Then the second person steps forward, hesitates, and asks for a Fiat.

Now it's your turn. What do you do? You should bear in mind that the first renter seemed sure of her choice, while the second didn't seem any more ignorant than you. Since both of them picked a Fiat, you probably should stifle your own hunch and pick a Fiat, too. So you say firmly, "I'll take a Fiat." Their choice has spilled over onto yours—and now yours will splash over onto the person behind you, triggering what's called an "information cascade" that can wash over everyone in the line and create a wave of demand for Fiats.

The cascade may well continue until someone who seems to be an expert finally steps up and demands a Hyundai. If the next person does the same, now everyone behind him who was leaning toward a Fiat will ask for a Hyundai. It doesn't take much new information to send a cascade sloshing in the opposite direction.

Informational cascades are not necessarily irrational. If you really don't know much about automobiles, you should try to figure out who does and take your cue from them, especially when you have to make a decision on the spot and have no time to learn more on your own. It's a simple guideline for making simple choices when you know you don't have all the information you need.

In fact, what ecologists call "public information"—the contagious spread of cues about risk and reward—is one of the most basic techniques that living things use to enhance their chances of survival. Incredible as it may seem, even some plants have developed the ability to share public information about the presence of risk: When sagebrush is damaged by animals, it releases aromatic chemicals, alerting nearby plants to step up their own production of defensive proteins that deter grasshoppers and other critters that munch on vegetation. And in the animal kingdom, creatures of all sorts band together—schools of fish, flocks of birds, herds of sheep, packs of wolves, pods of whales—to feed, migrate, fend off enemies, and learn from each other about risks and rewards.

Starlings, for example, rummage for food along the surface of the ground. They probe most painstakingly when they are alone. However, in the presence of at least one other starling, they quickly learn to skip over areas that are more difficult to evaluate—focusing on the patches where the behavior of the other birds suggests food is most likely to be found. Among fish, species of sticklebacks with smaller spines and softer bodies take more cues from the feeding patterns of other fish to locate the best places to forage. But stickleback species with tougher, spikier bodies explore much more independently for food. Animals seem more inclined to let others do some of their thinking for them when their own information is incomplete, outdated, or if they feel vulnerable.

Humans are animals, too. When people form investing herds, we crowd together to buy a stock, cheerlead its rise, and egg each other on. Just like starlings, we become less willing to venture off on our own—and merely being part of a group makes us less inclined to ask questions.

When investors pile into smaller stocks and newer industries, they become more willing to admit that someone else may know something they don't. Just like the run of Fiats or Hyundais in the rental car line, the result is an informational cascade, with everyone in the group buying the same stocks at the same time. "The trend is your friend," brokers have long told their clients. "Don't fight the tape." Being part of the herd is fun while it lasts, but it's seldom lucrative for very long, and it's impossible to predict when the herd will change its "mind." If you want to make more money than other people, you can't invest like other people.

Throwing a "Hail Mary"

How much money you made or lost on your last investments can transform how risky you think the next one is. The same bet can feel either dangerous or safe, depending on whether you are on a hot streak or in a slump. That's how your investing brain is designed.

Animals that are running low on food, water, or warmth have what ecologists call a "negative energy budget." Creatures that are hungry, thirsty, or cold can rarely afford to take the chance that small but steady gains will be enough to keep them alive. In effect, they need to try striking it rich. So animals in a state of deprivation tend to prefer more variable rewards: While that raises the risk of getting nothing, it's also the only feasible way to get the big boost they need to restore their depleted energy.

Biologist Thomas Caraco offered two different trays of millet seeds to yellow-eyed juncos, a bird native to Mexico and the southwestern U.S. One tray—let's call it the "risky" choice—provided either several millet seeds or none at all. The other tray—the "certain" option—always offered a constant amount of food. (If, for example, the first tray contained either zero or four seeds, then the second tray always provided two seeds; if the risky tray offered either zero or six seeds, then the certain tray always held three seeds; and so on.) When the juncos had recently been fed, they preferred the certain choice: the tray with the fixed, smaller amount of food. But when the birds had not eaten anything for a few hours, they became risk-takers, pecking at the tray that offered either double or nothing.

If birds run out of seeds, they might not survive. If humans run out of money, we also might not survive—because without it we might not be able to obtain the necessities of life. The less money people have, the more willing they often become to take on extra risk, just as a quarterback will throw a "Hail Mary" pass late in the fourth quarter or a basketball player will launch a desperate shot from half-court just before the final buzzer sounds. All too often, a "negative energy budget" makes people fling a Hail Mary with their money. Sometimes the results are heartbreaking, as in Manila in 2006, when seventy-nine impoverished Filipinos were trampled to death in a panic to get raffle tickets for a cash giveaway.

Even when no one dies, the sad fact is that those who can least afford to lose the little money they have are most prone to put it at high risk.

- In Virginia, residents who earn less than $15,000 per year spend 2.7% of their annual income on lottery tickets, while those who make more than $50,000 per year use only 0.11% of their income to play the lottery.
- When more than 1,000 Americans were asked to pick the most practical way to become wealthy, 21% said "win the lottery." Among those with incomes of $25,000 or less, nearly twice as many felt that their best chance at getting rich was a lottery ticket.
- Blacks and Hispanics are more reluctant than whites to take moderate financial risks—yet nonwhites are between 20% and 50% more willing to take *substantial* financial risks. Black and Hispanic households, on average, have roughly one-fourth the total net worth of the typical white household.
- In the second half of the year, mutual funds with below-average returns become up to 11% more volatile than those that had above-average results in the first half. Consciously or not, the managers of funds that lagged in the first six months buy riskier stocks in an attempt to salvage their returns by year-end.
- On average, when professional "market makers" at the Chicago Board of Trade lose money in the morning, they take extra risk that afternoon—making bigger bets and trading faster until it seems that their underpants are on fire.
- Poorer investors—those with an estimated net worth under $75,000—favor stocks that offer the same promise as lottery tickets: a low price and a small chance at striking it rich, plus a high risk of loss. By gambling on these long-shot stocks, the poorest investors fall behind the overall stock market by an average of roughly 5% of their annual income. Thus the very people who can least afford them take the worst risks. At the same time, they lack the resources to get advice that might help them invest more prudently.

Making Risk Work for You

Bobbi Bensman, the champion rock climber we met at the beginning of this chapter, knows that dangling off cliffs can be deadly to climbers who don't think clearly. But if you learn how to control the likeliest sources of danger, rock climbing becomes surprisingly safe. Investing is like that, too. Here are some reliable policies and procedures you can put in place ahead of time to help you manage your risks instead of letting them manage you.

➢ **TAKE A TIME-OUT.** Since even the most minor, momentary changes in your mood can make a huge difference in how you perceive risks, don't buy or sell an investment on the spur of the moment. You may be under the sway of temporary influences you're not even aware of. A road-rage encounter on your way to work, a fight with your spouse, background music that bothers (or pleases) you, a flash of red or a cool wash of blue—any of these things could skew the way you think about an investment. Sleep on it until tomorrow, and see whether you view it the same way.

➢ **STEP OUTSIDE YOURSELF.** In the mid-1980s, Intel Corp.'s main business, manufacturing memory chips, was collapsing under fierce competition from Japan. The company's profits fell by more than 90% in a year. Torn between the pain of the status quo and the fear of change, Intel's managers were paralyzed. At that point, recalls then-president Andy Grove,

> I looked out the window at the Ferris wheel of the Great America amusement park revolving in the distance when I turned back to [CEO Gordon Moore], and I asked, "If we got kicked out and the board brought in a new CEO, what do you think he would do?" Gordon answered without hesitation, "He would get us out of memories." I stared at him, numb, then said, "Why shouldn't you and I walk out the door, come back, and do it ourselves?"

By stepping outside themselves, Grove and Moore got the insight to see which risk was right—and the courage to take it. Intel moved out of memory chips and into microprocessors, a bold and brilliant leap that powered its growth for years to come.

Knowing, or even imagining, that someone else is relying on your advice can make you feel more accountable, forcing you to go beyond your gut feelings and fortify your opinions with factual evidence. After the terrorist attacks of September 11, I got dozens of e-mails from panicky readers asking me whether they should get out of the market. Stifling my own gut feelings of fear and rage, I answered as analytically as I could, gathering historical evidence on how the U.S. stock market had performed after earlier national tragedies. My conclusion: "The modern financial history of the U.S. holds no example of a physical disaster, or even an outright war, that wreaked lasting havoc on investment returns." Within a year, the bear market was over—and anyone who got out in September 2001 missed one of the best buying opportunities in a generation.

If you find it hard to imagine stepping outside yourself, try an exercise anyone can relate to. Before concluding that an investment decision is right for you, ask whether you'd be comfortable advising your mother to do the same thing. If you would tell her not to do it, then why should you do it? I call this the WWMD question (What Would Mom Do?).

➢**LOOK BACK.** If you have never lived through a bear market like 2000–2002 or 1973–74, it's easy to delude yourself into thinking you have nerves of steel. Every novice investor should study enough financial history to know that booms always end in busts—and that the cockiest traders die first. (Two good books are Edward Chancellor's *Devil Take the Hindmost* and Charles P. Kindleberger's *Manias, Panics, and Crashes.*)

➢ **WHEN THE PRICE DROPS, RISK GOES WITH IT.** Long ago, the financial analyst Benjamin Graham pointed out that most people measure what a stock is worth by looking at its *price,* while they judge what a business is worth by calculating its *value.* That leads to huge differences in how people think:

	STOCKS	BUSINESSES
Unit of measurement	price	value
Accuracy of measurement	precise (and often wrong)	approximate (but usually right)
Rate of change	every few seconds	a few times a year
Reason for change	different price is offered by somebody who doesn't own it	different amount of cash is produced for the people who do own it
How long people own	11 months, on average	up to several generations
Risk	temporary drop in stock price	permanent decline in business value

"The main reason investors struggle with how to react to bad news," says Oakmark Fund manager Bill Nygren, "is that they really haven't figured out why they own the stocks they own. If you buy a stock primarily because it has been going up, and then it falls on bad news, it shouldn't come as a surprise that your gut instinct is to sell."

This emotional mix-up arises because once a business sells shares to the public, nearly everyone focuses on the fast-moving price of the stock and forgets about the much more stable value of the business. Thus, when the stock price drops, it seems like bad news. As *BusinessWeek* put it during the heart of a bear market long ago, "For investors . . . low stock prices remain a disincentive to buy." But if the value of the business is solid, a declining stock price should be an incentive to buy, because it enables you to get more shares for less. And if the stock price drops below the business value, you have that rare opportunity that Graham called a "margin of safety"—the assurance that you can increase your stake for less than what it is actually worth.

I've often said that the problem with stocks is that little letter *T.* If you take the T out of *stocks,* you're left with *socks.* They have almost nothing in common:

SOCKS	STOCKS
You buy them when you need them.	You buy them when other people want them.
You buy more whenever they go on sale.	You buy more whenever they are not on sale.
You keep them for years.	You sell them as soon as possible.
If they develop holes, save them in a "rag bag."	If they go down in price, panic.

Would you ever pay $500 or $1,000 for a pair of socks without asking how they could possibly be worth so much? Would you be upset if your favorite store started selling socks at 50% off? Of course not. But when it's stocks instead of socks, people make those mistakes all the time.

Once the pain and fear of loss kick in, it can be nearly impossible to think calmly enough to use your reflective brain to figure out the right course of action. But in the face of falling share prices, you must systematically analyze whether bargains are being created. That's why it's so important to plan ahead. Buying a stock without first calculating the value of the underlying business is as irresponsible as buying a house without ever setting foot inside. And you should never sell a stock without first asking whether a falling share price has made it a better investment than ever.

➢ **WRITE YOURSELF A POLICY.** The best way to prevent yourself from being knocked off track by your emotions is to spell out your investing policies and procedures in advance, in what's called an "investment policy statement," or IPS. An IPS states what you, as an individual or an organization, are seeking to accomplish with your money and how you will get there. It lists your objectives down the road as well as your constraints along the way. (For a sample IPS, see Appendix 3.) Once you create an IPS, you must live by it; it is your contract with yourself or your organization. You can put extra "teeth" into your IPS by programming your Palm, BlackBerry, or desktop calendar software to send yourself periodic warnings not to violate your own investment policy.

➢ **GET REFRAMED.** The German mathematician Karl Jacobi had useful advice: "Invert, always invert." Explains the investor Charles Munger: "It is in the nature of things, as Jacobi knew, that many hard problems are best solved only when they are addressed backward." If someone tells you that the odds of success are 90%, invert the frame like this: That means there's a 10% chance of failure. Is that too high for you? Next, flip the percentage frame into a personal one: One out of every ten people who try this will fail. How do I know I won't be the one? If you are part of a large organization, break your researchers into groups and have each one report its findings in a different frame. Seeing risk estimates in both percentage and personal terms will help you make a more balanced decision.

You should also fit the evidence into the widest frames you can. Let's say you have a total of $24,000 invested in stocks. Now you're considering putting $1,000 into a stock that you believe has even odds of either doubling or going to zero. Thinking of it in a narrow frame—I could either make $1,000 or lose $1,000—turns the decision into a tug of war between greed and fear. Instead, put it in a broad frame: The total value of my portfolio could either go up to $25,000 or down to $23,000. That broader frame will take most of the emotion out of your choice.

➢ **TRY TO PROVE YOURSELF WRONG.** "The riskiest moment is when you're right," says investing sage Peter Bernstein. "That's when you're in the most trouble, because you tend to overstay the good decisions." That refusal to second-guess yourself can lead to huge losses and a crippling sense of regret. So it's a good idea to schedule an advance appointment with a devil's advocate. Make a rule: Whenever an investment doubles in price, find out who has the most negative view of it and give this devil's advocate a full hearing. Read or listen to the critique, take careful notes, then factor the devil's advocacy into a fresh comparison of price and value. If the price no longer makes sense, it's time to sell.

➢ **KNOW YOURSELF.** Most exercises designed to test your "risk tolerance" are a waste of time. As we have seen, your reaction to risk is not a single thing, fixed and unchangeable, like an isolated insect preserved in amber. Instead, you house a multitude of potential responses—ranging from rigid fear to feelings that stretch like taffy. Your attitudes toward financial risks, as we have learned, can differ drastically depending on how they are framed, whether you are alone or in a group, how your previous bets paid off, how easy it is for you to think about the risk, what mood you are in, even what the weather is like outside. The slightest change in any of those factors can raise or lower your tolerance for risk in a heartbeat. Even so, there's still plenty you can do to manage your attitudes toward risk.

- Bear in mind that when most people say they have "a high tolerance for risk," all they really mean is that they have a high tolerance for making money. It's easy to feel comfortable with the risks you took when they all seem to be paying off. But when the gains vanish and the losses swell, the harvest of risk is nothing but heartbreak. If you think a plunge in the value of your investments won't bother you, you are either wrong or abnormal.

- Economists used to say that "rational" people should pay the same amount for a chance to win $100 or to avoid losing $100. After all, either bet leaves you $100 better off. But experiments by Kahneman and Tversky proved that most people don't think like that. Try one of their ideas yourself: Imagine a coin toss in which you will lose $100 if tails comes up. How much would you have to win on heads before you would be willing to take this bet? Most people insist on a payoff of at least $200. What should that tell you? Losing money feels at least twice as painful as gaining the same amount feels good. So the thrill you feel when you make money will be dwarfed by the pain you feel when you lose it. If you've never yet experienced that pain, you have no idea how much it will hurt. You can minimize it by refusing to put more than 10% of your money into any single investment. That way, even if your hottest holding goes to zero, nearly all of your portfolio should still be intact.

▪ Most people run with the bulls and flee from the bears. Chances are, when the market is going up, you will end up being bullish, too, taking more risk than you normally would. Then, when the market turns back down, you will turn overly bearish, taking even less risk than you should. Being mauled once will make you afraid of getting clawed again. Simply knowing in advance that you are likely to act this way can help you plan for it. You could, for example, set a target range for how much money you will keep in foreign stocks. Let's say it's between 25% and 30%. You could trim back to 25% after a year in which foreign stocks go way up, then raise your allocation to 30% after a year in which they go way down. That way, you force yourself to take less risk when prices are dangerously high and more risk when they are attractively low.

▪ Since no one really has a fixed tolerance for risk, says psychologist Paul Slovic, it's more helpful to think in terms of "goals, objectives, and outcomes." How much money will you need down the road? How will you get there? What kind of result do you want to attain—or want to avoid? To answer these questions, you need to know your budget, calculate your current assets, and plan your future income and expenses. While those numbers aren't perfectly certain, either, they are a much more reliable basis for judgment than a squishy concept like "risk tolerance."

▪ When you're on fire with the hope of striking it rich on some investment, remember to consider not just how much you will make if you are right but how much you will lose if you are wrong. In what's known as "Pascal's Wager," the mathematician and theologian Blaise Pascal provided a model for how to think about this problem. Since God's existence is a matter of faith, not scientific proof, how should you live? Let's say you gamble that God exists, so you lead a virtuous life—but it turns out that there is no God. You miss enjoying a few sins while you are alive, but that's all your gamble costs you. Now let's say you gamble that there is no God and sin your way through life without a qualm—and it turns out that God does exist. The payoff on this gamble is a few decades of cheap thrills—then an eternity burning in Hell. In Peter Bernstein's words, Pascal's Wager shows that "whether you should take a risk depends not just on the probability that you are right but also on the consequences if you are wrong." To make reliably good decisions, you must *always* weigh how right you think you are against how sorry you will be if you turn out to be mistaken.

CHAPTER SEVEN

Fear

> Neither a man nor a crowd nor a nation can be trusted to act humanely or think sanely under the influence of a great fear. . . . To conquer fear is the beginning of wisdom.
>
> —*Bertrand Russell*

What Are You Afraid of?

Here are a few questions that might, at first, seem silly.

- Which is riskier: nuclear reactors or sunlight?
- Which animal is responsible for the greatest number of human deaths in the U.S.?
  - Alligator
  - Bear
  - Deer
  - Shark
  - Snake
- Match the causes of death (on the left) with the number of annual fatalities worldwide (on the right):

1. War	a. 310,000
2. Suicide	b. 815,000
3. Homicide	c. 520,000

Now let's look at the answers.

The worst nuclear accident in history occurred when the reactor at Chernobyl, Ukraine, melted down in 1986. According to early estimates, tens of thousands of people might be killed by radiation poisoning. By

2006, however, fewer than 100 had died. Meanwhile, nearly 8,000 Americans are killed every year by skin cancer, which is most commonly caused by overexposure to the sun.

In the typical year, deer are responsible for roughly 130 human fatalities—seven times more than alligators, bears, sharks, and snakes combined. How could gentle Bambi cause such bloodshed? Unlike those other, much more fearsome animals, deer don't attack with teeth or claw. Instead, they step in front of speeding cars, causing deadly collisions.

Finally, most people think war takes more lives than homicide—which they believe kills more people than suicide. In fact, in most years, war kills fewer people than conventional homicides do, and the number of people who take their own lives is almost twice the number of those who are murdered. (In the list on the previous page, the causes and the number of deaths that result from them are already matched correctly.) Homicide seems more common than suicide because it's a lot easier to imagine someone else dying than it is to imagine killing yourself.

None of this means that nuclear radiation is good for you, that rattlesnakes are harmless, or that the evils of war are overblown. What it does mean is that we are often most afraid of the least likely dangers, and frequently not worried enough about the risks that have the greatest chances of coming home to roost. It also reminds us that much of the world's misfortune is caused not by the things we are afraid of, but by being afraid. The most terrible devastation wrought by Chernobyl, for example, did not come out of its nuclear reactors. Instead, it came from the human mind. As panicky business owners fled the area, unemployment and poverty soared. Anxiety, depression, alcoholism, and suicide ran rampant among the residents who could not afford to leave. Fearing that their unborn babies had been poisoned, expectant mothers had more than 100,000 unnecessary abortions. The damage from radiation was dwarfed by the damage from the fear of radiation, as imaginary terrors led to real tragedies on a massive scale.

We're no different when it comes to money. Every investor's worst nightmare is a stock market collapse like the Crash of 1929 that ushered in the Great Depression. According to a recent survey of 1,000 investors, there's a 51% chance that in any given year, the U.S. stock market might drop by one-third. And yet, based on history, the odds that U.S. stocks will lose a third of their value in a given year are only around 2%. The real risk is not that the stock market will have a meltdown, but that inflation will raise your cost of living and erode your savings. Yet only 31% of the people surveyed were worried that they might run out of money during their first ten years of retirement. Riveted by the vivid fear of a market Chernobyl, they overlooked the more subtle but severe damage that can be dealt by the silent killer of inflation.

If we were strictly logical, we would judge the odds of a risk by asking how often something bad has actually happened under similar circumstances in the past. Instead, explains psychologist Daniel Kahneman, "we tend to judge the probability of an event by the ease with which we can call it to mind." The more recently an event has occurred, or the more vivid our memory of something like it in the past, the more "available" an event will be in our minds—and the more probable it will seem to happen again. But that's not the right way to assess risk. An event does not become more likely to recur merely because its last occurrence was recent or memorable.

Just say these words aloud: *airplane crash.* What do you see in your mind's eye? Chances are, you imagine a smoky cabin filling with screams, a bone-shattering crunch, a giant fireball pinwheeling down a runway. In principle, says Paul Slovic, a psychologist at the University of Oregon, "risk is brewed from an equal dose of two ingredients—probabilities and consequences." But in practice, when we perceive the risks around us, the doses of those two ingredients are not always equal. Since the consequences of a crash can be so horrific, while the probabilities of a crash evoke no imagery at all, we get zero comfort from the fact that the odds against dying in a U.S. plane crash are roughly 6,000,000 to one. Those images of death are scary, while "6,000,000 to one" is an abstraction that conveys no feeling at all. ("I don't have a fear of flying," the basketball player Toni Kukoc once said. "I have a fear of crashing.") Once again, the emotional force of the reflexive brain overwhelms the analytical powers of the reflective brain.

On the other hand, we feel perfectly safe—if not immortal—when we're behind the wheel of our own car. Many travelers think nothing of having a couple of beers, then climbing into their car and driving to the airport with a cell phone in one hand and a cigarette in the other. Many of them even worry about whether their plane might crash—and remain utterly blind to the ways their own behavior is riddled with risk. The numbers tell the story: Only 24 people died on commercial aircraft in the U.S. in 2003, while 42,643 people were killed in car accidents. Adjusting for the distance traveled, you're about 65 times more likely to die in your own car than in a plane. And yet it's air travel that frightens us. Over the twelve months after the terrorist attacks of September 11, 2001, the fear of flying put far more people onto U.S. roads, causing an estimated 1,500 extra deaths in car crashes.

The more vivid and easily imaginable a risk is, the scarier it feels. People will pay twice as much for an insurance policy that covers hospitalization for "any disease" than one that covers hospitalization for "any reason." Of course, by definition, "any reason" includes all diseases. But "any reason" is vague, while "any disease" is vivid. That vividness fills us with a fear that makes no economic sense. However, it makes perfect emotional sense.

The emotion generated in our reflexive system can shove our analytical abilities aside, so the presence of one risk can make other things seem riskier, too. In the wake of September 11, for example, the Conference Board's Consumer Confidence Index, a measure of how Americans feel about the economic outlook, slumped by 25%. And the number of people who said they planned to buy a car, a home, or a major appliance in the coming six months dropped by 10%.

When an intangible feeling of risk fills the air, you can catch other people's emotions as easily as you can catch a cold. Merely reading a brief newspaper story about crime or depression is enough to prompt people into more than doubling their estimates of the likelihood of unrelated risks like divorce, stroke, or exposure to toxic chemicals. Just as when you have a hangover the slightest sound can seem deafening, an upsetting bit of news can make you hypersensitive to anything else that reminds you of risk. As is so often the case with the reflexive brain, you may not realize that your decisions are driven by your feelings. Roughly 50% of people can recognize when they have been disturbed by a bit of negative news, but only 3% admit that being upset may influence how they react to other risks.

Our intuitive sense of risk is driven up or down by what Paul Slovic calls "dread" and "knowability." Those two factors, he explains, "infuse risk with feelings."

- Dread is determined by how vivid, controllable, or potentially catastrophic a risk seems to be. Repeated surveys have found that people consider handguns a bigger risk than smoking. Because we can choose not to smoke (or choose to quit if we do), the hazards of smoking seem to be under our control. But there's not much you can do to prevent some thug from putting a bullet through your head at any moment, and TV cop shows pump your living room full of gunshots every night—so handguns seem scarier. Yet smoking kills hundreds more people than handguns do.

- The "knowability" of a risk depends on how immediate, specific, or certain the consequences appear to be. Fast and finite dangers (fireworks, skydiving, train crashes, etc.) feel more "knowable" (and less worrisome) than vague, open-ended risks like genetically modified foods or global warming. Americans rate tornadoes as a much more frequent killer than asthma. Because asthma moves slowly and many of its victims survive, it seems less dangerous, even though it kills many more people. If the consequences of a risk are highly uncertain and poorly understood, any perceived problem can trigger a frenzy of publicity. Thus hedge funds, those giant investment pools that operate in almost complete secrecy, become front-page news whenever they lose money.

Dread and knowability come together to twist our perceptions of the world around us: We underestimate the likelihood and severity of common risks, and we overestimate the likelihood and severity of rare risks—especially if we have never personally experienced them. When we feel we are in charge and we understand the consequences, risks will seem lower than they truly are. When a risk feels out of our hands and less comprehensible, it will feel more dangerous than it actually is. It's as if we see the world through warped binoculars that not only magnify whatever is remote but shrink whatever is near.

That's why so many people buy flight insurance at the airport: The chance of dying in a plane crash is almost zero, and most passengers are already covered by life insurance anyway, but air travel still *feels* risky. Meanwhile, roughly three-quarters of all Americans living in vulnerable areas have no flood insurance. Because homeowners can readily see how high the water has risen in the past, and because they can easily invest in drainage systems and other techniques that seem to control the risk of flooding, they feel safer than they really are. Hurricane Katrina exposed how dangerous this feeling of safety can be.

In the stock market, these quirks of risk perception can be a big distraction. On March 22, 2005, a woman named Anna Ayala was eating at a Wendy's restaurant in San Jose, California. She spooned a helping of chili into her mouth, started to chew, and then spat out a human finger. When the news broke, Wendy's stock fell 1% on heavy trading volume, and by April 15, 2.4% had been chopped off the market value of the stock. Customers turned away, costing the company an estimated $10 million in revenues. But investigators soon found that Ayala had planted the finger (which one of her husband's coworkers had lost in an industrial accident) in the bowl of chili herself. Wendy's business recovered steadily, and anyone who sold the stock in the initial panic was left feeling like somebody with ten thumbs, as it nearly doubled over the coming year.

Much the same thing happened in June 1999, when eBay's website crashed and "went dark" for twenty-two hours. Trading in Beanie Babies and G.I. Joes ground to a halt, costing eBay about $4 million in lost fees and causing consternation among thousands of buyers and sellers. Over the next three trading days, eBay's shares fell 26%, a loss of more than $4 billion in market value. Because the Internet was still relatively young, many investors had no idea when eBay could fix the problem—so the consequences seemed highly uncertain, arousing enormous fear. But eBay's site was soon running smoothly, and the stock almost tripled over the next five years.

In short, overreacting to raw feelings—"blinking" in the face of risk—is often one of the riskiest things an investor can do.

The Hot Button of the Brain

Deep in your brain, level with the top of your ears, lies a small, almond-shaped knob of tissue called the amygdala. When you confront a potential risk, this part of your reflexive brain acts as an alarm system—generating hot, fast emotions like fear and anger that it shoots up to the reflective brain like warning flares. (There are actually two amygdalae, one on the left side of your brain and one on the right, just as office elevators often have one panic button on either side of the door.)

The amygdala helps focus your attention, in a flash, on anything that's new, out of place, changing fast, or just plain scary. That helps explain why we overreact to rare but vivid risks. After all, in the presence of danger, he who hesitates is lost; a fraction of a second can make the difference between life and death. Step near a snake, spot a spider, see a sharp object flying toward your face, and your amygdala will jolt you into jumping, ducking, or taking whatever evasive action should get you out of trouble in the least amount of time. This same fear reaction is triggered by losing money—or believing that you might.

While other parts of your brain also generate fear, the amygdala's role is probably the best understood so far. While it can fire up around pleasant stimuli, too, it seems to be custom-fit for fear. The amygdala links directly to areas that manipulate your facial muscles, control your breathing, and regulate your heart rate. Fibers emanating from the amygdala also signal other parts of the brain to release norepinephrine, a kind of starter fluid that prepares the delivery of energy to your muscles for instant action. And the amygdala helps infuse your bloodstream with cortisol, a stress hormone that assists the body in responding to an emergency.

Remarkably, the amygdala can flood your body with fear signals before you are consciously aware of being afraid. If you smell smoke in your home or office, your heart will hammer and your feet will start flying well before any fire alarm goes off. In the presence of real or potential danger, the amygdala waits for nothing. "You don't need to fall off a ten-story building in order to be afraid of falling off it," says neuroscientist Antoine Bechara of the University of Southern California. "Your brain doesn't need actual experience."

A rat born and bred in a laboratory, where it has never seen a cat, will nevertheless freeze instantly if it encounters one. The rat's amygdala senses danger and triggers an automatic fear response—even though the rat has no idea what a cat is. A rat with an injured amygdala, however, will not freeze; instead, it will scamper up to the cat, climb on its back, even nibble on its ear. (Fortunately for the rats, in these experiments the cat has been sedated.) When the amygdala is damaged, the sense of fear is broken.

"Emotion can be beneficial when it is triggered by a chain of prior experiences," explains Bechara. "Otherwise, you would take forever to decide." In speeches to investors, I sometimes reach into a sealed bag, pull out a rattlesnake, and throw it into the audience. In theory, "rational" people should sit there while the snake flies through the air. They should take a few moments to decide whether it's worth causing a ruckus by scrambling out of the way, and to calculate the odds that a writer would throw a live snake at them during a speech. Having weighed the potential costs against the possible benefits, "rational" people should conclude that there's no cause for alarm. Instead, they scream and bolt out of the chair. (Needless to say, the snake isn't real; it's a rubber toy.)

Does this lightning response of the amygdala make us "irrational"? Of course not. As it helped our remote ancestors survive, the fear reflex remains a vital survival tool in daily life today: It makes you look both ways before you cross the street and reminds you to hold the railing on high balconies. However, when a potential threat is financial instead of physical, reflexive fear will put you in danger more often than it will get you out of it. Selling your investments every time they take a sudden drop will make your broker rich, but it will just make you poor and jittery.

Social signals can set off the hot button of your brain as easily as physical dangers can. When photographs of fearful faces are flashed for 33 one-thousandths of a second—and immediately followed by longer exposures of emotionally neutral faces—your reflective mind has no time to become aware that you saw anything scary. But your reflexive brain will "know" it with lightning speed. The exposure to a fearful face for just a thirtieth of a second is enough to spark intense activation in the amygdala, priming your body for action just in case this subliminal threat turns out to be real.

The amygdala also enables us to spot fearful body language in a split second: The mere glimpse of someone standing hands-up makes us expect a mugging, and a hunched and cowering figure makes us anticipate a beating. If you were exposed for just a third of a second to images of anonymous actors making agitated gestures, your amygdala would instantly "catch" their fear, alerting the stress systems throughout your body in a flash.

Finally, the amygdala is sensitive to that uniquely human way of conveying threats—through language. Brain scans show that your amygdala will fire more intensely in response to words like *kill, danger, knife,* or *torture,* than to words like *towel, formation, number,* or *pen.* Researchers in France have recently shown that a frightening word can make you break out in a sweat even if it appears for only 12 one-thousandths of a second—roughly 25 times faster than the blink of a human eye! (No wonder you cringe when someone says, "I got killed on that fund" or "Buying that stock would be like trying to catch a falling knife.")

An alarming word or two can even be powerful enough to transform your memories. In a classic experiment by psychologist Elizabeth Loftus, people viewed video footage of car accidents. Some of the viewers were asked how fast the cars were going when "they hit each other." Others were asked how fast the cars were going when "they smashed into each other." Even though both groups saw the same videos, the people who were prompted by the words "smashed into" estimated that the cars were going 19% faster. "Hit" may not sound very scary, but "smashed into" does. That evidently switches on the amygdala, splashing emotion back onto your memory and changing your perceptions of the past.

What does all this tell us about investing? Humans are reflexively afraid not just of physical dangers, but also of any social signal that transmits an alarm. A television broadcast from the floor of the stock exchange on a bad trading day, for example, combines a multitude of cues that can fire up the amygdala: flashing lights, clanging bells, hollering voices, alarming words, people gesturing wildly. In a split second, you break out in a sweat, your breathing picks up, your heart races. This primal part of your brain is bracing you for a "fight or flight" response before you can even figure out whether you have lost any money yourself.

Both actual and imagined losses can flip this switch. Using brain scans, one study found that the more frequently people were told they were losing money, the more active the amygdala became. Other scanning experiments have shown that even the expectation of financial losses can switch on this fear center. Traumatic experiences activate genes in the amygdala, stimulating the production of proteins that strengthen the cells where memories are stored in several areas of the brain. A surge of signals from the amygdala can also trigger the release of adrenaline and other stress hormones, which have been found to "fuse" memories, making them more indelible. And an upsetting event can shock neurons in the amygdala into firing in synch for hours—even during sleep. (It is literally true that we can relive our financial losses in our nightmares.) Brain scans have shown that when you are on a financial losing streak, each new loss heats up the hippocampus, the memory bank near the amygdala that helps store your experiences of fear and anxiety.

What's so bad about that? A moment of panic can wreak havoc on your investing strategy. Because the amygdala is so attuned to big changes, a sudden drop in the market tends to be more upsetting than a longer, slower—or even a much bigger—decline. On October 19, 1987, the U.S. stock market plunged 23%—a deeper one-day drop than the Crash of 1929 that ushered in the Great Depression. Big, sudden, and inexplicable, the Crash of 1987 was exactly the kind of event that sparks the amygdala into flashing fear throughout every investor's brain and body. The memory was hard to erase: In 1988, U.S. investors sold $15 billion more shares in stock mutual funds than they bought, and their net purchases of stock funds did not recover to precrash levels until 1991. The "experts" were just as shell-shocked: The managers of stock funds kept at least 10% of their total assets in the safety of cash almost every month through the end of 1990, while the value of seats on the New York Stock Exchange did not regain their precrash level until 1994. A single drop in the stock market on one Monday in autumn disrupted the investing behavior of millions of people for at least the next three years.

The philosopher William James wrote that "an impression may be so exciting emotionally as almost to leave a *scar* upon the cerebral tissues." The amygdala seems to act like a branding iron that burns the memory of financial loss into your brain. That may help explain why a market crash, which makes stocks cheaper, also makes investors less willing to buy them for a long time to come.

Fright Makes Right

I learned how my own amygdala reacts to risk when I participated in an experiment at the University of Iowa. First I was wired up with electrodes and other monitoring devices—on my chest, my palms, my face—to track my breathing, heartbeat, perspiration, and muscle activity. Then I played a computer game designed by neuroscientists Antoine Bechara and Antonio Damasio. Starting with $2,000 in play money, I clicked a mouse to select a card from one of four decks displayed on the computer monitor in front of me. Each "draw" of a card made me either "richer" or "poorer." I soon learned that the two left decks were more likely to produce big gains but even bigger losses, while the two right decks blended more frequent but smaller gains with a lower chance of big losses. (The left decks were the rough equivalent of an aggressive growth fund that invests in risky small stocks, while the right decks resembled a balanced fund that mixes stocks and bonds for a smoother return.) Gradually, I began picking most of my cards from the decks on the right; by the end of the experiment I had drawn 24 cards in a row from those safer decks.

Afterward, I looked over the printout of my bodily responses with a profound sense of wonder. I could see that the paper was covered with jagged lines that traced my spiking heartbeat and panting breath as the red alert of risk swept through my body. But the reflective areas of my brain never had a clue that I was on edge. So far as I "knew," I was doing nothing more than calmly trying to make a few bucks by picking cards.

At first, the printout showed, my skin would sweat, my breath quicken, my heart race, and my facial muscles furrow immediately after I clicked on any card that cost me money. Early on, when I drew one card that lost me $1,140, my pulse rate shot from 75 to 145 in a split second. After three or four bad losses from the risky decks, my bodily responses began surging before I selected a card from either of those piles. Merely moving the cursor over the riskier decks, without even clicking on them, was enough to make my physiological functions go haywire—as if I had stepped toward a snarling lion. It took only a handful of losses for my amygdala to create an emotional memory that made my body tingle with apprehension at the very thought of losing money again.

My decisions, I now could see, had been driven by a subliminal fear that I sensed with my body even though the "thinking" part of my mind had no idea I was afraid. As anyone who has ever come upon a sudden danger knows, it's often only after the fact that you realize how keyed up you were in your moment of peril. My brain handled this danger the same way, even though it was a financial, not a physical, risk and even though it involved only play money, not real cash.

At least in the developed world, money has become an inherently desirable object. Current social pressures—plus centuries of tradition—lead us to equate money with safety and comfort. (Ironically, we even call stocks, bonds, and other investments "securities"!) So a financial loss or shortfall is a painful punishment that arouses an almost primitive fear.

"Money is a symbolic token of the problem of life," says neuroscientist Antonio Damasio. "Money represents the means of maintaining life and sustaining us as organisms in our world." Seen in this light, it's not surprising that losing money can ignite the same fundamental fears you would feel if you encountered a charging tiger, got caught in a burning forest, or stood on the crumbling edge of a cliff.

Ironically, this highly emotional part of our brain can sometimes help us act more rationally. When Bechara and Damasio run their card-picking game with people whose amygdalas have been injured, they find that these patients never learn to avoid choosing from the riskier decks. If amygdala patients are told that they have just lost money, their pulse, breathing, and other bodily responses show no change. With the amygdala knocked out, a financial loss no longer hurts.

The result is what Bechara calls "a disease of decision-making." With no emotional signal from the amygdala to alert the prefrontal cortex about how bad it will feel to lose money, these people sample cards from all the decks—good and bad—until they end up going broke. Normally, the amygdala plays a vital role as the alarm that signals "Don't go there!" But once the reflexive brain is impaired, then the reflective areas say, "Hmm, maybe I should try that one." Without the saving grace of fear, the analytical parts of the brain will keep trying to outsmart the odds, with disastrous results. "The process of deciding advantageously," says Damasio, "is not just logical but also emotional."

A team of researchers designed an even simpler game to test how fear affects our financial decisions. Starting off with $20, you could then risk $1 on a coin flip (or pass and risk nothing). If the coin came up heads, you would lose your $1; if it came up tails, you would win $2.50. The game ran for twenty rounds. The researchers tried the experiment on two groups: people with intact brains (or "normals") and people with injuries to emotional centers of the brain like the amygdala and the insula ("patients").

The "normals" were reluctant to bet. They gambled in only 58% of all the rounds (even though, on average, they could have come out ahead just by betting on every flip). And they proved the proverb "Once burned, twice shy": Immediately after a loss, the normals would bet only 41% of the time. The pain of losing $1 discouraged the normals from trying to win $2.50.

The people with damaged emotional circuits behaved very differently. They bet their dollar, on average, in 84% of all the rounds—and, even when the previous flip had lost them $1, the patients took the next bet 85% of the time. That's not all. The longer they played, the more willing the patients became to flip the coin again—regardless of how much they had lost. In their case, it's as if the pain circuits in the brain had been anesthetized, making it impossible for the patients to feel the anguish of loss. Therefore, they bet with abandon: Damn the consequences, full speed ahead!

The result? The people with emotionally impaired brains earned 13% more money than those whose brains were undamaged. With their fear circuits knocked offline, these people take chances that the rest of us are too scared to touch.

The lesson? It's not that you could raise your investing returns by whacking yourself upside the head with a hammer. It's that the fear of financial loss *always* lurks within the normal investing brain. When the market is flat or rising, your sense of fear may go into deep hibernation. But believing that you are fearless is very different from being fearless. During the peak of the bull market, investors bragged that they didn't mind taking big risks in the pursuit of bigger gains. But most of these people had never suffered a major financial loss—and the meltdown in the amygdala that goes along with it. That led all too many investors to the mistaken conclusion that big losses wouldn't bother them.

But you can't change the biological facts. Imagining that you can shrug off setbacks before you've ever suffered any is a disastrous delusion, since it leads you to take such high risks that huge losses become inevitable. When the bull market of the 1990s died, people lost trillions of dollars on stocks they never should have owned in the first place. These people paid a terrible price for their poor self-knowledge.

Is There Safety in Numbers?

Nowadays, investment herds often form in online chat rooms where intense peer pressure pulls each visitor toward the views of the most vocal and charismatic members. You look around and find a large support group all expressing similar views—so you feel "there's safety in numbers."

But groups of animals, points out UCLA ecologist Daniel Blumstein, "have more eyes, ears, and noses with which to detect predators." In general, animals in groups are *more* sensitive to risk than they are in isolation. The larger the group in which animals gather together, the sooner and faster they will tend to flee from danger. So there's safety in numbers only when there's nothing to be afraid of. The comfort of being part of the crowd can disappear in a heartbeat.

Of course, anyone who has ever been a teenager knows that peer pressure can make you do things as part of a group that you might never do on your own. But do you make a conscious choice to conform, or does the herd exert an automatic, almost magnetic, force? People were recently asked to judge whether three-dimensional objects were the same or different. Sometimes the folks being tested made these choices in isolation. Other times, they first saw the responses of either four "peers" or four computers. (The "peers" were, in fact, colluding with the researchers conducting the study.) When people made their own choices, they were right 84% of the time. When all four computers gave the wrong answer, people's accuracy dropped to 68%. But when the peer group all made the wrong choice, the individuals being tested chose correctly just 59% of the time. Brain scans showed that when people followed along with the peer group, activation in parts of their frontal cortex decreased, as if social pressure was somehow overpowering the reflective brain.

When people did take an independent view and guessed against the consensus of their peers, brain scans found intense firing in the amygdala. (There was no such pattern when they guessed independently of the computers, showing that it is human peer pressure that makes it so hard for us to think for ourselves.) Neuroeconomist Gregory Berns, who led the study, calls this flare-up in the amygdala a sign of "the emotional load associated with standing up for one's belief." Social isolation activates some of the same areas in the brain that are triggered by physical pain. In short, you go along with the herd not because you consciously choose to do so, but because it hurts not to.

Once you join the crowd, your feelings are no longer unique. A team of neuroscientists scanned the brains of people watching the classic spaghetti western *The Good, the Bad and the Ugly,* leaving the viewers free to daydream, get caught up in Ennio Morricone's eerie music, or wonder why Clint Eastwood can't stop squinting. Even so, a third of the surface of each viewer's cerebral cortex lit up in lockstep with the other viewers' brains—a striking phenomenon that the researchers call "ticking together." People's brains were especially prone to tick together at the most obvious turning points in the movie, like gunshots, explosions, or sudden plot twists. When emotions run high, individual brains converge to think almost as one. (If you have a DVD of *The Good, the Bad and the Ugly,* you can follow along on your computer, matching the footage of the movie with other people's brain activation patterns, at www.weizmann.ac.il/neurobiology/labs/malach/ReverseCorrelation/.)

"Ticking together" suggests that our own emotions tend to peak in synch with other people's reactions to the same stimuli. We move in herds partly because, although we *are* all individuals, our brains respond in common to common circumstances. When we face the same conditions, "ticking together" leads many of us to share the same emotions. If the financial news makes you feel anxious or afraid, surprised or elated, the chances are high that many other investors feel the same way.

Being part of a larger group of investors can make you feel safer when everything is going great. But once risk rears its ugly head, there is no safety in numbers: You may find that everyone in the herd is dumping your favorite stock and, in effect, running for their lives. One burst of bad news, and the support group can become a stampede. You will suddenly be all alone, just when nothing feels safe anymore.

When Nobody Knows the Odds

Military-intelligence scholar Daniel Ellsberg helped to bring down the presidency of Richard Nixon when, in 1971, he leaked the Pentagon Papers to the *New York Times.* That top-secret report documented systematic flaws of decision-making in the Vietnam War. Ellsberg was no stranger to the notion that people don't always have good judgment. A decade earlier, as an experimental psychologist at Harvard, he had published the results of a mind-bending little discovery that became known as the Ellsberg Paradox. Here's how it works. Imagine that you have two urns in front of you. They are open at the top so you can reach in, but you cannot see what is inside. The first—call it Urn A—contains exactly 50 red balls and 50 black balls. Urn B also contains exactly 100 balls; some are red and some are black, but you do not know how many there are of each. You will win $100 if you draw a red ball from either urn.

Which urn would you pick from? If you're like most people, you prefer Urn A.

Now let's repeat the game, but change the rules: This time, you win $100 if you draw a *black* ball from either urn. Which urn would you pick from now? Most people stick with Urn A. But that makes no logical sense! If you went with Urn A the first time, you obviously acted as if it contained more red balls than Urn B. Since you know Urn A has 50 red balls, your first choice implies that Urn B contains fewer than 50 red balls. Therefore, you should conclude that more than 50 balls in Urn B are black. Now that you are trying to draw a black ball, you should pick from Urn B.

Why, then, do people prefer Urn A in both the first and second rounds? In a press conference in 2002, U.S. Secretary of Defense Donald Rumsfeld made a widely mocked distinction between what he called "known knowns," "known unknowns," and "unknown unknowns." But—although he has less in common with Ellsberg than almost anyone else alive—Rumsfeld was right. "Known knowns," Rumsfeld explained, "are things we know that we know." In the case of known unknowns, he continued, "we know there are some things we do not know."

In those terms, Ellsberg's Urn A is a known known: You can be sure it has a 50/50 mix of red and black balls. Urn B, on the other hand, is a known unknown: You can be sure it contains both red and black balls, but you have no idea how many of each. Urn B is brimming with what Ellsberg called "ambiguity," and that feels scary. After all, what if 99 of the balls in Urn B somehow turn out to be red? Then you will stand a very high chance of winning nothing on the draw for black balls. The less sure we can be about the probabilities, the more we worry about the consequences. So we avoid Urn B, regardless of basic logic.

Ellsberg found that people persisted in choosing Urn A even after they realized it made no sense, and even if he asked them to bet money on whether they had picked the right urn. When Ellsberg tried his experiment on the leading economists and decision theorists of his time, many of them made the same mistake as the man in the street.

That's no surprise, since Ellsberg's Paradox is rooted in the same tension between thinking and feeling that drives so many of our investing decisions. A team of researchers recently scanned the brains of people who were asked to pick from a deck of 20 cards. Sometimes the players knew that the deck contained 10 red and 10 blue cards; at other times, all they knew was that the deck contained both red and blue cards. (They would miss out on a $3 gain if they picked the wrong card.) The first deck, like Ellsberg's Urn A, was a known known; the second, like Urn B, was a known unknown. When people considered picking from the ambiguous deck, the fear center in the amygdala went into overdrive. You can see this area sizzling with activity in Figure 7.1. What's more, thinking about an ambiguous bet dampened activity in the caudate, one of the brain's reward centers that, as we saw in Chapter Five, helps us trust someone and feel the pleasure of being in control of a situation. Not knowing the odds not only inflames our fears, but also strips us of the feeling that we are in charge.

Ellsberg's Paradox often shows up in the stock market. Even though the growth rate of every company is uncertain, some rates seem more predictable than others. When a company's growth seems reliable, Wall Street says it has "high visibility." Ellsberg might say it has "low ambiguity." Whatever you call it, investors pay a premium for this illusion of predictability:

- Stocks that are followed by more security analysts on Wall Street have higher trading volume, suggesting that investors prefer betting on companies that are eyeballed by more "experts."

- The more closely analysts agree about how much a company will earn over the coming year, the more investors will pay for the stock. (As we saw in Chapter Four, analysts are lousy at predicting corporate earnings; yet investors prefer a precise but wrong forecast over a vague but accurate one.)

- Among security analysts, 78% agree that ambiguity about future earnings "tends to make me less confident" investing in small stocks than large stocks.

- On average, the earnings of so-called "value" companies are more than twice as volatile as those of "growth" companies.

All this makes investing in value stocks or small stocks the equivalent of trying to pick a black ball from Urn B: The higher ambiguity makes your odds of success feel less certain. Picking from the "predictable" growth stocks in Urn A simply feels safer. So most investors steer clear of value companies and small stocks, driving their share prices down, and pile into big growth companies, sending their stocks soaring—at least in the short run. Over longer periods, however, growth stocks and the stocks most popular with analysts tend to earn lower returns than value stocks and underanalyzed companies. By avoiding stocks that are high in ambiguity, the investing public makes them underperform in the short run—creating bargains that go on to outperform over the long run.

Fighting Your Fears

When you confront risk, your reflexive brain, led by the amygdala, functions much like a gas pedal, revving up your emotions. Fortunately, your reflective brain, with the prefrontal cortex in charge, can act like a brake pedal, slowing you down until you are calm enough to make a more objective decision. The best investors make a habit of putting procedures in place, in advance, that help inhibit the hot reactions of the emotional brain. Here are some techniques that can help you keep your investing cool in the face of fear:

➢ **GET IT OFF YOUR MIND.** You'll never find the presence of mind to figure out what to do about a risk gone bad unless you step back and relax. Joe Montana, the great quarterback for the San Francisco 49ers, understood this perfectly. In the 1989 Super Bowl, the 49ers trailed the Cincinnati Bengals by three points with only three minutes left and 92 yards—almost the whole length of the field—to go. Offensive tackle Harris Barton felt "wild" with worry. But then Montana said to Barton, "Hey, check it out—there in the stands, standing near the exit ramp, there's John Candy." The players all turned to look at the comedian, a distraction that allowed their minds to tune out the stress and win the game in the nick of time. When you feel overwhelmed by a risk, create a John Candy moment. To break your anxiety, go for a walk, hit the gym, call a friend, play with your kids.

➢ **USE YOUR WORDS.** While vivid sights and sounds fire up the emotions in your reflexive brain, the more complex cues of language activate the prefrontal cortex and other areas of your reflective brain. By using words to counteract the stream of images the markets throw at you, you can put the hottest risks in cooler perspective.

In the 1960s, Berkeley psychologist Richard Lazarus found that showing a film of a ritual circumcision triggered instant revulsion in most viewers, but that this disgust could be "short-circuited" by introducing the footage with an announcement that the procedure was not as painful as it looked. Viewers exposed to the verbal commentary had lower heart rates, sweated less, and reported less anxiety than those who watched the film without a soundtrack. (The commentary wasn't true, by the way—but it worked.)

More recent, disgusting film clips—featuring burn victims being treated and closeups of an arm being amputated—have been shown to viewers by the aptly named psychologist James Gross. (Although I do not recommend watching it on a full stomach, you can view the amputation clip at www-psych.stanford.edu/~psyphy/Movs/surgery.mov.) He has found that viewers feel much less disgusted if they are given written instructions, in advance, to adopt a "detached and unemotional" attitude.

As we've learned, if you view a photograph of a scary face your amygdala will flare up, setting your heart racing, your breath quickening, your palms sweating. But if you view the same photo of a scary face accompanied by words like *angry* or *afraid,* activation in the amygdala is stifled and your body's alarm responses are reined in. As the prefrontal cortex goes to work trying to decide how accurately the word describes the situation, it overrides your original reflex of fear.

Taken together, these discoveries show that verbal information can act as a wet blanket flung over the amygdala's fiery reactions to sensory input. That's why using words to think about an investing decision becomes so important whenever bad news hits. To be sure, formerly great investments can go to zero in no time; once Enron and WorldCom started to drop, it didn't pay to think analytically about them. But for every stock that goes into a total meltdown, there are thousands of other investments that suffer only temporary setbacks—and selling too soon is often the worst thing you can do. To prevent your feelings from overwhelming the facts, use your words and ask questions like these:

Other than the price, what else has changed?

Are my original reasons to invest still valid?

If I liked this investment enough to buy it at a much higher price, shouldn't I like it even more now that the price is lower?

What other evidence do I need to evaluate in order to tell whether this is really bad news?

Has this investment ever gone down this much before? If so, would I have done better if I had sold out—or if I had bought more?

➢ **TRACK YOUR FEELINGS.** In Chapter Five, we learned the importance of keeping an investing diary. You should include what neuroscientist Antoine Bechara calls an "emotional registry," tracking the ups and downs of your moods alongside the ups and downs of your money. During the market's biggest peaks and valleys, go back and read your old entries from similar periods in the past. Chances are, your own emotional record will show you that you tend to become overenthusiastic when prices (and risk) are rising, and to sink into despair when prices (and risk) go down. So you need to train yourself to turn your investing emotions upside down. Many of the world's best investors have mastered the art of treating their own feelings as reverse indicators: Excitement becomes a cue that it's time to consider selling, while fear tells them that it may be time to buy. I once asked Brian Posner, a renowned fund manager at Fidelity and Legg Mason, how he sensed whether a stock would be a moneymaker. "If it makes me feel like I want to throw up," he answered, "I can be pretty sure it's a great investment." Likewise, Christopher Davis of the Davis Funds has learned to invest when he feels "scared to death." He explains, "A higher perception of risk can lower the actual risk by driving prices down. We like the prices that pessimism produces."

➢ **GET AWAY FROM THE HERD.** In the 1960s, psychologist Stanley Milgram carried out a series of astounding experiments. Let's imagine you are in his lab. You are offered $4 (about $27 in today's money) per hour to act as a "teacher" who will help guide a "learner" by penalizing him for wrong answers on a simple memory test. You sit in front of a machine with thirty toggle switches that are marked with escalating labels from "slight shock" at 15 volts, up to "DANGER: SEVERE SHOCK" at 375 volts, and beyond to 450 volts (marked ominously with "XXX"). The learner sits where you can hear but not see him. Each time the learner gets an answer wrong, the lab supervisor instructs you to flip the next switch, giving a higher shock. If you hesitate to increase the voltage, the lab supervisor politely but firmly instructs you to continue. The first few shocks are harmless. But at 75 volts, the learner grunts. "At 120 volts," Milgram wrote, "he complains verbally; at 150 he demands to be released from the experiment. His protests continue as the shocks escalate, growing increasingly vehement and emotional. . . . At 180 volts the victim cries out, 'I can't stand the pain' . . . At 285 volts his response can only be described as an agonized scream."

What would you do if you were one of Milgram's "teachers"? He surveyed more than 100 people outside his lab, describing the experiment and asking them at what point they thought they would stop administering the shocks. On average, they said they would quit between 120 and 135 volts. Not one predicted continuing beyond 300 volts.

However, inside Milgram's lab, 100% of the "teachers" willingly delivered shocks of up to 135 volts, regardless of the grunts of the learner; 80% administered shocks as high as 285 volts, despite the learner's agonized screams; and 62% went all the way up to the maximum ("XXX") shock of 450 volts. With money at stake, fearful of bucking the authority figure in the room, people did as they were told "with numbing regularity," wrote Milgram sadly. (By the way, the "learner" was a trained actor who was only pretending to be shocked by electric current; Milgram's machine was a harmless fake.)

Milgram found two ways to shatter the chains of conformity. One is "peer rebellion." Milgram paid two people to join the experiment as extra "teachers"—and to refuse to give any shocks beyond 210 volts. Seeing these peers stop, most people were emboldened to quit, too. Milgram's other solution was "disagreement between authorities." When he added a second supervisor who told the first that escalating the voltage was no longer necessary, nearly everyone stopped administering the shocks immediately.

Milgram's discoveries suggest how you can resist the pull of the herd:

■ Before entering an Internet chat room or a meeting with your colleagues, write down your views about the investment you are considering: why it is good or bad, what it is worth, and your reasons for those views. Be as specific as possible—and share your conclusions with someone you respect who is not part of the group. (That way, you know someone else will keep track of whether you change your opinions to conform with the crowd.)

■ Run the consensus of the herd past the person you respect the most who is not part of the group. Ask at least three questions: Do these people sound reasonable? Do their arguments seem sensible? If you were in my shoes, what else would you want to know before making this kind of decision?

■ If you are part of an investment organization, appoint an internal sniper. Base your analysts' bonus pay partly on how many times they can shoot down an idea that everyone else likes. (Rotate this role from meeting to meeting to prevent any single sniper from becoming universally disliked.)

■ Alfred P. Sloan Jr., the legendary chairman of General Motors, once abruptly adjourned a meeting this way: "Gentlemen, I take it we are all in complete agreement on the decision here. . . . Then I propose we postpone further discussion of this matter until our next meeting to give ourselves time to develop disagreement and perhaps gain some understanding of what the decision is all about." Peer pressure can leave you with what psychologist Irving Janis called "vague forebodings" that you are afraid to express. Meeting with the same group over drinks in everyone's favorite bar may loosen some of your inhibitions and enable you to dissent more confidently. Appoint one person as the "designated thinker," whose role is to track the flow of opinions set free as other people drink. According to the Roman historian Tacitus, the ancient Germans believed that drinking wine helped them "to disclose the most secret motions and purposes of their hearts," so they evaluated their important decisions twice: first when they were drunk and again when they were sober.

CHAPTER EIGHT

Surprise

> What can no longer be imagined must happen, for if one could imagine it, it would not happen.
>
> —*Karl Kraus*

Oops

Everyone knows the feeling: Out of habit, without looking, you sit down. And then you discover that the last person who used the toilet did not put the seat back down. With a shocking start just a few milliseconds and millimeters before you collapse backward into the bowl, you regain your balance and right yourself in midair—followed by a sharp scream or an angry curse.

What's striking about this experience is not only how common it is, but how fine and fast your response is. A toilet with the seat up is only about 8% shorter than a toilet with the seat down, but your brain is exquisitely sensitive to the slightest difference between what you expect and what you get.

It takes a while after such a near-miss for your heart to stop hammering. That's the simple power of surprise: Getting one thing when we expected another is one of the most potent ways we learn from our experience. Your pounding heart and rattled nerves ensure that it will be a good while before you sit down again without looking first. "There is little evolutionary value in accurately recording everything in our environment," says R. Douglas Fields, a neuroscientist at the National Institutes of Health. "But there is a premium on detecting novelty." Without the shock of recognition that something has gone wrong, you would keep making mistakes until the cumulative course of your actions led you into an irreversible mess.

For better or worse, your brain generates the same lightning-fast response to an unexpected change in an investment position as it does to a toilet seat in the wrong position. Of course, if you were constantly surprised by everything, you would live your entire life at fever pitch and die of nervous exhaustion. Fortunately, the more often you are exposed to something, the less intensely your brain tends to respond to it, a process known as adaptation. "As you become more practiced at or familiar with something, your brain activation declines," says neuroscientist Michael Gazzaniga of the University of California at Santa Barbara. "That reduces the metabolic load on your brain."

Your neurons conserve energy by downshifting, sending out fewer signals per second, as a potential danger becomes more familiar and less threatening. But another sudden change is all it takes to kick your neurons into high gear again. It's the feeling of surprise—the new and unexpected—that takes your brain off autopilot just in time to react to what has changed.

In the financial markets, even a minor surprise can upset you into making major changes, regardless of whether a sudden shift in strategy is really justified. So it's important to figure out how to limit the number of unexpected shocks you fall prey to, and how to minimize your own sense of panic when something does catch you unawares. This chapter will help make investing surprises a little less surprising for you.

"What's the Best Building to Jump Off Of?"

On January 31, 2006, Google Inc. announced its financial results for the fourth quarter of 2005: revenues up 97%, net profit up 82%. It's hard to imagine how such phenomenal growth could be bad news. But Wall Street's analysts had expected Google to do even better. The result: a "negative earnings surprise," or a shortfall between the market's expectations and the actual outcome. And the result of that surprise, in turn, was panic. As soon as the news of the negative surprise hit, Google's stock fell 16% in a matter of seconds, and the market in the shares had to be officially halted. When trading resumed, Google, whose stock had been at $432.66 just minutes earlier, was hammered down to $366. The bottom line was bizarre: Google earned about $65 million less than Wall Street had expected, and in response Wall Street bashed $20.3 billion off Google's market value.

In the online message boards dedicated to the stock, investors were in shock. "IT'S OUT . . . OH MY GOD, I CANT BELIEVE IT," gasped one Google fan. Another moaned, "here it comes i am about to throw up." Yet another lamented, "This is like the worst possible scenario that could have happened. I feel completely ill and sick. . . . this is a horrible and horrific day. I'm totally awe struck and feel completely shocked. . . . This is a dark dark day for Google." A post by a typographically challenged investor with the handle "bodjango2003" said it all: "What's the best Building to Jump Off of??????????"

Every investor should know that it's impossible to predict corporate earnings to the penny. Even so, people keep trying to do just that—and end up being surprised again and again. One study looked at more than 94,000 estimates of quarterly earnings over roughly twenty years and found that they resulted in negative surprises more than 29,000 times. More than 1,250 times in 2005 alone, companies announced quarterly earnings that fell one penny per share short of what Wall Street expected. According to Numeric Investors L.P. of Cambridge, Massachusetts, the shares of these companies immediately fell by an average of 2%.

Often, the results would be funny if they weren't so scary: In January 2006, when Juniper Networks Inc. announced quarterly earnings per share that were just one-tenth of a penny less than analysts expected—along with a slight slowdown in projected future growth—the market value of the stock almost instantly fell 21%, or $2.5 billion.

The least surprising aspect of earnings is that they are full of surprises; the most surprising aspect is that investors are continually taken by surprise. Why don't we learn more from experience? What makes negative earnings surprises so common? How come a $65 million booboo can set off a $20 billion bloodbath? The best way to answer these questions is to drill down into the brain and see what's going on in there.

The Belt of the Unexpected

In a typical day, you say something like "That's amazing!" or "Wow!" or "You've got to be kidding" over and over again. Our superabundant sense of surprise appears to be one of the main features that set humans apart from other animals. Where does it come from?

Humans and great apes—chimpanzees, gorillas, and orangutans—are the only land mammals that have specialized neurons called spindle cells in a central forward region of the brain known as the anterior cingulate cortex, or ACC. Humans have at least twice as many spindle cells in this region as the great apes do. The spindle cells, which resemble partially unwound corkscrews, may latch on to signals from other areas of the brain, helping the ACC to focus attention, perceive pain, and detect errors. The ACC also assists in generating the feeling of surprise when your normal expectations are shattered. (Some neuroscientists call it the "Oops!" center or the "Oh, s--t!" circuit.) *Cingulate* is from the Latin for "belt," and this part of the cortex lies directly above a strip of tissue that stretches like a band along the top of the limbic system. There's no doubt that other regions of the brain are also sensitive to surprise, but the ACC has been studied the most thoroughly so far.

The ACC must have given our ancestors an evolutionary edge. Because their ability to walk upright enabled them to range over a wider territory, the early hominids encountered a greater variety of both risks and rewards than other primates did. "This rudimentary system in monkeys probably expanded in humans to signal more general errors and violations of expectancy," says Harvard neuroscientist George Bush. "It's more important to know about mistakes sooner than it is to know about successes." The farther our ancestors ranged from "home," the more vital it became to make this kind of decision quickly and accurately. For these restless protohumans, the familiar was more often mixed with the unfamiliar; to give one possible example, a variety of berry that looked only slightly different from the usual kind might well be poisonous. The hominids that reacted with the greatest surprise to the least differences were the most likely to survive.

"These cells process the rapid integration of very large volumes of information across time and space on a second-by-second basis," says John Allman, a neuroscientist at the California Institute of Technology. "It's an intuitive system that's built for speed. In a state of nature, there's no luxury of working through all the logical steps to arrive at the ideal 'rational' solution. Where uncertainty is maximal, the importance of learning is maximal and attention is highly focused." Neurons in the

ACC respond to a variety of surprising or conflicting events in less than three-tenths of a second.

The ACC receives input both from dopamine neurons that carry reward signals and from neurons originating in the amygdala that fire in response to risk. It also is closely connected to the thalamus, at the center of the brain, which helps direct attention to the input from your senses (like sights or sounds or smells). And the ACC is linked to the hypothalamus, another part of the reflexive brain that acts much like a thermostat, regulating your pulse, blood pressure, body temperature, and blood chemistry to keep them all close to the proper setting. When a surprise jolts your ACC, it can hit your hypothalamus in turn, knocking your internal thermostat out of kilter. No wonder a one-penny shortfall can put Wall Street into a multibillion-dollar panic.

You can sense your own ACC at work in a simple experiment called the Stroop test. Devised in 1935 by a psychologist with a name out of a Marx Brothers movie, J. Ridley Stroop, this task asks you to name the color in which various words are printed. (It's harder than it sounds. You can try a basic version at www.jasonweig.com/stroop.ppt.) The first change will almost certainly catch you by surprise; even if you do get it right, you will probably feel yourself doing a double-take that slows you down. After a few stumbles, it will get easier. Neuroscientist Jonathan Cohen of Princeton University has shown that the more active your ACC is when you flub one step in the Stroop test, the faster you will succeed at the next one. By alerting you that something has gone wrong this time, the ACC enables you to adjust your behavior to get it right next time. Once burned, twice shy.

The Asymmetry of Surprise

Even in monkeys, the cingulate flares up when a choice leads to a lower reward than expected (for instance, getting less fruit juice). Among humans, money—that modern product of advanced civilization—triggers a peculiarly primitive response when surprise is involved. In one experiment, people had to identify the odd letter in a string as fast as possible (for example, the S in HHHSHHH). Each time they made a mistake, they could gain money, lose money, or suffer no financial consequence. When errors resulted in a monetary loss, the ACC fired much more intensely than it did when no money was at stake. And a wide swath of cells in the brainstem (a region atop the spinal cord that regulates basic bodily functions) also switched on whenever a mistake cost money. When an identical error did not produce a financial penalty, however, the brainstem barely responded at all. (See Figure 8.1.) That's especially striking because the brainstem is one of the most ancient parts of the human mind.

"People make mistakes all the time," says University of Michigan psychologist William Gehring, who helped run this study. "But the ones we really care about, and the ones we try hardest to avoid in the future, are the ones that have an important negative consequence"—like losing money.

Using tiny electrodes, researchers measured the activity of single neurons in the ACC while people tried to move a joystick in the right direction to earn a variable cash reward. It turns out that 38% of the neurons fired when the amount of money suddenly shrank, but only 13% were triggered when people got a larger than expected gain. And those neurons that did respond to a positive surprise sent out weaker signals. With these insights, we can finally understand why stocks that beat Wall Street's expectations go up an average of 1%, while those that come up short lose an average of 3.4%. At the most basic biological level, a positive surprise packs a lot less punch than a negative one.

Why Apple Got Bruised

The intensity of your surprise depends largely on how unexpected the surprise is. The longer a sequence has repeated, the more vehemently your ACC will respond when the pattern is broken. Batches of neurons throughout your reflexive brain will join in, especially in the insula, the caudate, and the putamen—areas that help generate intense emotions, including disgust, fear, and anxiety. Scott Huettel, a neuroeconomist at Duke University, has shown that the ACC reacts roughly three times more vigorously if a pattern reverses after eight repetitions than it does after a three-in-a-row pattern is broken.

The stock market provides uncanny real-world proof of Huettel's laboratory findings: The more times in a row a company has exceeded Wall Street's expectations, the worse its stock gets whacked when it finally misses the analysts' forecasts. While a shortfall after a run of three good earnings reports trims just 3.4% off the price of the typical growth stock, a miss after a streak of eight positive quarters hacks off 7.9%.

So there's a lot more at stake after a string of successes. When expectations are especially high, a miss hurts even worse; that's why a market superstar like Google gets hit extra hard after any shortfall. And "growth" stocks, with the accelerating profits that most investors prize, are more vulnerable to a negative surprise than slower-moving "value" stocks are. When firms warn in advance that they will be unable to meet Wall Street's lofty expectations, their shares lose an average of 14.7% in two days.

The more closely all the analysts agree on what a company's earnings will be, the worse the stock will get walloped if the actual numbers fall short. In addition, the higher the overall market rises, the harder any single stock will get hammered when it stands out with a negative earnings surprise. And Wall Street's analysts are twice as likely to downgrade a company after an unexpected decline in its stock price as they are after a surprise increase—even though a short-term fall in price makes a stock more attractive as a long-term value. There's nothing like a negative surprise to deepen Wall Street's chronic confusion between price and value.

How big and bad can a surprise be? On September 28, 2000, Apple Computer Inc. announced that its quarterly earnings would be about $55 million less than it had previously estimated. At about 27% better than the same period a year earlier, it was hardly a poor result. But Wall Street's analysts were expecting Apple to do even better. The next trading day, Apple stock fell by 52%, vaporizing $5 billion of the company's total market value. For each million dollars of Apple's shortfall in earnings, Wall Street slashed the market capitalization of the stock by more than $90 million.

Now that Apple's shares were available at more than 50% off, did analysts raise their ratings? Certainly not: Instead, they scurried to downgrade it, warning investors to stay away now that a "cloud" hung over the stock. But Apple's problems were only temporary; in no time, the iMac and iPod were powering the company to record profits. Investors who got scared out of Apple by that one negative surprise were in for a much nastier surprise: Over the next six years, the stock sextupled. Even people who bought Apple just before its "earnings warning" could have more than doubled their money if they had hung on instead of selling.

The Bad News Bears

Of course, the corporate world is well aware that investors hate negative surprises. In the 1990s, CEOs and other senior executives fixated on "making the numbers"—that is, ensuring that they would meet Wall Street's earnings expectations to the penny. In a survey at more than four hundred leading U.S. companies, fully 78% of senior financial managers said they would be willing to hurt the long-term value of their businesses in order to prevent a short-term dip in earnings. One chief financial officer admitted that the company would delay routine maintenance to meet quarterly earnings expectations, even if putting off this vital upkeep would cost far more money down the road. Another executive revealed that when his company earned a $400 million profit by selling off one of its businesses, he used a complex financing technique to spread the gain in equal $40 million increments over the next ten financial quarters. While this gimmick created the illusion of "smoother" earnings in the future, it incurred very real (and unnecessary) investment banking fees in the present.

Companies that consistently give "earnings guidance" to Wall Street—by setting a target that they attempt to hit to the penny—end up spending much less on research and development. Their fear of giving Wall Street a current surprise hurts the future profitability of these companies, since today's research and development budget can be a leading source of tomorrow's growth. Companies that cannot bear the pain of a surprise in the present become afraid of investing in their own future.

Over time, thousands more companies beat earnings forecasts by one or two pennies than fall short by the same amount—a difference far too great to be a coincidence. As Harvard economist Richard Zeckhauser explains, when earnings are in danger of "falling just short of thresholds," they are "managed upward" by company insiders. But these short-run fixes fall apart in the long run. Firms that barely meet their earnings targets—implying that they had to pull out all the stops to get there—end up with lower earnings growth (and stock performance) than companies that either beat their targets by a wide margin or fall short.

A few years ago, Joseph Nacchio, the CEO of Qwest Communications International Inc., was obsessed with hitting Wall Street's earnings targets. "The most important thing we do is meet our numbers," Nacchio declared at a companywide meeting in January 2001. "It's more important than any individual product, it's more important than any individual philosophy. . . . We stop everything else when we don't make the numbers." And if Qwest couldn't make the numbers, it could make up the numbers. To avoid the terrible penalty that Wall Street deals out for negative surprises, Qwest relied on accounting gimmicks, delaying current expenses into the future and moving revenues from the future into the present. Employees were ordered so often to use these tricks that they nicknamed them "heroin." But Qwest overdosed: The stock fell 65% in 2001 and another 65% in 2002 as the company was forced to erase more than $2.5 billion in phantom earnings. Desperate to avoid even the tiniest surprise, Qwest ended up creating a monster.

Breaking the Cycle of Surprise

After writing about Wall Street since 1987 and studying centuries' worth of financial history, I have become convinced that the prevailing view of what the future holds is almost always wrong. In fact, the only incontrovertible evidence that the past offers about the financial markets is that they will surprise us in the future. The corollary to this historical law is that the future will most brutally surprise those who are the most certain they understand it. Sooner or later, sometimes slowly and sometimes suddenly—but with a diabolical ability to root out everyone who has ever gazed into a crystal ball—the financial markets always humiliate whoever thinks he knows what's coming. So the best way to immunize yourself against being surprised is to *expect* to be surprised. In an axiom proposed by Edgar Allan Poe and perfected by G. K. Chesterton, "wisdom should reckon on the unforeseen." Here are some specific steps you can take to embrace the unexpected.

➢ **EVERYBODY KNOWS NOTHING.** Investors often justify their decisions by saying, "Everybody knows that . . ." In 1999, for example, everybody knew that the Internet would change the world. In 2006, everybody knew that energy prices would keep going up. But whatever "everybody knows" is already embedded in the price of a stock or the expectation of an entire market. If what "everybody knows" does come perfectly true, the price won't change. If it comes only partly true, however, then the price will collapse. Unless you can come up with insights of your own that have nothing to do with what "everybody knows," you don't know anything that the market hasn't already built into the price. Whenever you are tempted to do what everybody else is doing, don't. Instead, try to find a less obvious investment opportunity that most people have somehow overlooked.

➢ **HIGH HOPES CAUSE BIG TROUBLE.** As money manager David Dreman has pointed out, expectations are usually so high for glamorous growth stocks that a positive surprise will barely make a ripple, while even the smallest negative surprise can capsize them. On average, a growth company that misses earnings by as little as three cents per share will see its stock drop two to three times farther than a value company with a similar shortfall. Conversely, expectations are so low for cheap value stocks that a negative surprise won't hurt much; but a positive surprise can send a value stock shooting straight up.

In the short term, value stocks tend to have lumpier, less linear earnings than growth stocks do. As the profits and share prices of value stocks bounce around, our brains probably seek to interpret them as an alternating pattern ("up, down, up, down") rather than the simple repetition that growth stocks often show in the short run ("up, up, up"). Alternation requires extra mental energy to understand. Experiments in neuroeconomics have shown that the brain takes longer to grasp an alternating pattern than a repeating one—about six iterations, as opposed to two.

That may explain why value stocks are so consistently underpriced: Because the path of their earnings growth is more erratic in the short run, your reflective brain has to work harder to predict what's coming next. The simple upward sweep of a growth stock, on the other hand, automatically lights up the emotional circuits in your reflexive brain. A steadily rising price automatically *feels* more predictable. The longer a company keeps beating Wall Street's expectations, the more "buy" orders outnumber "sell" orders for its stock; buying demand is over five times higher after six straight positive surprises than it is after two in a row. That helps drive growth stocks up, at least for a while. A string of positive surprises makes investors more confident that a stock is "predictable"—but also jacks up the risk of a horrible loss if the next surprise turns out to be negative.

Because repetition is so much easier for our brains to interpret than alternation, these short-term patterns lead us to overlook the longer-term truth: Over time, value investing is at least as lucrative as a growth-only strategy.

➢ **TRACK THE "WHYS" OF SURPRISE.** In a classic experiment, psychologists mixed carefully faked cards into a deck of regular playing cards. For example, they added an ace of diamonds printed in black, a six of clubs printed in red. This simple switch triggered several results. First of all, it took the average person four times longer to name the suit and value of the altered cards than the normal ones. Secondly, people were flummoxed: They were sure that something was wrong but unsure what. Shown a red club or a black diamond, people often insisted it was printed in "purple" or "brown." One person sputtered at the sight of a red spade, "I'll be damned if I know now whether it's red or what!" On the other hand, the more times people had previously seen an altered card, the easier it became for them to identify it correctly and quickly.

The obvious lesson is that the more often you see things that surprise you, the less likely you are to be flustered. Keeping careful mental notes—or better yet, written notes in an "emotional registry" or investing diary—will help you learn from your past surprises. Keep track of what surprised you, how you felt, and what you did in response. Pay special attention to describing what had happened earlier to make the surprise so unexpected. Flesh out your explanation by filling in the blank in a sentence like "This outcome took me by surprise because ______________________________." Instead of writing a generalization like "I just wasn't expecting it," think of specific factors, like "the stock had doubled since I bought it" or "all the news about the company was positive."

➢ **STAY AWAY FROM GUIDED MISSILES.** Many companies give "earnings guidance," doing a nudge-nudge-wink-wink with Wall Street's analysts to hint at what the next quarter's earnings will be. If an analyst estimates earnings of $1.43 per share, the company's chief financial officer may say, "We think that's a tad aggressive," leading the analyst to cut his forecast to $1.42. Then, when the company earns $1.43, the analyst and the company alike can boast that it "beat the forecast." Calling such a cynical manipulation a "positive surprise" is just silly. But the negative surprises that inevitably follow are very real, as the messy reality of business life comes barging in out of left field. Interest rates or oil prices suddenly rise or fall; a hurricane or earthquake hits; workers go on strike; a competitor invents a great new product. The result is a guided missile: a stock whose earnings guidance blows up precisely when it can do the worst damage.

For sensible investors, there is only one logical conclusion: You should ignore the entire farce. Increasing numbers of brave companies refuse to give any earnings guidance; among them are Berkshire Hathaway, Citigroup, Coca-Cola, Google, InterActive Corp., Mattel, Progressive Corp., and Sears. Instead of wasting precious energy trying to make their results match their guidance in the short run, these companies have the luxury of focusing on improving their performance over the long run. On average, firms that give less guidance should be more rewarding over time.

➢ **LOOK UNDER THE STATISTICAL HOOD.** The financial industry teems with people who are experts at twisting numbers to separate investors from their cash. Here are some of their statistical gimmicks, and tips on how you can keep them from taking you by surprise.

Incubation: Mutual fund companies create private portfolios, then track which of them succeed. The failures are terminated in secret, while the successes are trumpeted in advertisements hyping their market-beating track records. Across different fund categories, this fancy footwork can add between 0.2 and 1.9 percentage points annually to the average reported return—even though no one outside of the fund company ever earned it. "Incubated" funds typically go on to underperform in the long run. To avoid being surprised by one of these babies, always read the fine print in the prospectus of a new fund. Among the red flags: footnotes under "inception date" indicating that the fund was not available to the public early in its life, total net assets that stayed at or below $1 million in the fund's first year, and total return that was inexplicably high in year one.

Bait and switch: Brokers and fund companies wait until one of their investments—any of their investments—is doing well over some time period, and then advertise that track record. As soon as that investment stops doing well, they hype the results from whatever time period happens to look better. The simplest defense is to look at returns over several periods that begin and end on different dates.

In another bait and switch, fund companies run ads bragging about how many of their portfolios have earned "five stars," the highest rating from the Morningstar research firm. But, as a matter of policy, Morningstar awards five stars to 10% of all funds. If a company boasts that three of its funds have five-star ratings, but it offers a total of 60 different portfolios, that's not something to brag about; it's something to be embarrassed about. The company has only half as many top-rated funds as the average firm. There's an easy way to keep from being surprised by this kind of statistical shell game: Whenever someone throws a number at you, always remember to ask, "Compared to what?"

Burying the dead: Often, the track records of losing investments are erased from history as if they had never existed at all. The reported returns of technology mutual funds in 2000, for example, no longer include the results of dozens of failed portfolios that have since gone out of business. Once funds close down, they drop out of the averages that are compiled by firms like Morningstar and Lipper. Including only the funds that survived, and ignoring those that died, the average tech fund lost 30.9% in 2000. But all the tech funds then in existence actually lost an average of 33.1% in 2000.

Counting the winners and skipping the losers can even distort numbers that are taken as gospel truth by investors. It's almost universally claimed that since 1802, U.S. stocks have returned an annual average of 7% after inflation. What no one ever tells you is that the reported track record of U.S. stocks in the early 19th century includes just a tiny fraction of all the companies then in existence. The only stocks included are the winners. Hundreds of companies, in such doomed industries as canals, wooden turnpikes, the pony express, and bird-manure fertilizer, went broke and took their investors with them. If these losers were included in the track record, the average performance of stocks in the early years would drop by roughly two percentage points annually. That doesn't mean that stocks aren't worth owning today, but it does mean that their outperformance is not a certainty—so diversifying with bonds and cash is a necessity. Anyone who claims that history "proves" that young people should put all their money in stocks doesn't know much about history.

The technical term for counting the winners and dropping the losers is "survivorship bias" (because the resulting averages are biased by the success of the surviving companies). Whenever anyone tries to sell you an investment based on long-term "average" performance, ask whether the average has been adjusted for survivorship bias. If the person doesn't know what you mean or cannot clearly explain the result, hang on to your wallet.

➢ **DOES IT HAVE LEGS?** Rather than focusing only on the fact that something has changed, concentrate on what has changed. Five years from now, will investors look back on it as a watershed, or will no one even remember it? Take some time to evaluate the event that caught you by surprise. How much less should you be willing to pay for this stock or that asset, as a long-term investment, now that something surprising has occurred? Why might the fundamental future health of the business have been damaged? Are there any reasons to believe that this news will impair the earning power of the company? Forget about how the surprise affected the market price of the stock. Instead, focus on what—if anything—it means for the underlying business. Years from now, will customers and suppliers really care whether a company disappointed Wall Street's analysts by one penny per share over the past three months? Now *that* would be the biggest surprise of all.

CHAPTER NINE

Regret

Clear, unscaleable, ahead
Rise the Mountains of Instead.
—*W. H. Auden*

The Dog in the Rain

To this day, Dan Robertson has an indelible memory of the pain he felt on July 22, 2002, when his portfolio hit rock bottom. Thanks to a series of disastrous bets on Internet stocks and technology funds, Robertson had lost almost exactly $1 million, turning $1,457,000 into $468,000 in less than two and a half years. "I felt like this dog I'd seen once on an L.A. freeway in the rain," recalls Robertson, a retired teacher who lives just north of Los Angeles. "It had been grazed by a car, and it was limping, and the cars kept coming, and then the dog just stopped and looked into the oncoming traffic with this kind of pained smile on its face, like it was thinking, 'It doesn't matter to me whether you hit me or whether you miss me, all I know is I can't run anymore. I can't run anymore.' And I said to myself, 'That's you, Dan. You're just like that dog you saw in the rain.'"

As Robertson knows, losing money on investment mistakes can make you feel almost unimaginably miserable. We saw in Chapter Eight that a negative financial surprise blasts through the reflexive brain like a startle response—quick and astonished, panicky and raw. But then, in the aftermath of a nasty surprise, your reflective brain may take over, generating regret—bitter and contemplative, cold and slow. A surprise makes you exclaim in the heat of the moment, "Uh-oh!" or "Oops!" or "Oh no!" Regret comes in the chilly light of the morning after, when you kick yourself with thoughts like "What on earth was I thinking?" or "How could I ever have believed such baloney?"

Why do some investing decisions fill you with more remorse than others? At least in the short run, your regret is likely to be hotter, sharper, and more painful when:

- the outcome appears to have been caused directly by your own actions, not by circumstances that seem beyond your control;
- you could have chosen other options;
- you came close to your goal, making a "near-miss";
- your mistake was the result of what you did rather than what you did not do, or an error of commission rather than omission;
- and the action you took was a departure from your normal or routine behavior.

One of your hardest tasks as an investor is properly anticipating how much regret you will feel over your mistakes. You can eat yourself up in an endless cycle of "woulda shoulda coulda," or you can let go of your regrets and learn from your errors. It's impossible to invest without making mistakes, but it is possible to stop kicking yourself when you make them. This chapter will show you how.

Are You Well Endowed?

Imagine that I show you something cheap and simple—a plain coffee mug, perhaps. It's not worth a lot of money, it has no sentimental value, and you already have several at home. Would you be interested in buying it? If so, how much would you be willing to pay?

Now imagine that instead of offering to sell it to you, I just give you the mug. It's not a present or a reward, but it's all yours now. Would you be willing to sell it to someone else? If so, how much would you ask for it?

In theory, your buying and selling prices should be identical. It's the same mug, and you are the same person, regardless of whether you happen to be buying or selling. But countless experiments have shown that people demand a price two to three times higher to sell a mug they have just gotten than they are willing to pay for one they do not yet have. And that includes people who—just like you, probably—insist beforehand that they would never do such a silly thing.

Researchers call this the "endowment effect." What makes you reluctant to sell something you may not have wanted to buy in the first place? Before you buy the mug, you focus on how much money you must give up to get it (and on what else you could do with the money). Thinking about those questions will tend to lower the price you are willing to pay. But once you own the mug and are asked to sell it, you focus more on being asked to surrender something that has become yours (and on what you like about having it). Those questions typically make you jack up the price you are willing to accept for selling. What's more, buying a mug you do not own feels like an act of commission; deciding not to sell one you do own feels like an act of omission. Your instincts tell you that you will feel more regret over a mistake of action than a mistake of inaction. So it's natural to offer less money to buy a mug you don't yet have than you will accept to sell one you do have.

How different are stocks from coffee mugs? Not very. The investments we own tend to seem better to us than those we don't own, even though that isn't always true.

- One firm began enrolling new employees in its 401(k) automatically, with a 3% contribution that went into a money-market fund. Before the change, employees had put roughly 70% of their savings into stock funds. Now that the company plunked all their contributions into the money-market fund, however, the new participants left more than 80% of their assets sitting in that low-return vehicle.

- To encourage employees to save in a 401(k), many firms offer matching contributions. Some make the "match" go automatically into their own stock. Even when employees are free to shift the matching money out of company stock, they rarely do. The companies keep a big slug of their shares out of the grasp of corporate raiders, which makes the financial future of the senior managers more secure. Meanwhile, the financial future of the regular employees who are counting on this money for retirement becomes less secure as they forgo the greater safety of a more diversified portfolio.

- In Sweden, workers who are not ready to choose their own pension investments can have the money placed automatically into a "default" fund, a low-cost index portfolio that blends stocks and bonds. In recent years, roughly 97% of eligible workers have left their money in the default fund, even though they were free to switch at any time to any of more than four hundred other funds. (Luckily, in this case, that's not a bad choice.)

Once you make an investment, you can't help regarding it as yours. You have invested part of yourself in it. The word *invest* literally means to clothe yourself in something. When you buy a stock, you wrap it around yourself, and it becomes a part of you. From that moment forward, the prospect of having to get rid of it becomes a wrenching thought.

In a psychology lab in Haifa, Israel, 61 people were given lottery tickets, each with an equal chance of winning a prize worth about $25. Before the prize drawing, the participants could trade their ticket for someone else's; if they did, they got a gourmet chocolate truffle. Only 80% of the people actually believed that every ticket had the same chance of winning; 10% thought their own ticket was more likely than others to win, while another 10% felt they held a ticket with less than equal odds. Not surprising, 5 out of 6 people who thought they held a superior chance of winning refused to exchange their tickets. But then came two surprises. First, among the folks who agreed that every ticket was equal, 55% refused to exchange their own for someone else's. And among those who thought their own ticket was less likely than others to win, 67% still refused to trade for another person's ticket!

What makes people act so strangely? If you trade away your own ticket for another, and then your original one turns out to be the winner, you will feel like a loser and an idiot. On the other hand, if you keep your original ticket and someone else's turns out to win instead, you can just shrug it off. (After all, every other ticket you might have traded for was a loser.) When you imagine your future feelings, doing something that results in a loss feels intensely real and painful. But not doing something—and thereby missing out on a gain—feels much more vague.

Sometimes, investing inertia is a form of avoidance. A lot of people are too intimidated, too busy, or just can't be bothered to devote time and effort to their finances.

- A recent study of the accounts held by 1.2 million 401(k) investors found that 79% never shifted a dime from one fund to another in 2003 or 2004 even as the stock market gained more than 40%.
- A 1986 study examined the decisions of 850,000 people saving for retirement through the TIAA-CREF pension system. Over the course of their investing lifetimes, 72% of the participants never made a single asset-allocation change; they always invested in exactly the same funds they chose at the beginning.
- A later look at more than 16,000 TIAA-CREF accounts found that 73% of the participants made zero changes in asset allocation over a ten-year period; 47% never once changed the proportion of money they put into any fund.

In the financial version of Sir Isaac Newton's First Law of Motion, investors at rest tend to stay at rest unless an outside force acts upon them. Instead of taking action whenever necessary, we do nothing whenever possible. We invest by inertia.

Nobody Loves a Loser

Imagine that you can choose between winning $3,000 for sure, on the one hand, or a gamble with an 80% chance of winning $4,000 and a 20% chance of winning nothing. If you're like most people, you will pick the sure thing.

Next, imagine that you can choose between losing $3,000 for sure, or a gamble with 80% odds of losing $4,000 and 20% odds of losing nothing. What would you do now? In this case, people reject the sure thing and take the gamble 92% of the time.

You would be better off taking the gamble in the first example and the sure thing in the second one—the opposite of what you probably chose. On average, an 80% chance of winning $4,000 is worth $3,200 (.80 x 4,000 = 3,200). So, in the first example, the gamble has an "expected value" $200 higher than the sure thing. By the same rule, an 80% chance of losing $4,000 leaves you $3,200 poorer. And in the second case, you logically should favor the sure loss of $3,000; on average, it will leave you with $200 more.

But it's hard to be strictly logical in these choices, because the idea of losing money triggers potential regret in your emotional brain. If you take the 80% gamble of winning $4,000 and win nothing instead, you will kick yourself for missing out on the $3,000 sure thing. And a 100% chance of losing everything feels a lot worse than the risk of an even bigger loss coupled with a small shot at losing nothing. Doing anything—or even *thinking* about doing anything—that could lead to an inescapable loss is extremely painful.

That's why football coaches almost always punt the ball on fourth down, even though statistics prove they would usually be better off trying for a first down. Punting is the "sure thing," since it typically backs up the opposing team deep in its own territory. If you don't punt and then turn the ball over, your opponents start off where they can really hurt you. So the fact that you will succeed *most* of the time on fourth down matters less than the regret you will feel, and the blame you will take, if you fail *this* time. Likewise, a baseball manager typically saves his best relief pitcher (the "closer") for the ninth inning. A manager knows that if his best reliever is no longer available in the last inning, and the winning run then scores, he will kick himself (and look like a fool to the fans). But it makes more sense to use the ace in any late inning of a close game, especially when the opposing team's best hitters are coming up to bat. The logical goal is to prevent the winning run from scoring in any inning—but, emotionally, the regret of losing at the last minute is much worse.

The same instinct drives millions of retirement investors to keep their money buried in cash and bonds; they are afraid of the regret they will feel if they invest in stocks right before a market crash. Yet adding stocks to a portfolio of cash and bonds will almost certainly boost returns over the long term.

Avoiding whatever felt like a loss probably helped keep our ancestors alive. Researchers at Yale University trained five capuchin monkeys to trade metal tokens for goodies like apples, grapes, or jello. Two humans—let's call them Seller 1 and Seller 2—traded with the monkeys. For the price of one token, the monkeys could choose to buy either from Seller 1, who offered one apple piece for sure and a 50/50 chance of getting a second one, or from Seller 2, who offered two apple pieces at first, then added a 50/50 chance of losing one of them. The monkeys "traded" dozens of times, long enough to learn that they would always get at least one apple piece and that the average outcome was identical in either case. Nevertheless, the monkeys preferred to deal with Seller 1 fully 71% of the time. Their choice was evidently driven by a desire to avoid the pain of having a bigger reward taken away. The capuchin behavior suggests that loss aversion is very ancient; our most recent common ancestor dates back about 40 million years.

Breaking Up Is Hard to Do

What happens to investors when they are racked with regret? Most people don't panic; instead, they freeze. To understand that paralysis, consider this example:

> Paul owns shares in Company A. During the past year he considered switching to stock in Company B, but he decided against it. He now finds that he would have been better off by $2,500 if he had switched to the stock of Company B.

> George owned shares in Company B. During the past year he switched to stock in Company A. He now finds that he would have been better off by $2,500 if he had kept his stock in Company B.

Who feels worse? In surveys, an overwhelming 92% of people say George feels more regret than Paul. Almost everyone shares the same powerful intuition: In the here and now, a mistake that stems from an action hurts worse than a mistake that results from inaction. "You were at the secure trunk of the tree," says psychologist Thomas Gilovich of Cornell University, "and then you went out on a limb, and when it gives way you feel like a fool because you didn't need to go out on that limb." And that's why errors so often paralyze investors. After you screw up once, you become afraid of taking another action that could make things even worse. And the only thing worse than losing is having to admit that you're a loser. So, while most investors are only too eager to cash in on their gains, they hate to sell an investment when it's down, since selling turns a "paper loss" into a real one.

Under U.S. tax law, that makes no sense. When you sell a winner, you turn a paper profit into a taxable capital gain (and, if you hold it for less than twelve months, it is taxable at up to 35%). Meanwhile, clinging to a loser keeps you from taking it as a tax deduction. Hanging on makes sense *only* if you have determined that the value of the business is greater than the price of the stock.

As is so often the case, however, what makes no economic sense makes perfect emotional sense. "When you sell a loser," explains psychologist Daniel Kahneman, "you don't just take a financial loss; you take a psychological loss from admitting you made a mistake. You are punishing yourself when you sell." On the other hand, says Kahneman, "Selling a winner is a form of rewarding yourself."

Holding losers too long and selling winners too soon is no way to get rich, but it's how nearly everyone invests:

- An analysis of roughly 2 million transactions found that Finnish investors are 32% less likely to sell a stock after a sharp fall in price than after a rise. Professional money managers in Israel cling to their losing stocks for an average of fifty-five days, more than twice as long as they hold winners.

- A look at more than 97,000 trades found that individual investors cashed in on 51% more of their gains than their losses—even though they could have raised their average annual returns by 3.4 percentage points (and cut their tax bills) if they had held on to the winners and dumped the losers.

- Among more than 450,000 trades in 8,000 accounts at a discount brokerage, 21.5% of the clients never sold a single stock that had dropped in price!

- In a study of mutual funds that got new managers, researchers ranked the funds' stock holdings from best to worst return. On average, the new managers sold 100% of the stocks ranked at the bottom—implying that their predecessors must have been so paralyzed by their own mistakes that only a new broom could sweep the portfolio clean. The funds that cling most desperately to their losers underperform by up to five percentage points annually.

- People trying to sell their house will hold out longer when they are facing a loss—and will often take the house off the market rather than lose money on it.

As Hamlet said, we would "rather bear those ills we have than fly to others that we know not of." Once you make a mistake, your gut tells you that another sin of commission is sure to hurt more than a sin of omission. Almost everyone believes that even when you think your answer on a test might be wrong, you still should "stick with your first instinct" rather than change your answer. But test-takers who switch an answer are twice as likely to go from wrong to right as from right to wrong! On average, you can significantly raise your test score by changing the answers you have second thoughts about. But any difference from the status quo sticks out in your memory, so you will tend to overestimate how many times you changed your original right answers to wrong ones—and to underestimate how often you should have switched from your first answers to ones that actually would have been right. So you expect to feel foolish if you change an answer, even though that anticipated regret is based on a mistaken view of mistakes.

For investors, there is always a chance that this dog of a stock or fund might turn into a star in the future. Your instincts tell you that if you sell it now, you could end up kicking yourself twice: once for buying it in the first place, and once for selling it right before it bounced back. "Maybe if I leave it alone it will go back up," you tell yourself. "If it just got back to what I paid for it, then I could sell it and at least break even."

Your potential regret is clear: If you sell a favorite stock that you know well, you take an immediate hit to your self-esteem, and you might replace it with a less familiar one that loses you even more money. The result is portfolio paralysis: a petrified state in which you know you made one mistake but are afraid to do anything else, lest you make another.

Found Money

You've just received $10,000. Would you spend it differently depending on how you got it? Your initial answer might be "Of course not." But consider three scenarios:

1. Imagine you earn a year-end bonus of $10,000. You would probably:
   a. spend it on luxuries
   b. spend it on necessities

c. invest it
d. save it in an absolutely safe account

2. Imagine your favorite aunt dies and leaves you $10,000. You would probably:
    a. spend it on luxuries
    b. spend it on necessities
    c. invest it
    d. save it in an absolutely safe account

3. Imagine you win $10,000 on a lottery ticket. You would probably:
    a. spend it on luxuries
    b. spend it on necessities
    c. invest it
    d. save it in an absolutely safe account

While your answers may vary, people typically respond like this: 1, b or c; 2, c or d; 3, a or b. While the amounts of money are identical, they could not feel less alike. Each comes wrapped in different emotions and images. The bonus carries a message of pride: "I made it the old-fashioned way—I earned it." The inheritance evokes the image of your aunt hovering over you, watching what you do with the money. And the lottery carries the thrill of a once-in-a-lifetime chance to splurge with money you never thought you would have.

"Found money" can do funny things to your head. Let's say you go searching in a department store for a $100 item and unexpectedly find it on sale for $50. You buy it—and then take the $50 you just "saved" and spend it on things you would never have bought otherwise. When supermarket shoppers come across "instant coupons" in the store, they spend roughly 12% more on spontaneous purchases than other shoppers do—as if they feel compelled to reward themselves for saving money.

As part of President Bush's tax reform in 2001, every U.S. taxpayer received a rebate of up to $600. People who thought of the rebate as a windfall from the government spent more than three times as much as those who regarded it as a return of their own money. And an unexpected windfall makes you more eager to spend than an expected one. One group of college students was told they would get $5 to spend the next day at a basketball game; a second group showed up at the game and was handed $5 each without warning. The students who got the surprise windfall spent more than twice as much as those who knew it was coming.

For years, Benjamin Franklin used an asbestos purse, reputedly so his money would never burn a hole in his pocket. Many of us could also benefit from a fireproof wallet. In 1988, I was hurrying through New York City's Greenwich Village to meet some friends for lunch when, out of the corner of my eye, I spotted a roll of cash on the sidewalk. I stepped on it and stood there for several minutes to give the owner a chance to claim it. When no one came, I scooped up the bills and raced to meet my friends. Inside the restaurant, I counted the money: It was $300, a lucky strike for a struggling young writer. What did I do with it? First I treated my four friends to lunch. (They felt that would only be fair.) Then I bought myself some books, a few records, a couple of nice neckties. I took my girlfriend out for an expensive dinner. By the time I was done, out of the $300 I had found, I spent about $430. But I had no regrets.

On the other hand, when a windfall is very big and you blow it all, you are nearly certain to look back and kick yourself for wasting the chance of a lifetime. That's why, unfortunately, many lottery winners end up both poor and depressed.

A windfall that you believe you deserve can create yet another feeling. About a year before I found that $300, the top investor at one of Wall Street's leading firms—let's call him Mr. X—earned one of the biggest cash bonuses then on record: roughly $100 million. Mr. X, who came from a poor immigrant background, suddenly found himself sitting on a greater mountain of wealth than he had ever dreamed of. Mr. X was used to moving around millions of dollars of other people's money as casually as most of us shuffle a deck of cards. So how did this expert stock picker invest the $100 million? He put the entire sum into a money-market fund, the investment with the lowest possible return and the least amount of risk. He kept it there for years, always meaning to move it into the stock market but never quite taking the plunge. Mind you, Mr. X remains a very rich man, but his fear of future regrets cost him a fortune. One of his former colleagues later recalled, "If he had put that money into stocks, he'd be a billionaire by now."

The lesson: How good a windfall feels depends on whether you control it or it controls you.

The Chains of Choice

"It's good to have choices."

We all believe it, and it seems to be a fundamental truth about living in a democracy—or being alive at all. Even pigeons prefer to have more than one way of getting food. Knowing you have plenty of choices gives you a feeling of freedom and power. If you want a cup of plain black coffee you can go to the nearest convenience store, but when you really crave a grande decaf skinny iced latte with a shot of almond, there's no place like Starbucks. It seems obvious that the more options you have, the better the choice you will end up making, and the happier you will be.

Like a lot of notions that seem obviously true, it's largely false. Believe it or not, once you have a handful of options, adding even more choices will lower your odds of making a good decision and increase your chances of regretting whatever decision you do make. Having a few choices may be good, but too much choice is nothing but trouble.

In a classic experiment, shoppers in Draeger's Market, a gourmet grocery in Menlo Park, California, dropped by a "tasting booth" where fancy jams were displayed. Sometimes 24 different kinds of jam were available; other times, only 6 were offered. Shoppers were 50% more likely to drop in for a taste when the full variety of 24 jams was displayed.

Then choice started to show its dark side. Among the people who were offered 24 choices, a paltry 3% actually bought at least one jar of jam. Among those who had only 6 choices, however, 30% ended up buying at least one. When a similar test was done with Godiva chocolates, the people who had selected one piece of candy from a display of 6 were much happier with their choice than those who picked one out of an array of 30 different candies. The more choices they left on the table, the more they worried that the one they picked was not the best. Too much of a good thing creates "choice overload," releasing a swarm of potential regrets.

The same thing happens when we invest. A study of hundreds of 401(k) plans found that the more funds a plan offers on its "menu," the more reluctant people are to sign up and save for their retirement. The harder the choice feels, the less people want to choose.

Yet the threat of having less choice almost always disturbs us. If the public address system announces, "The store will close in five minutes," you may be more inclined to grab something off the shelf that you might not otherwise have picked. When mutual funds announce that they will soon "close to new investors," meaning that people who do not yet own shares will no longer be allowed to buy in, millions of dollars can come rushing in over a matter of days. The mere warning that you may miss out on a choice can be enough to make it seem worth choosing—even if you might barely have given it a thought otherwise.

What Might Have Been

True or false: You would rather receive $150 than $100.

If you answered "true," consider this scenario:

When Ralph gets to the box office window of the Roxy Cinema, he's told that he's just won $100 for being the theater's 100,000th customer. Meanwhile, at the Bijou Theater, Bill wins $150 for being the 1,000,001st customer. Who would you rather be: Bill or Ralph?

A moment later, however, Bill learns that the man who was immediately in front of him was awarded $10,000 for being the millionth patron at the Bijou Theater.

Now who would you rather be: Bill or Ralph?

As this example shows, how much financial regret you feel is determined not just by what did happen, but by what you think could have happened. If you're like most people, you'd rather be Bill at first, because he made even more money than Ralph. But you also sense that Bill's delight turned to ashes in his mouth the instant he learned that he could have won $10,000 instead. That's what psychologists call "counterfactual thinking"—imagining what might have been.

Counterfactual thoughts often begin with "If only I had . . ." or "If only I hadn't . . ." Bill, for example, might torment himself with thoughts like, "If only I hadn't stopped to tie my shoe, then I would have won the $10,000 myself."

Counterfactual thinking creates an alternative universe in which the outcomes are always knowable and the right thing to do is always obvious. The more easily you can transport yourself into one of these imaginary worlds, the more regret you will feel over the mistake you made in the real world.

In Holland, the government runs a "postcode" lottery. If you play, the number of your ticket is not chosen randomly, nor do you get to choose your own number. Instead, your ticket is marked with the postcode of your home address (with four digits and two letters, this is the Dutch equivalent of a U.S. zip code). One postcode is selected randomly, and anyone with a winning ticket wins between 12,500 and 14,000,000 euros (up to $18 million). In conventional lotteries, it's hard to know whether you would have won if you had played; on any given day, you might pick a different number. But there is no doubt in the Dutch lottery: If your postcode comes up, you would have been a winner if you had played. When Dutch citizens are asked how they feel when they did not buy a ticket and their postcode won, they report emotions like jealousy, anger, sadness and, of course, regret. The more regret they feel, the more likely they are to play the postcode lottery again—as if taking another chance can make up for having missed the last one.

In the 2006 Winter Olympics in Torino, Italy, American snowboarder Lindsey Jacobellis was far in front and making a beeline for the finish. The gold medal was just a few dozen meters away. Jacobellis hit the next-to-last jump and, elated with how she had dominated the race, grabbed her board in midair. She lost her balance and crashed. After Jacobellis recovered and finally crossed the finish line in second place, the camera zeroed in on the anguish in her eyes. "This is not the face of someone who just won a silver medal," exclaimed NBC broadcasters Pat Parnell and Todd Richards. "This is the face of someone who lost a gold."

That's the pain of the near-miss. Winning a medal in the Olympics is one of the biggest thrills anyone could ever have. But being better off can make you feel worse if you know you could—or should—have done even better. The closer you came to hitting your target, the more regret you are apt to feel if you miss it. A study of dozens of Olympic athletes showed that, on average, bronze medalists were happier than silver medalists. After all, the silver medalists just missed the gold, while those who won the bronze just missed earning no medal at all.

An athlete may never get another chance at Olympic gold. But other fields, like gambling or investing, offer repeated shots at winning. Here, a near-miss may feel partly painful but mostly pleasant. It enables you to tell yourself: "I didn't really lose—I almost won." Coming so close can make your odds of winning feel higher in hindsight than they actually were at the time. And narrowly escaping a big loss can make you feel unusually lucky, as if a guardian angel were watching over you. Either feeling can compel you to come back for more.

Students in Scranton, Pennsylvania, chose between two stocks in an investing game. Then the players saw the returns on the stock they picked and the one they passed over. Sometimes, their chosen stock finished far behind the one they did not pick; other times, their choice earned only 1% less. Given the chance to reinvest, people were more willing to buy back their previous pick if it had nearly won than if it had clearly lost. Asked why, they said things like, "I'm feeling pretty good right now because I came so close to winning." But, of course, a stock doesn't know whether you almost made money on it, and a company's future prosperity has nothing to do with when you bought or sold its shares.

Out of two million trades by retail investors over a six-year period, 15% of the buy orders were repurchases of stocks that the investors had sold within the previous twelve months. And which stocks did they buy right back? Investors were twice as likely to repurchase the stocks they had sold at a gain than those they had sold at a loss. Knowing these stocks were winners once, investors eagerly bought them back, especially when they dropped below the price at which they first sold them. People seem to be telling themselves, "I came this close to striking it rich, and I'm not missing out again!" (Alas, the stocks they buy back do not perform better than those they sell or hold.)

So it pays to be on guard against your own regrets. The stock market breeds counterfactual thoughts the way a swamp breeds mosquitoes. Every second, you can compare the value of what you do own against the value of what you did own or could have owned. No matter what you do, something else can seem like what you should have done.

The Crucible of Regret

The human brain is a brilliant machine for comparing the reality of what is against the imagination of what might have been. If you had no way of knowing how things might otherwise have turned out, many of your bad decisions would never torment you. Knowing (or believing) that you could have done better is what makes you feel bad when something goes wrong. The ache of regret encourages you to play out in your mind what else could have happened and focuses your attention on what you should have done instead. And that, in turn, motivates you to do better in the future. Regretting our mistakes keeps us from rushing to commit them again. This function probably evolved to help our ancestors plan how to produce and consume limited resources in a hardscrabble world, where risks and rewards followed more predictable rules than today's financial markets do.

At least one region of the prefrontal cortex (known technically as "Brodmann's Area 10") is much larger in humans than in the brains of any other primate. If you clap the palm of your hand to your forehead just above your eyebrows, as though you were smacking yourself for a stupid mistake, you'll mark approximately where Area 10 lies on the surface of your brain. This patch of tissue is roughly twice as large in humans as it is in any of the great apes and is nearly four times more densely packed with neurons. Area 10 also appears to be more finely interconnected with the rest of the brain in humans than it is in apes.

Area 10 is in the same neural neighborhood as the orbitofrontal cortex, or OFC. Along with the nearby ventromedial prefrontal cortex, or VMPFC, this appears to be one of the main regions in the brain where we evaluate the gains we actually receive against what we hoped to get. The OFC is tied especially tightly to other parts of the brain that process memories, emotions, and our impressions of taste, smell, and touch—which may explain why our sense of regret often feels so tantalizing and tangible. ("Selling Google stock too soon left me with a bittersweet feeling," you might say, or "I was so close to making money I could taste it.") Neurons in the OFC predict whether the potential outcomes of your actions should produce reward or punishment, and then monitor any mismatch with reality. If you think a stock is going to go up but it goes down instead, the regret you feel is generated largely in the OFC.

People with injuries to these areas of the brain can become impulsive and incapable of planning for the future. Neuroscientist Jordan Grafman of the National Institutes of Health studied a group of military veterans who had suffered head injuries during the Vietnam War. With damage to the OFC and VMPFC, these patients spent less than half as much time as uninjured people in planning to pay for their children's college education, and barely any time on figuring out how to save for their own retirement. They also had a hard time imagining new ways to increase income, as if they were stuck in the reality of what they were given. Other people with similar brain damage have frittered away their money on tacky jewelry, plunged into risky business partnerships without doing their homework, or blown their entire insurance settlement on an expensive car.

When asked to recall past experiences that made them feel sad or afraid, VMPFC patients can conjure up clear memories of what happened, but, unlike normal people given the same challenge, they do not break out in a sweat, nor does their pulse quicken. It's as if they can remember these feelings without feeling them. In a similar way, they often say they "know" their behavior is "wrong" and are frequently heard telling themselves "No!"—right before they knuckle under to the latest whim.

Gambling experiments at the University of Iowa have shown that half the time, VMPFC patients can identify which bets are more likely to lose. Unfortunately, more than half the time, the patients go ahead and play the bad bets anyway, out of curiosity or caprice. With their regret circuits knocked out, they can't stop themselves.

For the rest of us, the anticipation of regret is a kind of emergency brake that keeps us from speeding after every greedy impulse that pops into our heads. Unfortunately, it can keep us from pursuing some good investments, too.

Compare and Contrast

Neuroeconomics is now explaining how the orbitofrontal cortex reacts not just to what is, but to what might have been. In one recent experiment, Slot Machine A offered 50/50 odds of winning either $20 or nothing, while Slot Machine B had a 25% chance of winning $20 and a 75% chance of no gain. Brain scans showed that when people played B and got nothing, neurons in the OFC barely reacted; after all, these folks knew that winning was a long shot compared to breaking even. However, when people played A and missed out on the $20, their OFC neurons fired furiously. When the odds of a gain were 50/50, missing it suddenly flipped on the regret circuits. That suggests a basic rule of how your investing brain is built: The higher you think the odds are of making money, the more regret you will feel if you don't.

In the Human Reward Learning Lab at the California Institute of

Technology, neuroscientist John O'Doherty used an MRI machine to scan people's brains while he gave them the chance to win $1 or break even, and to lose $1 or break even. He measured the activity in their brains when they won or lost, as well as when they broke even instead. Neurons in the orbitofrontal cortex surged within about four seconds of whenever people won $1, but also became nearly as active when people avoided losing $1. On the other hand, the outright loss of $1 slashed the activity of the same neurons, much the way turning off a light switch plunges a room into darkness. And missing out on a $1 gain dimmed the output of these neurons almost as dramatically.

So your OFC reacts both to what happens and to what could have happened. Avoiding a loss will create a neural thrill more than half as intense as earning an outright gain. And when you try making money but only break even instead, that will depress your OFC neurons about half as much as an actual loss.

The obvious conclusion is that avoiding a loss is a milder form of gain, and missing out on a gain is a diluted form of loss. When you think about what might have been, you are creating imaginary outcomes—but real emotions.

In Bron, France, people played a simple gambling game in which they could either win or lose 50 to 200 French francs (about $9 and $36 at the time). Some of the players had injuries in their OFC or VMPFC, while others had fully intact brains. When the players picked between either of two gambles and could see only the outcome of the one they chose, they all felt roughly the same mild sense of disappointment. But when they could see not just the result of their own gamble, but also the result of the one they could have chosen instead, there was suddenly a difference. People with normal brains said they felt "extremely sad" (and instantly broke out in a sweat) if they found out they could have won 200 francs instead of the 50 they had just gained or lost. Meanwhile, the players with a damaged OFC or VMPFC felt no more regret (and never broke a sweat) when they learned how much better they could have done. The knowledge that they would have been better off if they had taken the other bet caused them no pain. Like the French singer Edith Piaf, who warbled *"Je ne regrette rien,"* they regretted nothing.

A follow-up study took MRI scans as people with intact brains played the same simple gambles. The OFC fired up in response to a gain or loss, but only after people learned how much better or worse they could have done on the gamble they did not choose. The more active the OFC was, the more regret people felt over the choice they did not make. The next time they faced a choice, they anticipated the regret they would feel if they made a mistake—and changed their behavior accordingly.

Thus these areas of the brain seem to have a compare-and-contrast function that triangulates between what you expected to happen, what did happen, and what else could have happened. When these circuits are knocked out, that function breaks down. Asked how they would decide which of several apartments they would choose to rent, people with normal brains tended to compare information about factors like size, location, and noise levels across all the apartments at once. But people with VMPFC injuries looked at information on one apartment at a time; often, when they found one that seemed desirable, they stopped searching and picked it on the spot. Unable to anticipate their own regret if they made a bad decision, they didn't care about making the best of all possible choices. They just wanted one that was good enough.

Murphy Was an Investor

On June 13, 2006, I got an e-mail from a distraught investor I will call Michael Buchanan. A retired social-studies teacher, Buchanan could not believe his bad luck. "For years, I've been meaning to put some of my money in an emerging markets fund," he recalled. "I knew they would win big, and they did. And I knew they would keep winning big, and they did." (The average emerging-markets fund gained 55.4% in 2003, 23.7% in 2004, and 31.7% in 2005.) "It got to the point where I couldn't sit on my hands anymore, so on May 13th, I put $10,000 into an emerging markets stock fund." But then rising interest rates and geopolitical worries hammered investments in places like Brazil, Russia, India, and China, and Buchanan lost 22% of his money in four weeks.

"Believe it or not, this wouldn't actually bother me so much," Buchanan continued, "if I hadn't bought Jacob Internet Fund in January 2000. I got my guts ripped out by that fund." (Jacob Internet lost 79.1% in 2000, another 56.4% in 2001, and 13% more in 2002.) "So I sold it at the end of 2002. As soon as I got out, the damn thing turned into a superstar." (Jacob went up 101.3% in 2003 and 32.3% in 2004.)

"Why does this keep happening to me?" asked Buchanan plaintively. "I know—I don't think, I KNOW—that the second I sell my emerging markets fund, it will take off. But if I keep it, it will keep losing money! What's wrong with me? What should I do? Is it Murphy's Law of mutual funds?"

Buchanan e-mailed me because of a column I had written in 2002 called "Murphy Was an Investor." In our daily lives, we all shake our heads over the apparent workings of Murphy's Law ("Whatever can go wrong will go wrong") and its corollary (". . . in the worst possible way at the worst possible time"). We tend to believe that it will rain if we forget our umbrella and be sunny if we lug it along, or that whichever checkout line we stand in will turn out to be slowest, or that whenever we change lanes on the highway the other lanes will speed up. But does the perverse logic of Murphy's Law govern investing, too? And is the whole concept merely a cleverly expressed superstition, or does it have some basis in fact?

The maven of Murphy's Law is an Oxford-trained physicist named Robert A. J. Matthews. A few years ago, Matthews set out to investigate one of the oldest examples of Murphy's Law: Why does bread always seem to land butter-side-down when it falls on the floor? You might think it's because the buttered side is heavier; a psychologist might say we are more apt to recall a wet landing than a dry one; a skeptic might simply insist that which way the bread lands is random. It turns out all those views are wrong.

"Like most people, I guess," says Matthews, "I thought it's a 50/50 chance, unless you've got a pound of jam on one side." With the uniquely British gift of taking essentially silly things very seriously, in 2001 Matthews enlisted 10,000 schoolchildren across England to tip buttered toast off plates. Just over 62% of the time, the bread landed butter first—a percentage much too high, across so many trials, to be the result of chance. Matthews easily ruled out the weight of the butter as a cause: When unbuttered toast was inscribed with the letter B in magic marker, then placed face-up on a plate and tipped off the table, it landed B-side down most of the time.

So why does toast tend to go splat on the wrong side? "The universe is designed against us," Matthews says flatly. Given the width and velocity of falling bread and the typical height of tabletops (29 to 30 inches), there isn't enough room for a tipped piece of toast to make a full rotation before it hits the floor. And tabletops are so low because humans average less than six feet in height. And why is that? If we were much taller, says Matthews, "our heads would hit the ground with so much energy that it would break the chemical bonds forming the linkages in our skull," so that people would constantly be killed by tripping and falling.

That's what engineers call a fundamental design constraint. Does investing have its own design constraints? Of course it does. From the beginning of 2003 through the end of 2005, emerging markets gained an average of 36.3% annually. But decades—in fact, centuries—of history show that economic growth of greater than 2.5% to 3.5%, after inflation, is not sustainable. In the short run, stock markets can perform better than the economies they represent and the companies that make them up. In the long run, it's impossible. A period of unusually high returns must be followed by more normal returns. That's why the Japanese stock market, after its record-setting returns in the 1970s and 1980s, lost roughly two-thirds of its value in the 1990s. It's why the U.S., after the boom of the late 1990s, suffered the bust of 2000 to 2002. And it's why emerging markets, after years of scorching gains, were not a good choice to throw money at in early 2006. At that point, the only question was not whether they would lose money, but when. (I told Michael Buchanan to sit tight and, in fact, emerging markets went on to have a good year overall in 2006. But I reached Buchanan too late; he had already sold.)

The pursuit of extreme growth carries within it the seeds of its own destruction. As Warren Buffett quips, "Nothing recedes like success." That brings us to Murphy's Law of Investing: Sooner or later, a stock or fund return much higher than average almost always fades back toward average. By the same token, a badly below-average return is also liable to reverse.

This tendency for trends to flip with the passage of time is called *regression to the mean.* Without it, giraffes would get taller with each passing generation until their hearts and hips burst under the strain. Big oak trees would drop bigger acorns, yielding larger and larger saplings until full-grown trees collapsed of their own height and weight. Tall people would always have even taller offspring, and so would their kids, and so on, until no one could get through a nine-foot-tall doorway without ducking. (And, as Matthews points out, they would crack their heads open if they fell.)

Regression to the mean is nature's way of leveling the playing field, in almost every game, including investing. So, whenever you gamble that a very high (or low) investment return will continue, the odds are overwhelmingly against you. Michael Buchanan should have been betting *on* regression to the mean; instead, he bet *against* it. By constantly grabbing the hottest returns he could find, he virtually guaranteed that he would get scalded sooner or later.

Other aspects of Murphy's Law apply to investing. Robert Matthews points out that a great Cambridge mathematician, G. H. Hardy, believed in Murphy's Law of Umbrellas. "Hardy was convinced that there is a malevolent rain god," says Matthews, "so he would send an assistant outside carrying an umbrella to trick the god and ensure that it wouldn't rain on Hardy's cricket match that day." Even in soggy England, however, the odds that it will rain during any given hour of the day are only about 10%. So, even when the forecast is for a 100% chance of rain that day, the odds of rain at any particular hour are much lower. Therefore, most of the times you lug an umbrella because of a rainy forecast, you will end up never opening it. And the more often you tote an umbrella around under a sunny sky, the more likely it is to stick in your selective memory. You will be much less inclined to remember the less common cases when you brought your umbrella and it did rain. As a result, you will tend to overestimate how often you carried an umbrella in vain, and to underestimate how often you didn't bring it when you should have.

Likewise, whenever one sector of the stock market is hot, diversifying your money across other assets will always feel like a waste of effort—an umbrella you never seem to need. As Michael Buchanan's story shows, however, it is a mistake to think you don't have to be diversified. No matter how many times you carry an umbrella without needing it, you will be very glad indeed to be carrying one when a downpour finally hits.

Your apparent tendency to pick the wrong checkout line holds an investing lesson, too. If three cash registers are open, the odds that you will pick the fastest line are only 33%. Two-thirds of the time (assuming the same number of people are waiting and the checkout clerks are about equally efficient), one of the other lines will move faster. With four lines open, your odds drop to one in four. So the raw math is always against you: No matter which line you pick, it will usually be the wrong choice. You may think your success rate is a function of how well you size up lines, but in fact it's predetermined.

Now consider mutual funds. On average, over time, half the funds will do better than the market and half will do worse—before expenses like trading costs, management fees, and taxes. After expenses, the odds of sustained outperformance go from 50/50 to about one in three. Thus, if you try picking mutual funds that will beat the market on the basis of their past returns alone, you will end up wrong about two-thirds of the time. That's why intelligent investors don't make that mistake.

The regret you feel from chasing a hot fund or stock becomes even more painful when you hear strangers boasting about their successes—on television, online, at the next party you go to. You screwed up, but somehow *they* keep making money. It's that same uncanny feeling you get after you switch lanes on the highway: As soon as you move out of the "slow" lane into the "fast" lane, the fast lane turns into a parking lot. Whichever lane you are in is the wrong one—or so it seems. The truth is more subtle: When the other lane is slow, you can pass many vehicles in a blur, so you have only a vague sense of how many you have passed. But when your own lane is slow, one car after another passes you in a discrete whoosh. What's more, safe driving requires you to focus more of your attention on the road ahead than on what is in your rearview mirror. So you get a much better and longer look at the cars that have passed you than at the ones that you yourself have passed.

With investing, too, your losers, and other people's winners, can often feel more visible than your own good decisions. At a cocktail party or a barbecue, it might seem as if everyone but you has a great investing move to brag about. As you sheepishly excuse yourself to get a refill, it might not occur to you that all these folks made investing mistakes, too, and that a party is the last place they would ever discuss them. This mistaken feeling of being the only one with investing regrets can tempt you into taking risks you normally would avoid. It's important to remember that everyone makes mistakes, and that everyone who makes mistakes has regrets.

The Island of Disgust

Why is it so hard for us, at least in the short run, to live with our losses? What gives Michael Buchanan, and the rest of us, such a hot sense of regret over a knuckleheaded investing move?

On the inner edges of your upper brain, underneath flaps of several other parts of your cortex, lies a region called the insula (pronounced "IN-syu-la" from the Latin word for "island"). The insula is one of your brain's main centers for evaluating events that arouse negative emotions like pain, disgust, and guilt—just what you feel when you lose money. Like the anterior cingulate cortex that we learned about in Chapter Eight, the forward part of the insula is packed with unusual neurons called spindle cells. These neurons may specialize in helping us adjust our behavior when circumstances change. They are unique to humans and great apes, and the fronto-insular cortex in the average human brain contains nearly thirty times more spindle cells than a chimpanzee's insula does.

Strikingly, these cells carry a molecule that is rare in the human brain but plentiful in the digestive system, especially the colon, where it helps trigger the contractions that propel food through your pipes. People with the digestive ailment Crohn's disease respond more intensely to video footage of scary, sad, or disgusting scenes; it's literally a visceral response. When you get a "gut feeling" that an investment has gone sour, you might not be imagining. The spindle cells in your insula may be firing in sync with your churning stomach.

The abundance of these neurons in the human insula may explain why we are so keenly disgusted by foul smells and tastes even though our sense of smell is much duller than that of other animals. (Just think of all the things you have seen your dog sniffing, or eating, that turned your stomach regardless of whether you got a whiff of them.) Decades ago, scientists found that stimulating the insula directly with electrical current triggers intense nausea and an overpowering sense of revolting tastes. The insula also appears to be one of the key places where your brain turns flashes of emotion into conscious feelings; when you become aware of how your heart is racing, it's your insula that puts you in touch with what your body is doing.

Despite its name, no insula is an island. Your insula is intimately connected to the hypothalamus, which helps regulate your heart and lungs;

to the thalamus, where sensory impressions are sorted out and basic rewards are compared; to the amygdala, where fear is processed; to the motor areas of the cortex, which prepare your muscles for action; to the anterior cingulate cortex, where surprise and conflict are registered; and to the orbitofrontal cortex, the area that seems to evaluate "what might have been."

You don't need direct contact with something disgusting for your insula to light up. When people had their brains scanned while they sniffed butyric acid, a chemical that makes vomit smell bad, their insula burst into action. The same part of the brain kicked in when people saw a photograph of someone else reacting to a revolting smell. (We learn what's disgusting partly by observing what disgusts other people.) It takes about a quarter of a second for the insula to generate this "Eeeww!" reaction.

Just a glimpse of something gross, like cockroaches or rotting food, switches on the insula. In a series of tests, a patient with damage to the insula and a related area of the brain, the putamen, was unable to identify which face, in a series of photos, had a disgusted expression. Listening to a recording of someone retching, he couldn't explain why a person would make such a noise. When he filled out a questionnaire that asked, "If you were hungry would you eat a bowl of soup that had been stirred with a washed fly swatter?" he said yes. Nor was he put off by a description of a piece of chocolate in the shape of a lump of feces. People with an intact insula, however, are readily grossed out by such things. (Admit it: Does poop-shaped chocolate sound tasty to you?)

There's something else that sets your insula on fire: losing money. In one experiment, the insula was roughly three times as active after people lost money as it was after they won money. Meanwhile, whenever people chose a bet that (based on their recent experience) might lead to a loss, the insula lit up more than four times as intensely as it did when they made money. And the more furiously the insula fired during one of those risky bets, the more likely the person was to pick a lower-risk option the next time. New research also shows that when people shop for consumer goods, the insula flares up if products are overpriced; the thought of paying too much may literally be painful.

I felt my own insula at work in an experiment at Scott Huettel's neuroeconomics lab at Duke University. Inside an MRI, I was shown images of three slot machines: a black one that always broke even, a blue one that paid out a mix of small gains and small losses, and a red one that yielded either big gains or big losses. The right forward part of my insula became active whenever I decided to play either blue or red, but its level of activity surged if I selected the riskier red machine. And my insula kept sputtering, generating a butterflies-in-my-stomach feeling that did not die down until I made my next choice. Not surprising, I ended up shifting toward the safer machine, picking it 70% of the time, which helped minimize my losses. (See Figure 9.1.)

In another experiment, people with stroke damage to the insula played a simple investing game. They got $20 of play money at the start. On each of 20 rounds, they could invest $1 or stand pat. If a patient invested the dollar, the experimenter flipped a coin. If heads came up, the patient lost his dollar; if it was tails, he won $2.50. When normal people play this game, 60% of the time they will decline to invest if they lost their dollar on the previous coin toss. But the insula patients who lost $1 on the last round chose to put more money down an astounding 97% of the time. With their disgust circuitry knocked out, they felt no pain from past—or future—losses.

Knowing that something painful might happen is nearly as bad as the pain itself; the brain flares up almost as fiercely in anticipation of expected pain as it does in response to actual pain. The insula generates feelings of disgust not only when you lose money, but when you think you might, just as you are grossed out not only when you step in dog poop, but whenever you see it. After all, that's *why* you don't step in it. And anticipating that you will feel disgusted with yourself if you lose money drives you away from riskier investments.

At Stanford University, neuroeconomists Brian Knutson and Camelia Kuhnen put people into an MRI scanner and gave them a simple choice: invest in either of two stocks or a bond. Here's how it worked. First you were told that one of the stocks was "good" and the other was "bad"—but you had to figure out which was which. The low-risk stock had a 50% chance of gaining $10, a 25% chance of breaking even, and a 25% chance of losing $10. The high-risk stock, on the other hand, would go up $10 a quarter of the time, break even a quarter of the time, and lose $10 half the time. Midway through, the good and bad stocks would be switched. The bond, however, would always pay $1.

Each time you made a choice, you not only learned what you gained or lost on your pick, but what you could have gotten on this round from the other two choices instead. The bigger the difference between how you did and how you could have done, the more active your insula became. Furthermore, the hotter your insula got as you chose a stock, the more likely you were to pick the safe bond on the next round. As losses fired up the disgust center of your brain, you retreated from further risk.

So when you make a big investing mistake, your insula generates much the same response to your own action as it would to a pile of rotting fish or a bag of garbage that's been sitting in the sun. You move away from the stench. You try to get it off your mind. Above all, you don't want to go anywhere near it again. In effect, these specialized neurons scream inside your head, "You make me sick." This is regret at its hottest, a feeling that makes you want to wash your hands of your foul mistake.

When investors are disgusted with their own blunders, their natural aversion to taking a loss finally breaks. Instead of grimly hanging on as usual, they now become desperate to get rid of what they own. That makes them willing to incur much higher trading costs, slashing the net proceeds of the sale. When you buy or sell a stock, you don't just pay a commission; you also pay a variety of invisible costs ranging from the "spread" (or gap between the buying and selling price) to "market impact" (how much your own order pushes the price up or down) and "delay" (the cost of waiting for your trade to be completed).

On average, across all markets, sellers pay total trading costs up to six times higher than buyers do. Desperate people do desperate things. Take the first quarter of 2001, when the NASDAQ index lost 25%. In that market, according to brokerage expert Wayne Wagner, traders incurred average total costs of 3.52% when they sold stocks that were falling fast. Meanwhile, it cost only 0.21% to buy stocks whose prices held steady. In other words, panic selling was roughly 17 times more costly than patient buying. In the first quarter of 2005, when NASDAQ fell 8%, the average total trading cost was 0.52% for patient buying but 1.8% for panic selling—nearly three and a half times higher. That may be because people tend to buy in dribs and drabs but sell in one fell swoop, a fact that traditional economics cannot account for, but that the neuroeconomics of disgust helps explain.

Your disgust helps make other people rich. After all, selling a stock after it drops is a confession that you were wrong. Kicking yourself for being such an idiot, you just want out of the damn thing so you can go back to believing that you know what you're doing. The sooner you can wash your hands of any evidence to the contrary, the better. Believe me, whoever gets to take the other side of *that* trade will be delighted that he found you.

As Time Goes By

Imagine that you buy 100 shares of Schmegeggy Corp. and, the day after you buy it, it plunges 29%. You will probably exclaim, "How could I have been so stupid? I knew I shouldn't have bought that lousy stock!" Your regret is centered on what you did that you now realize you should not have done. But as time passes and you look back from a greater distance, your perspective widens. With a more panoramic view of your decision, what you will see more clearly is all the better choices you should have made. In the long run, you may come to have more regrets over your inaction: what you did not do that you now realize you should have done.

As memories decay over time, it gets harder to recall what you were actually thinking when you made a decision. Thus it becomes easier to convince yourself, with hindsight, that any number of alternatives seemed equally attractive when you were deciding which stock to buy—even though, at the time, you were convinced that Schmegeggy was going to be a bigger gold mine than Google. Looking back, there appears to have been a clearly better choice than the one you made. Now that you know that shares in Noodnick Corp., Amalgamated Schnooks, and Schmendrick Inc. all outperformed Schmegeggy, it seems obvious that you should have invested in one—or all—of them. At the time, however, none of these alternatives had even been on your rader screen.

In a marriage that goes bad, one spouse might first focus on angry thoughts like "I never should have married him (or her)!" But as time passes, regrets tend to shift from what went wrong in the narrow confines of the real world to what could have gone right in the unbounded world of the imagination. In the long run, an unhappy spouse will often think, "Why didn't I marry Boris instead of him?" or "If only I had married Natasha." Over time, hot regrets of commission tend to fade into cooler regrets of omission—a wistful longing, a soft despair over what might have been.

And there's no end to the roads not taken. Who knows how things might have worked out with that person you kissed twice in high school?

So your regrets over what you did not do tend to be more "open" than your regrets over what you did do. Every time you see another stock price, your set of missed opportunities may appear to grow larger ("I knew I should have bought that one!"). And the passage of time can keep pumping up the value of each stock you imagine that you should have bought instead.

As time goes by, almost everybody regrets something. In a study of 176 individual investors, 175 said they felt regret over at least one financial decision. Among them, 59% said they were sorry they had kept a losing stock too long, while only 41% said they felt worse about selling a winner too early.

Looking back, it was their errors of omission, not commission, that bothered these investors more. And yet, oddly enough, their actions almost certainly cost more money than their inactions. Investing returns are not symmetrical: You can't lose more than 100% when a stock goes down (unless you buy it with borrowed money), but there is no limit to how much you can make if goes up. So, on average, you will take a bigger bite out of your return by selling winners too soon than you will by holding losers too long.

But when you look back, you will keenly remember a deep loss that you could have limited by selling earlier; you are the one who took the hit on it. It's a lot harder to put a dollar amount on how much more you could have made if you had hung on longer to a good stock; someone else made that extra money. Even though it may do less financial damage, a loss is more psychologically painful than a foregone gain. Thus, with hindsight, you will kick yourself more over what you did not do than over what you did, even though your errors of commission probably cost you more money. Because of the funny way regrets unfold over time, your feelings blind you to the financial reality of the situation.

Reducing Your Regrets

It might not quite be true that "time heals all wounds," but most investing scrapes do not leave lasting scars. In general, we do a much better job of reducing regret than we realize. It can take a while, but people adapt and adjust; they make do and move on. Even a paralyzed portfolio can make money if enough time passes, and it takes only one big winner to make up for a multitude of losing investments. In an odd paradox, *expecting* regret often hurts worse than *experiencing* it. Thus, investors probably hurt themselves more by avoiding risks they imagine they might regret than by taking risks they really do end up regretting. People who avoid actions that they think might later hurt, says Harvard psychologist Daniel Gilbert, often are "buying emotional insurance that they do not actually need."

In investing, there are two basic kinds of mistakes. The first is instantaneous and infuriating: You buy and the price tanks, or you sell and the price soars. You instantly know you did something wrong, and you immediately kick yourself.

The second kind of mistake is not obvious at first. While you're lying on a towel at the beach, there is no single moment when you can look at your skin and see it turn from a healthy bronze glow to the neon red of a painful sunburn. A burn occurs so gradually that the transition is invisible. An investment mistake is often like a sunburn: It results from forgetfulness, carelessness, or creeping commitment to a choice that you may never have been happy about. But after the fact there's no mistaking it, and it can burn like hell, and you're sorry you did it.

The more an outcome appears to be the result of your own choice and the more readily you can imagine having done something different, the more painful your regret is likely to be. So, whenever possible, do as little as possible. Instead of making judgments one at a time, you should follow policies and procedures that put your investing decisions on autopilot. Think of it as cruise control for your portfolio. In 1995, I got a speeding ticket driving my in-laws' car—and was so mortified that I swore I would never get another. Ever since, whenever I get on a highway, I check the speed limit and set my cruise control—eliminating all worries that I will get careless or emotional and end up speeding. "The more you can automate your investing," says psychologist Thomas Gilovich of Cornell University, "the easier it should be to control your emotions." Here are several forms of investing cruise control, along with some other lessons from neuroeconomics.

➢ **FACE IT AND FESS UP.** At the start of this chapter, we met Dan Robertson, whose losses on tech stocks left him feeling like a "dog in the rain." But Robertson did not roll over. At first he and his partner, Steve Schullo, were in denial over their losses. Once about 40% of their money was gone, however, they forced themselves to face the pain. "When you put it out there by talking about it," says Robertson, "that somehow changes the nature of it so you can change your behavior." Robertson felt they had to take action before the rest of their money disappeared, too. "I kept saying, 'Do we have a loss or a lesson?' Since we didn't take advantage of the market high to pay off our debts, we took advantage of the lesson." Dumping all the high-tech investments they now realized they never understood in the first place, Robertson and Schullo used the proceeds to pay off their mortgage and build a new portfolio of conservative stocks, bonds, and index mutual funds. At last count their financial assets, which had peaked at $1.46 million and bottomed at $468,000, were pushing $1 million.

One of the best ways, then, to cure portfolio paralysis is to talk the situation over with someone you trust. A friend or parent, spouse or partner, can help you push aside your shame and blame. You should never sell an investment purely because it has gone down, but if a sudden drop makes you aware that you never knew what you were doing, then talking it over can help. In order to learn from a mistake, you must first admit you made a mistake. It's much healthier to do that out loud than to kick yourself in private shame.

➢ **HAVE RULES FOR RULING THINGS OUT.** It was easy in 2006 to be angry that you didn't put all your money into energy stocks right before oil prices blew sky-high: "I knew it!" But you are less likely to feel regrets later if, at the time, you followed rules for ruling investments out. Sticking to a few simple guidelines for why not to buy enables you to look back and say, "I didn't put all my money into energy stocks because I would have had to break my own investing rules. That wouldn't have felt right. Sooner or later, it's bound to be a mistake." This way, you make an impulsive decision feel like a bigger departure from your normal behavior, so you are less likely, when you look back, to regret not having acted on that impulse. (See Appendix 2 for rules on ruling investments out.)

➢ **GET HELP PULLING THE TRIGGER.** Because it can be so hard to sell a hopeless loser, you may need to get used to the idea. If you've reexamined your original reasons (see "Use Your Words," Chapter Seven, p. 172) and concluded that an investment truly was a mistake—but you still can't face getting rid of it—then you need a push. Psychologist Robin Hogarth of Pompeu Fabra University in Barcelona suggests changing the log-on password for your brokerage account to something like "dumpmylosers." Typing that reminder every time you check on your account, puts you in the position of a musician who practices constantly. The idea of selling losers will become "second nature" to you; as you internalize it, you will become more comfortable with the need for action.

Writers, engineers, and graphic designers all know that the best way to spot their errors is to have someone else look over their work. A few money-management firms make it mandatory for each investment holding to be reviewed by someone other than the person who bought it; banks can reduce their losses by having bad loans reevaluated by someone other than the executive who first authorized them. It is a lot easier to admit that investing in a stock was a mistake if you are not the person who made the mistake. Get a second opinion whenever you can.

➢ **SEEK THE SILVER LINING WHEN YOU SELL.** The logic of selling is crystal-clear: What you paid for an investment should not determine whether you bail out. If you think a stock is more valuable than its current price, you should keep it. If its current price is higher than what you think it is worth, you should sell it. And if you desperately need cash, of course that justifies a sale. But how much you paid for it is beside the point. And yet, says economist Terrance Odean of the University of California at Berkeley, "For most people, the decision to sell depends more on what a stock has done than on what it is likely to do."

That's because emotion overwhelms reason. Regret makes people focus on what has changed (the price of the stock) without analyzing what may or may not have changed (the value of the business). A recent survey of individual investors found that only 17% felt that buying a stock is harder than selling. However, 62% of them devoted more time to buying decisions than to selling decisions. People know perfectly well that a good selling decision takes more work and more thought; therefore, they sweep it under the rug.

It helps to think of your losing investments not as liabilities but as the assets they are. Taking a tax loss is one of the only attractive loopholes left for individuals in the U.S. tax code. If you let your loss fester, it has no value to you. If, instead, you sell and lock in the loss, then you generate cash you can put to work elsewhere, and you can write off the loss and cut your tax bill. When you sell, you turn the dead money of a "paper loss" into a realized loss—and a gift from the government in the form of a tax break with real value. Each year, you can use up to $3,000 of realized losses to offset your capital gains and reduce your taxable income. (Consult a tax professional and www.irs.gov/pub/irs-pdf/p544.pdf before taking a tax loss.)

And stop focusing on whether the stock you sell might bounce back as soon as you sell it. As money manager Whitney Tilson of T2 Partners likes to say, "You don't have to make it back the same way you lost it." If a stock or fund was really a mistake, you should get rid of it and find a better use for the money.

Selling may be easier if you can make it feel like buying. First, find another stock or fund you would love to own, making sure you identify it with the help of an investing checklist. Do you have the cash on hand to buy it? If not, tell yourself that the easiest way to fund the purchase of this potential winner is by dumping an actual loser.

Another way to smooth your way to selling is by not investing the proceeds in something radically different. Psychologists gave people various trinkets—candies, crayons, pencils—and then offered them five cents if they would exchange any of their new stuff. People were much more willing to trade for another trinket of the same kind (a crayon for a crayon) than for a different kind (a crayon for a candy). So if you are not sure you can find something better to invest in, then invest in something comfortably similar to what you just sold. That will keep you from feeling that you will be left behind, or that you are plunging into the unknown with both feet. (If, for instance, you are afraid that Dell Computer might soar after you bail out at a loss, you could sell it and put the proceeds into a similar stock like Hewlett-Packard or, better yet, an exchange-traded fund of computer stocks.)

➢ **CUT YOUR LOSSES—BUT NOT TOO MUCH.** Market pundits often suggest using a "stop-loss," an advance order that instructs your broker to sell a stock automatically if it drops below a certain price. But if you set the stop-loss too close to the current price, you will constantly get sold out of good stocks taking short-term dips on their way to long-term greatness. Every time you get "stopped out" like that, your broker gets a commission. You, however, may get nothing but the regret of watching the stock go right back up after you get out. Some pundits recommend setting a stop-loss as close as 5% below the current price. That's crazy. Unless you are a day trader, a stop-loss that isn't at least 25% below the current price is much too narrow. You want to enhance your own wealth, not your stockbroker's.

Another approach is to set "gut-check points" in advance—levels of loss at which you commit yourself to review the original rationale for the investment. If, for example, you pay $50 a share for a stock, you could do a new round of research at $45 (a 10% loss), $40 (a 20% loss), and $37.50 (a 25% loss). You should sell not because the stock is dropping, but because your research tells you something is wrong with the underlying business. Ask yourself three questions:

> If I didn't already own this stock, would I want to buy it at this price?
>
> If I had no idea what the stock price was, would I want to own a piece of this business?
>
> Now that the stock is even cheaper, hasn't my margin of safety (see Chapter Six, p. 149) gotten even wider?

➢ **PUT INERTIA TO WORK.** If you have a hard time saving, save out of your "found money." Millions of Americans receive their federal income tax refund through direct deposit, automatically putting money into their bank account. But once the money lands there, most people spend it. If, instead, you have the refund direct-deposited into an investment account, your own inertia will almost certainly lead you to leave it there, where it should earn a much higher return than it would in the bank. Most major mutual fund companies can provide you with the routing number and account number that the IRS requires to direct-deposit a tax refund. If you already have an Individual Retirement Account at a fund company, you can pump it up with your tax refund every year. By putting this process on autopilot, you will never miss the money or be tempted to spend it.

➢ **DON'T LET TOO MUCH CASH PILE UP.** Mr. X, the Wall Street financier who let his $100 million bonus languish in a money-market fund, was paralyzed by the regret he was afraid he would feel if he moved his bonus into the market before a big drop. You can avert portfolio paralysis by investing gradually and automatically over time with a dollar-cost averaging plan (see Chapter Four, p. 76). That's the best way to control a windfall before it controls you.

➢ **CHANGE YOUR FRAMES.** If an investment takes a 25% plunge in two days, that can be terrifying—especially if you view it on a market website, tumbling in real time, trade after downward trade. In your panic, you keep comparing the current price to the peak price (or to its price just before it collapsed). Counterfactual thoughts can eat you alive: "If only I had sold at the top . . . If only I had sold three days ago . . . if only I had gotten out while I still could . . . If only I had listened to my gut."

But you can also use a website like Yahoo!Finance to retrace how a stock has performed ever since you bought it. Focusing on how much you made from your starting point, rather than on how much you lost from the peak, creates an opposite kind of "if only" that may keep you from feeling like a fool for not selling. After all, if you had never bought the stock in the first place, you would have missed out on all those gains you earned before the drop. Reminding yourself of how far the stock had risen before it fell, and how much you may still be ahead of the game even after the fall, should take out some of the sting.

Harvard economist Richard Zeckhauser has lived this lesson. In 1996, a startup technology company in which he was a private investor sold out to America Online, giving Zeckhauser a big block of shares in AOL at an effective cost of less than two cents per share. AOL stock peaked in early 2000 around $95 per share—but Zeckhauser did not sell out. AOL (now part of TimeWarner) fell all the way to $16, where it froze for years. Zeckhauser knows he could have made a fortune by cashing out at the peak. But instead of kicking himself for not selling, he pats himself on the back for investing. "If your AOL stock goes from two cents to almost $100 and then back to $16," he says wryly, "you should refer to the two cents, not to the $100." Zeckhauser concludes, "Happy is the investor who can control his frame of reference."

➢ **WHAT YOU DON'T KNOW CAN'T HURT YOU.** "If you have a truly diversified portfolio, then you guarantee yourself—by definition—that some of your assets will do well while some others will do badly," says psychologist Elke Weber of Columbia University. "If you regard risk as the probability of experiencing some loss somewhere in any given time period, then a diversified portfolio probably is riskier than a single, concentrated investment. But that's not what risk should mean—and, if you don't go looking for losses one by one, you don't have to feel them."

To numb the sting of loss, choose a portfolio that doesn't let you look at each investment in isolation. Many 401(k) plans now offer "lifecycle" or "target" mutual funds that combine a variety of assets—U.S. and foreign stocks, bonds, and sometimes a cash cushion—in a single package. Because it reports the returns of all its separate holdings as a single overall number, a lifecycle fund can spare you the regret you would feel if you fixated on losses one at a time. May 2006, for example, was a market nightmare around the world. An investor in these funds would have racked up alarming losses that month:

Vanguard Total Stock Market Index Fund	–3.2%
Vanguard European Stock Index Fund	–2.4%
Vanguard Total Bond Market Index Fund	–0.1%
Vanguard Pacific Stock Index Fund	–5.3%
Vanguard Emerging Markets Stock Index Fund	–10.7%

It's hard to imagine not feeling regret over owning a fund that dropped 10.7% in a month. But Vanguard Target Retirement 2035, a lifecycle portfolio that owns a mix of these five funds, lost "only" 2.8% in May 2006. By making it harder for you to view each of the five returns in isolation, the target fund makes the loss seem more bearable, and triggers much less regret. The emerging-markets fund is in there, and it still dropped like a rock, but all you feel is the much milder loss of the overall portfolio. As anyone who has ever gotten a vaccination knows, how much something hurts often depends on how closely you watch.

➢ **KEEP YOUR BALANCE.** Someone who put $10,000 into the typical U.S. stock fund at the beginning of 1984 and left it there undisturbed until the end of 2003 would have finished with $50,308. But the average investor in U.S. stock funds did not leave his money undisturbed; instead, he constantly fiddled with it, buying more whenever the market was hot and freezing (or selling) whenever it was not. As a result, the average investor saw $10,000 grow to $46,578—nearly 10% less than he could have earned if he had just left his money completely alone! The French philosopher Blaise Pascal hit the heart of the truth when he wrote, "All the misfortunes of men come from only one thing: not knowing how to remain at rest in a room."

The cure for chasing what's hot and ending up with what's not is called *rebalancing*. Decide what percentage of your money you want in each of several kinds of investments, spreading your eggs across those baskets in precise proportions. Let's imagine that your original target allocation was:

Total U.S. stock market index fund	50%
International stock index fund	25%
Emerging-markets stock index fund	5%
Total U.S. bond index fund	20%

Now let's say that over the next year, U.S. and international stocks lose a fifth of their value, emerging markets stocks shed a quarter of their value, and bonds stand still. That would leave you with roughly 48% in U.S. stocks, 24% in international stocks, 4% in emerging markets, and a ballooning 24% in bonds.

To rebalance those numbers back to your original targets, you have to sell bonds and buy stocks. In a retirement account like a 401(k) or IRA, you can rebalance without even triggering a tax bill. Make it mandatory twice a year, every year—on two easily remembered days roughly six months apart, like your birthday and a holiday, or on your semiannual dentist appointment. (Use calendar software or a website like www.backpackit.com to send yourself a reminder, and to get someone close to you to check whether you followed through.) Over the long run, rebalancing is all but certain to raise your return and lower your risk. The more your investments fluctuate, and the less they tend to move in sync when they do, the greater the benefit you will get from rebalancing.

This way, instead of letting your reflexive brain yank you into chasing winners and clinging to losers, you compel yourself, through a reflective commitment made in advance, to sell a bit of whatever has gone up the most and buy a little of whatever has gone down the worst. "Instead of just investing in specific securities," says Cornell's Thomas Gilovich, "you are also investing in the more abstract idea of 'buy low, sell high.' That should take out some of the emotion."

Most people don't rebalance because they are afraid of the regrets that swirl around buying and selling. It's a shame they don't realize they can automate the process. One recent survey of 1,000 investors found that 61% would rather admit they were wrong to a spouse or loved one than sell a winning stock. Yet only 34% of these investors said they rebalanced their portfolios "on a regular or set schedule," even though half of them admitted checking their account values at least once a month.

Like someone who refuses to trade one lottery ticket for another, investors are afraid they will jinx their portfolio performance if they make a change. What they overlook is that they are at least as likely to jinx themselves if they *don't* make a change. Over time, putting more money into investments that have dropped in price and trimming those that have risen—the essence of rebalancing—is the best way to lock yourself into buying low and selling high. If you stay stuck in the status quo, you miss out on the chance to raise your returns. "It's better for people to be forced to make decisions at some point," says John Ameriks, an analyst at the Vanguard fund company, "than to let inertia completely rule the day."

The beauty of automatic rebalancing is that you don't have to make decisions one at a time over and over again. Instead, a growing number of 401(k) providers allow you to set targets for what percentage of your money you want in each of your funds. Then, between once and four times a year, the fund company will automatically sell enough of whatever has gone up and buy enough of whatever has gone down to get your balances back to the target you set. You never have to think about it—taking away most of your potential regrets. This is investing cruise control at its best. If your company offers automatic rebalancing, sign up for it. If not, ask for it. There are very few other ways you can raise your returns and bolster your peace of mind at the same time.

CHAPTER TEN

Happiness

> There is nothing, Sir, too little for so little a creature as man. It is by studying little things that we attain the great art of having as little misery and as much happiness as possible.
>
> —*Dr. Samuel Johnson*

Money (That's What I Want)

Can money buy happiness?

Traditionally, when Americans are asked what would improve their quality of life, their most common answer has been "more money." Even though most people consider themselves mostly happy most of the time, nearly everyone wants to be rich, and increasing numbers of people say they would rather "be very well off financially" than live a meaningful life. The modern American dream, says the psychologist David Myers, has become "life, liberty, and the purchase of happiness." And, of course, the Beatles (and several other musical acts) made a bundle singing "Money (That's What I Want)."

Unfortunately, if you already earn enough cash to live on, the odds that merely having more money will make you happier are pretty close to zero. Fortunately, there is a lot more to the story than that. How much money you make is less important than how much you want—and how you spend it. Furthermore, no matter how much or how little money you have, you can use it to lead a happier life if you understand the limits of what it can do for you and the power you can exert over it with self-control.

While it's doubtful that money can buy happiness, happiness can buy money—meaning that many of us go through life getting it exactly backwards. The harder we work to earn more money, the less time we have for exercise and vacations, for our hobbies, for our charities or religion, and for creating memories with our friends and families. It is activities like these, rather than merely making more money, that create lasting happiness. Instead of laboring under the delusion that we would be happy if we just had a little more money, we should recognize the reality that we might well end up with more money if we just took a little more time to be happy.

If I Were a Rich Man

As any parent knows, one of the first words a baby says is *more.* Money is not much different from milk or applesauce; once we taste it, it makes us say "more." A survey of 800 people with a net worth of at least $500,000 found that 19% of them agreed with the statement, "Having enough money is a constant worry in my life." But among those who were worth at least $10 million, 33% felt that way. Somehow, as wealth grows, worry grows even faster. Fewer than half of these rich people felt that "as I have accumulated more money in my life, I have become happier."

In 1957, the average American earned about $10,000 (adjusted for inflation) and lived without a dishwasher, clothes dryer, television, or air conditioner. But 35% of people surveyed then said they were "very happy" with their lives. By 2004, personal income had nearly tripled after inflation, and the typical house was bursting with consumer goods. Yet just 34% of people now said they were "very happy." Somehow, almost tripling our wealth has made Americans a little less happy—and still we want more.

As the philosopher Arthur Schopenhauer warned, wealth is "like seawater: the more you drink, the thirstier you become." Desperate to quench that thirst, many of us have thought at some point, "If only I had as much money as Bill Gates, then all my problems would be solved." Is there any truth to this belief?

In an affluent society like the United States, the rich are certainly happier than the poor. Poverty subjects people to violent crime and strips them of the comfort that wealthier people derive from feeling that they are in control of their circumstances. The poor are much more liable to suffer from hypertension and chronic heart disease, as well as to die in chronic pain. Growing up in poverty may even reduce the level of activity in the left prefrontal cortex, one of the brain's centers for generating happiness, thus making the poor more prone to chronic depression. The poor also suffer from less stable households—a finding so basic that it has been documented even among bluebirds, whose families break up prematurely when food runs low. Overall, an American earning less than $20,000 is roughly 3.5 times more likely to die in middle age than someone who makes at least $70,000. When you have very low levels of income and wealth, every extra dollar can make a big difference in your quality of life.

But are the rich a lot happier than those of us who are fortunate enough to be above the poverty line? The surprising answer is no. For years, psychologists have been presenting standardized questions to people the world over: "Taking all things together, how would you say things are these days? Are you very happy, pretty happy, or not too happy?" The answers are usually ranked from 1 (not at all happy) to 7 (extremely happy). On average, members of the Maasai ethnic group, who herd livestock on the arid high plains of Kenya and Tanzania, score 5.7 on this scale. The Inuit, who live in the frigid wilds of northern Greenland, average 5.8. The Amish, with their antiquated rural lifestyle, also score 5.8. When members of the Forbes 400, the famed "Rich List" of the wealthiest people in America, took a similar test, their average response was 5.8.

In other words, having a vast fortune in America—with mansion and Mercedes, chef and trainer, yacht and private jet—makes you only marginally happier than the typical Maasai sipping cow's blood mixed with milk in a hut made of dried dung.

The minimum net worth to qualify for the Forbes 400 at the time of this survey was $125 million; the members had an estimated average annual income of more than $10 million. The richest people in America said they were happy 77% of the time on average; a sample of middle-class Americans answering the same survey said they felt happy 62% of the time. That is not a tiny difference, but it is not very big, either, considering that the people on the Forbes 400 earned an average annual income roughly 300 times greater than the average middle-class person in the survey. Furthermore, although most of the rich reported being marginally happier, 37% of them rated their feelings *lower* than average Americans did.

The truth, then, is not that money can't buy happiness. It's that once you have enough to meet your basic needs, more money buys much less extra happiness than you think it will.

If you had passed through Pittsburgh International Airport in November 1995, you might have bumped into students from Carnegie Mellon University handing out candy bars as a prize for participating in a survey. They wanted to know how great an impact people expected a future change in salary to have on the quality of their life—and then to measure the difference in living standards reported by folks whose household income had already changed. The research team surveyed dozens of travelers and found that they predicted that a change in their income would be roughly three times as important to their future quality of life as it actually turned out to be for the average person.

Why do we think money will matter so much more than it really does? It's how the brain is built. As we saw in Chapter Three ("The Wi-Fi Network of the Brain," p. 37), the nucleus accumbens in your reflexive brain becomes intensely aroused when you anticipate a financial gain. But that hot state of anticipation cools down as soon as you actually earn the money, yielding a lukewarm satisfaction in the reflective brain that pales by comparison. Imagining your future wealth often makes you happier than actually having it. Put another way, the pleasure you *expect* tends to be more intense than the pleasure you *experience.* That sets you up for chronic disappointment until you learn the truth.

Out of Focus

In shock after the assassination of John F. Kennedy, the journalist Mary McGrory turned to Daniel Patrick Moynihan, then the assistant secretary of labor, and said grimly, "We'll never laugh again." "Heavens, Mary," answered Moynihan, "we'll laugh again. It's just that we'll never be young again."

Moynihan sensed that humans bounce back from adversity much more quickly than we give ourselves credit for. We come equipped with what behavioral scientist Daniel Gilbert calls a "psychological immune system," which makes us expect bad things to be worse than they generally turn out to be. Because we imagine that our reactions to bad events will never fade, our own powers of recuperation take us by surprise. On the flip side, we also adjust to good events much faster than we anticipate.

Close your eyes for a moment and imagine something wonderful happening to you—winning a $250 million lottery jackpot, for instance. What do you imagine the rest of your life would be like after that?

Now imagine something terrible—for example, a car crash that paralyzes you from the neck down. How do you feel about the life that now would lie ahead of you?

Chances are, your intuitive response to the idea of becoming an instant zillionaire was something along the lines of "I'd never have another worry" or "I'd be able to live out all my dreams." Your response to the thought of becoming paralyzed, on the other hand, was probably something like "I couldn't bear it" or "I'd be better off dead."

Something strange happens when we predict how happy or unhappy something will make us feel. We focus on the wrong thing. When you imagine the impact of a dramatic event on your future quality of life, it seems like a bolt of lightning or the smack of a hammer onto an anvil—a sudden, sharp change that rivets your attention and monopolizes your emotions. And at the moment they happen, the big events of our lives often do feel just as wonderful or terrible as we hoped or dreaded they would. But after the moment of change passes, what remains is the aftermath of change and the process of adapting to it. That process is subtle, sporadic, and spread out over time. Because the adjustment to change is so much less vivid than the change itself, it's much harder to imagine in advance how this phase will make you feel. Your imagination focuses on the moment of *becoming* rich or paralyzed, not the state of *being* rich or paralyzed.

Being is very different from becoming. When you imagine winning the lottery, you focus on the incredible thrill of getting tens of millions of dollars in a heartbeat. You see yourself shedding every financial worry, taking the vacation of a lifetime, moving into a mansion, buying a Bentley. The images swarm through your head in a rush, and you picture all the wonders of wealth cascading onto you at once—as if time stood frozen in the wake of this enormous change.

But time does not stop. Becoming a lottery winner takes only an instant; being one lasts the rest of your life. People who actually win the lottery are often shocked by the aftermath of their lucky draw. There are plenty of thrills from suddenly making a fortune, just as the winners expected. But there are less obvious and less predictable consequences, too. The phone rings off the hook with calls from crooks and desperately friendly acquaintances. Ensconced in your new mansion, you no longer see your old neighbors as often; instead, you are besieged by long-lost relatives who should have stayed lost. Everyone you ever rubbed the wrong way files a lawsuit against you. Quit your job, and you miss your friends and go crazy with boredom; keep it, and your co-workers all seem to hate you or hit you up for money. It becomes hard to tell who your real friends are, so you spend more time alone. At home, you bicker constantly with your spouse over what to do with the money.

Your sudden wealth may end up feeling like a mocking reminder of all the things money can't buy: youth, time, self-control, self-respect, friendship, love. That frustration, in turn, may make you spend like mad on all the things money can buy. By one estimate, 70% of the people who earn a sudden windfall squander it all. No wonder the typical lottery winner, a few years after hitting the jackpot, is barely happier than before—and many are miserable. New York State lottery winner Curtis Sharp Jr. voiced a remarkably common sentiment when he said: "The lottery brought me a false joy. If you abuse it like I did, when it's over, you ain't going to have nothing."

Just as we mistakenly think that being rich is as full of joy as becoming rich, so we incorrectly imagine that being paralyzed is as horrible as becoming paralyzed. When you picture the fate of a quadriplegic, you focus on the shock and terror of a catastrophic injury, the loss of your mobility and freedom, the end of your working life, the temptation to end it all. But after someone is paralyzed, a new routine replaces the old one. Typically, after a terrible period of denial, shock, anger, and depression, the trauma fades and the condition becomes bearable. You concentrate your energy and attention on making the best of your situation. For everything you knew you would never be able to do again, there is something you can do that you never would have anticipated.

Thus, although no one would ever volunteer to become paralyzed, being paralyzed turns out to be more bearable than most people—even experts in the field—can imagine. Out of more than 150 nurses, emergency medical technicians, and doctors at three trauma centers, only 18% thought they would be glad to be alive if they suffered a spinal-cord injury; just 17% felt their quality of life would be average or above-average after paralysis. But among patients who actually were paralyzed by spinal cord injuries, 92% said they were glad to be alive, and 86% felt their quality of life was average or better. Incredible as it may seem, in the second year after being injured, 1 out of 4 spinal-cord patients will already agree with the statement, "In most ways my life is close to my ideal." That's partly because they have begun to make the best of their situation, and partly because one of the greatest sources of happiness for anyone is the social support of family and friends—and these relationships can grow deeper when people are disabled.

Jack Hurst is an enthusiastic investor who manages his stock and mutual fund portfolio online even though amyotrophic lateral sclerosis (ALS, or Lou Gehrig's disease) has left him almost completely paralyzed. Hurst breathes through a ventilator, eats through a feeding tube, and needs his lungs suctioned dozens of times a day. In his entire body, the only area he can still move is a small part of the right side of his face; a device taped to his cheek converts the electrical activity in his facial muscles into signals that enable him to operate a laptop computer. I first met Hurst in November 2004, and we have kept up a lively e-mail correspondence ever since. (Using his facial muscles, Hurst can type up to ten words a minute.) He is one of the most contented and charismatic people I know; the twinkle in his eye is undimmed after two decades of paralysis. This man who cannot move can still count his blessings—and he does. "The general attitude and love demonstrated by my wife makes me happy," Hurst has e-mailed me. "The number of friends who help and support us enhances my happiness. I have always been an optimist, which prevents me from being negative. I don't have a lot of regrets, because I've always performed to the best of my abilities in whatever task I faced." Hurst has not been able to walk since 1988; nevertheless, he insists, "A person in my position doesn't have much to complain about."

Wouldn't It Be Nice?

Even though our predictions about our future emotions often turn out to be wrong, we are usually blind to our own blunders. "In our lives there are literally thousands of things happening over long periods of time," says psychologist Ed Diener of the University of Illinois, "so it is hard to sort out the mistakes we make in forecasting our own feelings."

"If only the Red Sox win the World Series," you tell yourself, "then my life will finally be complete." Then they do, and in a few days the glow is gone. "If I can just get this new job," you swear, "then I'll be able to stop and smell the roses." Then you get the job, and you end up more stressed out than ever. "If only she'll agree to marry me," you pray, "then I'll be happy every day for the rest of my life." Then she does, and you are not happy every day for the rest of your life, even if your marriage survives. Furthermore, when you prayed for her to marry you, you managed to forget that not long before, you had made a solemn vow—over your own broken heart—that you would never fall in love again.

When we imagine the future, we exaggerate both how intense our emotions will be and how long they will last. That leads to what Daniel Gilbert calls "miswanting," the desire for possessions or experiences that we think will make us happy in the future, but turn out not to. Unless you learn to conquer this illusion, you are likely to waste your money buying things that seem to promise bountiful happiness but turn out to be little more than empty husks. The variety of ways in which humans mispredict their own future emotions is dazzling:

- College football fans were asked how happy they would be, and how long they would feel that way, if their favorite team won a big game. They predicted they would be elated for days but, forty-eight hours later, it was as if nothing had happened; many fans turned out to be happier even though their team lost.

- Asked whether they were hypothetically willing to accept $5 in return for performing a mime routine in public, many of those asked said yes. But when they were actually put in front of an audience and asked to do it, suddenly only half as many were willing to go through with it—even though they still could earn $5 just by imitating an elephant or a washing machine.

- At art and antique auctions, people regularly swear they will not bid above a certain price—but then, in the heat of the moment, find themselves the startled owners of something at two or three times the maximum price they "wanted" to pay. Even experienced CEOs suffer the same "buyer's remorse" or "winner's curse," often finding that the company they were thrilled to buy is nothing but a headache to run.

We base our goals upon what we believe to be our desires; in fact, one of the most basic principles of economics is that people know their own likes and dislikes. But we often find out that what we thought we wanted before we got it is no longer what we really want once we have it. Just as people adapt rapidly to drastic events like winning the lottery or becoming paralyzed, we get used to almost anything we are frequently exposed to. That's why the money we spend on big purchases gives us such a perishable pleasure. To put a twist on the Rolling Stones, you can't always want what you get.

Take that new SUV. When you first drive it off the dealer's lot, it glistens like a gigantic jewel and feels just as fast and safe and soft and roomy as you had dreamed. Becoming an owner is even better than you imagined. But being one turns out to be worse. In a couple of weeks the last trace of new-car smell is gone. In a month or two, the exterior is scarred with dings and scratches, while the interior is splattered with coffee, Gatorade, and goodness knows what else. It's hard to park, and every time you pull out of a gas station you leave at least $50 behind. With each passing day, the contrast between your vision of what ownership would be and the reality of what it has turned out to be will become more glaring. Therefore, while you might not quite be willing to renounce your new SUV, it will probably give you much less pleasure than you were sure you would get from it.

The same goes for your beautiful new suit or shoes (they will get stained or go out of style all too soon) or the kitchen you just remodeled (the countertop gets chipped, the floor tiles get scratched, and somehow the fridge still isn't big enough). Big capital expenditures can give us a great deal of pleasure when we imagine the results ahead of time. Unfortunately, the vision pales by comparison when it collides with reality. You are left weighing what you got against what you dreamed you would have, and the results feel flawed and dirty compared with your dreams.

Because those original dreams remain so vivid, you usually come to the wrong conclusion. Instead of realizing that big spending will probably never make you happy, you conclude that you simply spent your bundle on the wrong thing: "Next time, I'm getting a Lexus instead of this damn Acura." Then if the Lexus lets you down, you wish for a BMW, and so on—condemning yourself to an unending cycle of buildups and letdowns. If you spend like this, money might not only fail to make you happy, it could make you miserable. George Bernard Shaw was right when he wrote, "There are two tragedies in life. One is to lose your heart's desire. The other is to gain it."

How Was It for You?

If it seems strange that our predictions of what will make us happy are so unreliable, consider this: Our memories of what *did* make us happy are not much better. Because no one likes admitting a mistake, the past often gets polished in our memories—making us feel it wasn't so bad after all. That, in turn, may make us more willing to repeat experiences that we did not always enjoy at the time.

- College students were equipped with handheld computers as they went on spring break. Seven times a day, the students filled out an electronic survey rating how intensely they were feeling joyful, calm, friendly, pleasant, or happy. On average, they ranked their emotions right smack in the middle, reflecting the reality that fun in the sun also comes with bug bites, sunburn, sand-filled bathing suits, and bad hangovers. A month after they got back, the students were asked to recall their emotions during the vacation. The feelings they now remembered were 24% more positive than they had been in real time.

- On a tour of California, bicyclists got plenty of exercise, fresh air, and dazzling views—just as they expected. They also got bored, soaked by rain, and exhausted by heat and distance. During the trip, 61% of the bicyclists said that at least one aspect of the trip was worse than they had expected. Only one month later, however, just 11% remembered feeling that way, as if the recent past were reflected in a rose-colored rearview mirror.

- The vacation in your photo album is more pleasant than the one you actually had, and that may skew your memory. After all, you had to put the camera down when your husband fell off the boat ride, there was nothing to photograph while you waited forever in 95-degree heat to use the restroom, and you certainly didn't memorialize the moment when your six-year-old threw up his French fries. Instead, you posed everybody, smiling, in front of Cinderella's castle.

- Tracked over time, dozens of people were asked how happy their childhood was. Looking back from around age 30, just 40% felt they had been "generally happy" as children. By their late 60s, however, 57% decided that their childhood had been mostly sunny. And when they reminisced in their 70s, 83% now felt they had been generally happy as children.

- Half the patients undergoing colonoscopy at a hospital in Toronto got a conventional procedure lasting about twenty-seven minutes. For the other half, the colonoscope instrument was left motionless in place (yes, *that* place) for up to three extra minutes at the end of the exam. In real time, both groups rated the procedure almost equally painful. But when they looked back on it later, the second group felt the exam had been considerably less painful. These patients had been in pain for longer, but because their final minutes involved a declining level of discomfort, they now had a more positive memory. How long a feeling lasts is less important than how it ends.

Your memories, then, are not just recollections. They are also reconstructions. That helps explain why you learn so scantily from your own experience. Your memory of what *was* is shaped largely by what *is;* as you become happier in the present, it seems that you must have been happier in the past. And as you forget what actually happened, it starts to seem more like what you originally expected to happen. Your vacation was supposed to have been fun, so maybe it really was. After you kick yourself for doing something stupid, it may occur to you that if it had really been such a dumb thing to do, you would not have done it in the first place. These tricks of memory reveal a valuable lesson: While your new SUV or your renovated kitchen tend to depreciate as you grow used to them, past experiences often become more positive as you look back at them. While the money you spend on *acquisitions* tends to feel more and more like a mistake as time passes, the money you spend on *experiences* is apt to grow in value as your memories grow warmer.

The Elusive Butterfly

We've seen that people are often wrong about what will make them happy in the future, or even what did make them happy in the past. But surely we must know what makes us happy in the present.

Unfortunately, it turns out that isn't really true, either.

- How satisfied married people say they are with their lives is determined largely by how happy they are with their spouse—but only if you ask about married life *before* you ask how happy they are in general. Likewise, how happy college kids say they are is unrelated to how often they have been dating—if you ask them first about their overall happiness. If, instead, you first ask them about their social life and then about their happiness, you will get very different answers.

- Nearly 1,000 working women in Texas kept a detailed journal of one day in their lives, ranking sixteen common activities by how satisfying they were. It turned out that being with their children ranked near the bottom, barely ahead of surfing the Internet and slightly behind taking a nap. The only activity that made them less happy than being at work was commuting. They preferred praying to shopping, while socializing with friends made them much happier than relaxing alone. Yet, if someone asked what makes you most happy, you might think of achieving something at work, spending time with your kids, shopping, or just relaxing. It might not cross your mind that you could enjoy *thinking* about those things more than you enjoy *doing* them.

- Once a day for two weeks, college students who had recently begun dating rated how satisfied they were with their relationship in general, their sex life, and life overall. Then, at the end, these lovebirds were asked to look back and rate how they felt about their life and love over the preceding two weeks. When the researchers checked six months later to learn which romances had survived and which had broken up, they found that how happy people say they *are* each day is not a good way to forecast how long the relationship will last—but how happy people say they *were* is an excellent predictor of whether they will stay together.

Strange as it seems, thinking about the ups and downs of daily life as you go through them can make you feel less happy. Stranger still, the passage of time can smooth out the highs and lows, making you feel more satisfied with your life when you look back than you actually were while you were living it.

So Happy Together

Measuring happiness, it seems, can be like trying to trap a rainbow in a jar. Fortunately, a few findings hold water. How satisfied you are with your life depends, above all else, on how connected you feel to other people. Psychologists Ed Diener and Martin Seligman studied more than 200 people for months on end, administering numerous tests to determine who was genuinely happy. The happiest people were much more satisfied with the general state of their lives much more of the time. They seldom were sad for long when things went wrong, nor were they swept away by euphoria when things went right. Above all, happy people had more friends and spent less time alone.

Researchers had more than 100 folks wear wristwatches that beeped as a reminder to fill out a survey on happiness. Those who had scored high on a test for extroversion were happier than introverts. In fact, people who say they have more than five friends are nearly 50% more inclined to feel they are "very happy" than those who have fewer than five friends. Solitude produces so little happiness, on average, that most people would rather hang out with their boss than be alone.

We also know that when people are happy, the brain lights up with a special glow. Using brain scans and electrical recordings from the scalp, neuroscientist Richard Davidson of the University of Wisconsin has found that happier people have a lot more going on in the left prefrontal cortex. Neurons firing in the left side of the PFC help you recover from an upsetting event, focus on achieving a positive goal under trying circumstances, and suppress the negative emotions generated in the amygdala. People with a higher average level of activity in the left PFC get less upset by scary or gross movies. What's more, investors with stronger activation in this part of the brain have a better memory for how they earned their recent profits. It's almost as if this area is a source of internal sunshine for the mind: Autopsies have shown that people who suffered from chronic depression often have severe shrinkage of cells in the left PFC.

Some people may be born with better wiring in the left prefrontal cortex. Davidson studied ten-month-old babies as their mothers stepped out of the room for a minute. Comparing the infants who cried against those who didn't, he found that calm babies had much more activity on the left side of the PFC than crybabies did. Fortunately, it seems that at least some people who are not born with better wiring in the left PFC might be able to develop it. Buddhist monks, who spend years practicing meditation to generate inner calm, seem to have remarkably high levels of activity in the left PFC that persist even when they are not meditating. Deliberately dispelling negative emotion seems to create positive emotion.

Knowing that money is a means, not an end in itself, is another key to satisfaction. People who place less priority on having more money tend to be happier than those who are more materialistic. Pursuing wealth for its own sake, says Diener, has "toxic effects" on happiness. People who believe that money is the most important thing in life are more likely to suffer mental health disorders (unless they are already rich). Unfortunately, insecurity and uncertainty can bring out the inner materialist in almost anyone. Children who are raised with inconsistent rules or in a harsh or hostile atmosphere are more liable to grow up looking for money to provide psychic compensation. That sets up a vicious circle: Pursuing money as an end in itself can cause depression, anxiety, stress, and family tension, which in turn prompt a more urgent chase for more money in hopes of relief.

But you cannot slow down a treadmill by running faster. In a recent survey, working women estimated that people with annual incomes no greater than $20,000 are in a bad mood more than twice as often as those who make more than $100,000 a year. In fact, the low earners report being in a bad mood only a smidgen more often. After all, the more time you spend at a high-paying job chasing even more money in the pursuit of imaginary happiness, the less time you will have for the things that can add real happiness to your life.

Keeping Up with the Joneses

One reason it can be so hard for you to get off the treadmill is that your neighbors are pounding away on it, too. How good your money makes you feel depends partly on how much money the people around you have.

To envy is human. But humans are not the only creatures that care about how they rank against their peers. Many species form their own social ladders, with those on the bottom rungs kowtowing to those that are higher up. A dominant animal might have its fur groomed by others in the group, or the others might defer to its authority over food or choices of mates. Animals often show high levels of stress hormones in their blood when they are low on the totem pole in their own species. When a mouse is intimidated by a more powerful member of the group, its brain produces extra amounts of a protein that enhances memory, encouraging future "social defeat"—not just around an alpha mouse immediately, but around any of its peers for weeks to come. Rats that are low on their group's pecking order tend to lose their appetite, turn lethargic, and sleep fitfully, and their adrenal glands, which produce stress hormones, become enlarged. When their territory is dominated by another member of the same school, fish shut down production of a protein that enhances fertility; the less "real estate" they "own," the less likely they are to reproduce.

Trained by neurobiologist Michael Platt to use fruit juice as currency, monkeys pay up to view images of other monkeys that are higher in the pecking order. Conversely, they will spend units of fruit juice to avoid viewing pictures of monkeys lower on the social ladder. After as little as three months of socially dominating their neighbors, monkeys show a 20% increase in the volume of special molecules that sponge up dopamine; being high on the totem pole may actually beef up the reward system in the brain.

Our modern minds still retain these primitive responses to social status. When young men in Germany viewed photographs of automobiles, they strongly preferred pictures of flashy sports cars to either small cars or stodgy limousines. No surprise there. But when the men made their choices inside an MRI scanner, it turned out that just looking at the images of the hot sports cars fired up the reward centers in the reflexive brain. A mere glimpse of the kind of car that most people envy was enough to flood the viewer's brain with dopamine, in the same areas that light up when a man sees a picture of a beautiful woman's face. The craving called "car lust" is probably driven by an ancient, elemental desire to show off and move up the social ladder.

The roots of envy and social comparison run so deep that they are an instinctive part of our biological makeup. In a hunter-gatherer society, dominant individuals dominated precisely because they were better at getting and keeping scarce resources. Social comparison probably served early man well; by observing those who had more, our ancestors learned how to get more themselves. For example, envying—then imitating—the group member who was best at gathering fruit would have helped the others learn how to gather more, too. In a primeval world, envy helped everyone survive.

Having it ingrained in our modern genes, however, is a mixed blessing. For most people in today's Westernized societies, life is no longer a struggle to fend off starvation. Failing to imitate your successful peers will not normally condemn you to a childless life or early death. Suffering a mild case of the "comparison complex" can be beneficial; it motivates you to work hard, gives you hope for the future, keeps you from being a total miser, and prompts you to clean up the house before visitors arrive.

There is a good reason, however, that the Ten Commandments close with a thundering inventory of all the things that "Thou shalt not covet." While a secret pinch of envy is a positive motivator, a chronic comparison complex can ruin your life. If you cannot control the ancient urge to measure your success against that of your peers, your happiness will always depend less on how much money you have than on how much money they have. And that's something you will never have any control over. Always wanting more, in order to keep up with whoever has more, makes millions of people perennially unhappy.

Even if you think you don't care about "keeping up with the Joneses," you may suffer from a comparison complex without realizing it. Imagine two situations: In one, you buy the biggest house in a middle-class neighborhood; in the other, you buy the smallest house in a rich part of town. In both cases, you earn an upper-middle-class income and pay the same price for each house. Which neighborhood will you be happier in?

Financially, the house in the wealthy enclave might turn out to be a better investment. Psychologically, however, you are much more apt to be happy in the middle-class area, where your house will not seem smaller with every passing year and where your neighbors will not constantly outspend you. A study of more than 7,000 people in over 300 towns and cities found that, on average, the more money the richest person in your community makes, and the greater the number of neighbors who earn more than you, the less satisfied you will probably feel with your life.

Thousands of people in Switzerland were asked how much income they felt would be enough to meet all their needs. Over time, the survey showed that for every 10% increase in their income, people wished they earned still another 4% more. The more you have, the more you want more than you have.

Almost 5,200 British workers were asked to rate how happy they felt about their jobs and income. The less the typical person earned relative to those with similar jobs, the less happy he felt about his own work and wages—even if he was rich. In other words, someone earning a few farthings more than the typical pittance for his job may well be happier than someone earning a fortune in a field where his competitors make even more.

That appears to be why many people in poor countries are happier than an American or European might expect. When nearly everyone you know is about as poor as you, you are less likely to suffer from a comparison complex. So the Maasai and Inuit, who live largely without electricity or televisions, among other things, very seldom worry about how much richer someone else might be in Nairobi or Nanortalik—let alone Beverly Hills, Dallas, or Zurich.

How happy your money makes you even depends on how you stack up against your own family. The journalist H. L. Mencken once quipped that wealth is "any income that is at least $100 more a year than the income of one's wife's sister's husband," and economists have recently shown that he was right: Women are much more satisfied with their household income when their husbands earn more than their sisters' husbands. Furthermore, people who earn less than their parents did, even if they themselves are rich, are considerably less likely to feel "very happy" than those who make more money than their parents did.

Comparing your own wealth to that of your friends can also hurt. A survey done for *Money* magazine in 2002 found that 63% of affluent

Americans agreed with the statement "It's harder for me to be friends with people who have a lot more money than I do"—triple the proportion who felt that it is harder to have friends who are *less* well off. That's yet another reason that the race for riches can leave you feeling poorer. Your newfound wealth may cut you off from your old friends. Since an active social life is one of the keys to happiness, ending up richer but lonelier is no way to live a more rewarding life.

Can Happiness Buy Money?

"What good is happiness?" someone once quipped. "You can't buy money with it." It's a clever remark, but it probably isn't true. People who are in a good mood are more inclined to try learning new skills, to see things in a broader context, to think of creative solutions to problems, to work well with other people, and to persist instead of giving up. If you were writing a recipe for how to make more money, those are among the first ingredients you would include. *Happen* and *happiness* come from the same Old English root word, and happy people seem to make good things happen more often.

Neuroscientist Richard Davidson has found that people with greater activity in the left prefrontal cortex—one of the main areas where the brain seems to generate happiness—produce more antibodies after a flu shot, suggesting that their immune systems are stronger. In both humans and monkeys, greater activity in this part of the brain is associated with lower levels of stress hormones in the blood, helping prevent us from overreacting to the ups and downs of daily life. Happier women start the day with lower levels of stress hormones in their bodies—and their levels stay lower throughout the day. Among nearly 1,000 elderly Dutch men and women, those who laughed more often, looked forward to the future, and strove to fulfill their goals incurred a 29% lower risk of mortality than those who were less optimistic. Being extroverted is also good for your health; people who are more outgoing have lower levels of glycosylated hemoglobin in their blood, indicating that they are at lower risk for diabetes and related diseases.

In 1976, thousands of college freshmen rated how cheerful they were. Almost twenty years later, those who had ranked at the top for good cheer were earning an average income 31% higher than those with bottom scores. Happier professional cricket players have higher batting averages than gloomier players, and employees who are more consistently in a good mood miss fewer days at work—even in Russia. A study of nearly 300 workers at three U.S. companies found that the happier they were, the higher their salaries were eighteen months later. CEOs who are more cheerful are more likely to have more productive workers who generate higher profits. And day traders who stay in a good mood regardless of whether they made or lost money tend to earn higher returns over time.

It's very simple: The happier you are, the longer and healthier your life will probably be—and the more money you are likely to have.

Getting Lucky

On a lovely morning in May 1994, Barnett Helzberg Jr. was walking past the Plaza Hotel in New York City when he heard someone yell, "Mr. Buffett!" Helzberg turned and saw a woman in a red dress talking to a man he recognized as Warren Buffett. Recalls Helzberg: "I walked up to him and said, 'I'm Barnett Helzberg of Helzberg Diamonds in Kansas City. I'm a shareholder in Berkshire Hathaway, I really enjoy your annual meetings, and I believe that my company fits your criteria for investments.'"

Within weeks, Buffett had bought the company from Helzberg and his family for an undisclosed price. "My luck is uncanny," says Helzberg. "The more you believe that you're lucky, the luckier you are."

As Helzberg's story shows, luck is more than just being in the right place at the right time. Luck is making the most of being in the right place at the right time. In 1970, U.S. Navy Lieutenant Robert U. Woodward delivered a package to the office of the chief of naval operations. No one was there to sign for it, so the lieutenant had to sit and wait. Soon, an older man came into the waiting area and sat down silently. The lieutenant forced himself to make eye contact, introduced himself, and was soon chattering away to the stranger about the worries of a young naval officer who did not know what to do with the rest of his life.

It turned out that both the lieutenant and the stranger had taken graduate classes at George Washington University—and with that in common, they hit it off. The older man, whose name was Mark Felt, volunteered some career advice, and the two stayed in touch. Before long, Bob Woodward was out of the Navy and working as a reporter at the *Washington Post,* and Felt was a top official at the Federal Bureau of Investigation. Felt later became Woodward and Carl Bernstein's confidential source "Deep Throat."

Woodward had no idea who Felt was before they shook hands, nor any inkling that he himself would become an investigative journalist. If Woodward had sat there in the typical awkward silence of people in waiting rooms, he and Bernstein might never have been able to expose the full story of the Watergate scandal. By converting a stranger into a friend, Woodward ended up changing American history.

British psychologist Richard Wiseman has studied hundreds of folks who describe themselves as very lucky or unlucky. He found that some are indeed luckier than others—and that being lucky is a kind of skill. Wiseman identified several characteristics that lucky people tend to have in common and that help bring them good fortune.

➢ **THEY DON'T GIVE UP.** "Unlucky people collapse under bad luck," says Wiseman. "Lucky people treat bad luck as a learning experience." One painstaking trial at a time, Thomas Edison tested thousands of materials, including boxwood and bamboo, as potential filaments for his "incandescent lamp." (When he said that genius is 1% inspiration and 99% perspiration, he knew firsthand what he was talking about.) Finally, Edison discovered that carbonized cotton thread would do the trick. If he had been a quitter, there would be no lightbulb switching on over your head when you get a good idea.

➢ **THEY LOOK OUTWARD.** Lucky people are curious and observant, eager to engage and explore the world around them. In 1946, Percy Spencer, an engineer at Raytheon Corp., walked into a lab where magnetrons, the power tubes at the heart of shortwave radar, were being tested. After a few moments, he had the odd sensation that the candy bar he had in his pocket was melting. Sure enough, it had become a little bag of chocolate-and-peanut soup. Spencer wasn't the first engineer this had happened to, but he was the first one who did anything about it—even though he had had only a grammar-school education. Spencer promptly got a bag of popcorn and put it in front of the magnetron; after that, he got an egg. Both soon exploded. By seizing on serendipity, Spencer had invented the "radar range," which later became known as the microwave oven. If you focus too narrowly on the task at hand, you may never use your peripheral vision—and thus miss out on what the sociologist Robert K. Merton called "the importance of letting accidents happen," or "structured uncertainty."

Fund manager Bill Miller of Legg Mason Value Trust forces his mind open in much the same way; he built his superb investing record by straying far afield to buy stocks like America Online and Dell Computer that more conventional "value" investors wouldn't touch. A couple of decades earlier, Peter Lynch of the Fidelity Magellan Fund got investing ideas from everywhere, including his wife's new pantyhose; he had no qualms about buying assets other stock-pickers never looked at, like Swedish auto companies or long-term Treasury bonds. Intelligent investors do not just tunnel in one place with blinders on; they open their eyes to everything and throw themselves in serendipity's path whenever possible.

➢ **THEY LOOK ON THE BRIGHT SIDE.** Picture two cars, each occupied by only the driver, colliding head-on at high speed. Both cars are totaled. Both drivers walk away without a scratch. The unlucky one wails, "Oh God, look at my car!" The lucky one exclaims, "Thank God I'm alive!" One sees only what went wrong; the other, what went right. Both are holding the same glass, but the one who considers himself lucky sees it as half full, while the one who thinks he is unlucky is convinced it is half empty.

Now imagine what happens next. The driver who feels lucky is so happy to be alive that he shrugs off any setbacks for the rest of the day. He tells everyone that simply surviving is a miracle. Other people are infected by his enthusiasm and congratulate him on his good fortune; their approval makes him feel even luckier.

The driver who considers himself unlucky fixates on anything else that goes wrong, no matter how minor, as yet another sign that the universe is against him. Picking up on his negativity, waitresses dawdle, security guards ask for several forms of ID, checkout clerks don't even tell him to "have a good one." By the end of the day, he is sure that he is cursed.

Which driver will be a bigger success in life? The question answers itself. Starting from the same point, these two people have driven themselves in opposite directions on the road of luck. As Louis Pasteur said, chance favors the prepared mind.

The Time of Your Life

Would you rather get $10 today, or $11 tomorrow? If you are like most people, you would rather get $10 today than wait twenty-four hours for an extra dollar.

Next, consider this: Would you rather get $10 a year from now, or $11 in exactly a year and a day? When the question is phrased this way, most people change their answer to $11, even though the two rewards are still separated by an identical twenty-four-hour delay. Somehow, it just seems easier to be patient in the future than in the present.

As this thought exercise shows, your investing brain often gets tangled when you make decisions that unfold over time. Even our proverbs about time are a mess of contradiction. A bird in the hand is worth two in the bush, but don't count your chickens before they are hatched; the early bird catches the worm, but slow and steady wins the race; he who hesitates is lost, but all good things come to those who wait; devil take the hindmost, but fools rush in where angels fear to tread. We are literally of two minds when it comes to time. On the one hand, we are impetuous and impatient, obsessed with the short term, eager to spend now and get rich quick. On the other, we are capable of setting money aside for goals that may be decades away, like our children's college tuition or our own retirement.

Neuroscientist Jonathan Cohen of Princeton University invokes Aesop's fable about the ant and the grasshopper, in which the ant busily gathers up food for the winter, while the grasshopper basks in the sun and mocks the ant for working on one of the best days of the summer. Aesop's grasshopper lives for today; the ant plans for tomorrow. Each of us has a grasshopper and an ant battling within our brains for dominance over our decisions about time. These are the same two systems—the emotional grasshopper symbolizing the reflexive brain, the analytical ant representing the reflective brain—that we learned about in Chapter Two. You cannot be a successful investor or a completely happy person, unless you can control the impulsive power of your inner grasshopper.

The itch to spend, borrow, and trade as soon as possible originates in the emotional circuits of the reflexive brain that we share with most other animals. Experiments on many species of birds and rodents have shown that they will generally not take extra risk to increase the amount of food or other rewards they can get—but they usually are willing to take a bigger risk if it might shorten the time they have to wait for a reward. In a state of nature, it often pays to be impatient. Millions of years ago, when life was short, food was perishable, and territory was difficult to defend, it made sense to live for today.

Getting rich quick fills us with a special thrill because it is the modern version of an impulse that kept our ancestors alive. Today, however, we also have the luxury of living for tomorrow. The focus on the long term is generated in the analytical centers of the reflective brain that are unique to humans. Other animals can also plan ahead, and all creatures have to estimate the tradeoff between how large a reward may be and how long they might have to wait for it. But no other species is capable of laying such complex plans that reach so deep into the distant future. However, we don't always get the planning right. One way or another, time trips up almost everyone:

- When Americans leave their jobs, they can "roll over" their 401(k) into another retirement savings account or pay a tax penalty and get immediate access to the cash. Fewer than half choose the rollover into future savings; instead, 12% spend it on consumer goods and everyday expenses, while another 22% use it to buy a house, start a business, or pay down debt. For many people, the temptation to spend now overwhelms the resolve to save for later.

- Investors would much rather buy a mutual fund that charges an extra 0.75% in expenses every year than one that carries a one-time, 5.75% sales charge up front—even though they would earn higher returns the other way around.

- Thousands of credit card companies compete furiously for business. Why, then, do the interest rates they charge never drop much below 20%? Roughly one out of every two credit-card users think they "nearly always pay in full"—yet three out of four accounts incur monthly interest charges! Applying for a card in the here and now, people aren't bothered by a high interest rate, because they don't think they will ever go into debt; in the future, they almost always do.

- If you begin taking Social Security at age 62, your monthly benefits will be only 75% of what you could earn by waiting a few more years. And yet roughly 70% of recent retirees have chosen to start taking their Social Security payments before the full retirement age of 65. Their impatience gets them some money sooner—at the cost of missing out on a much larger stream of income later.

- Most people who join a gym pay for it up front, monthly or annually, as a way to force themselves to use it. They typically go so infrequently that they could save money by paying per visit—but are afraid to change their payment plan, lest they stop going to the gym entirely.

- To encourage military personnel to retire early, in the early 1990s the Pentagon began offering buyouts. Officers and soldiers could choose between an upfront lump-sum cash payment or a long-term annuity (a stream of monthly checks guaranteed for many years into the future). Over 90% of enlisted personnel took the lump sum, choosing more money up front—but missing out on an average of $33,000 in foregone benefits down the road.

- Offered a single, 50% chance of either winning $20 or losing $10, two out of three customers in a coffee shop in Westwood, Los Angeles, were willing to take the gamble. Given the chance to play the same bet 100 times, only 43% would take it; the near certainty of greater gains in the future was not vivid enough to counteract the pain of possible losses in the present.

A research team led by Jonathan Cohen of Princeton recently explored how the ant and the grasshopper wage their war over time inside your head. The researchers offered people a series of choices between two gift certificates at Amazon.com. One was for a smaller amount sooner; the other, for a larger amount later. (A typical choice was between getting a gift certificate worth $20.28 today or waiting a month to get one worth $23.32.) What Cohen's team found was simple but stunning: Reflective areas of your prefrontal and parietal cortex will be activated regardless of whether you go for immediate or delayed gratification. But your reflexive brain will kick in only when you pick the earlier reward—setting off a fiery surge of activity around the nucleus accumbens and nearby areas. Thus, choosing an immediate gain gives you a jolt of dopamine that you cannot get from choosing a delayed reward—unless the future profit is much larger. "It takes a really juicy future reward," says Cohen, "to get this system as excited as it is about an early or immediate reward."

That helps explain why most people prefer $10 today over $11 tomorrow but would rather wait for the extra dollar a month from now. When you can get the $10 immediately, just thinking about it generates a dopamine burst that crowds out your ability to wait a day for the larger amount; the grasshopper wins out over the ant. But if you can't get either gain until much later, then your reflective brain is not swamped by emotion and you can make a less impulsive choice; the ant can prevail over the grasshopper.

Long ago, psychologist Walter Mischel gave preschoolers a series of grasshopper-vs.-ant choices, for example, one marshmallow now, or two marshmallows 15 minutes from now. He found that the four-year-olds who were best able to defer their gratification—often by distracting themselves with a toy like a Slinky—turned out, as young adults, to have better social skills, greater self-confidence, and higher SAT scores.

As with marshmallows, so with money. Researchers recently asked people to imagine that they had won ten gift certificates good for dinner at a fine restaurant of their choice at any time over the next two years. The researchers asked who would be tempted to use so many certificates in the first year that they might kick themselves in Year Two; 25% admitted they would. How many people were willing to accept certificates that were usable only in the second year? Only 7% said they would. Strikingly, those 7% had already saved much more money in their retirement accounts. Saving comes naturally only to people with better self-control—something the government should keep in mind as it considers privatizing Social Security.

Because gains and losses have so much emotional power in the present, but fade when they are delayed into the future, we suffer chronic confusion between the price of buying something and the cost of owning it. We are very sensitive to the former because it is now, and much less sensitive to the latter because it is later. A century ago, a man named King Camp Gillette became rich by exploiting this quirk in our brains: He coaxed men into paying next to nothing for a safety razor in the first place, then hooked them on buying new blades every few days *ad infinitum.* That's also why today's consumers get such a thrill out of buying inkjet printers at amazingly low prices, only to realize later that replacement ink cartridges will cost them a small fortune indefinitely. It's your inner grasshopper that does the splurging, leaving the ant to find a way to pay for it.

I Love Ya, Tomorrow

"I'll do it later." We all say those words, and they seem pretty harmless. But they can make a huge difference in your financial life.

Procrastination, says Harvard economist David Laibson, is simply "delaying an unpleasant task for longer than you yourself think you should." Of course, unpleasant tasks often lead to pleasant results down the road. We often procrastinate the worst on things that are good for us: saving more in our 401(k), quitting smoking, paying down our credit card debt, exercising, prepaying the mortgage, going on a diet, raising our insurance deductibles, signing up for online banking and bill-paying, shopping around for cheaper phone service, filing our expense accounts.

Joseph Ferrari, a psychologist at DePaul University, estimates that 25% of Americans are chronic procrastinators who show "habitual delays in meeting deadlines." (Men and women, he says, procrastinate more or less equally, regardless of age.) In 2005, according to the Internal Revenue Service, 27% of America's 120 million individual tax returns were filed at the very last minute (or just beyond); 32 million returns came in between April 9 and April 22. Almost 40% of those returns were owed refunds. Even the thrill of getting money back from Uncle Sam isn't enough to keep people from procrastinating.

At one large U.S. company, 68% of employees felt their 401(k) savings rate was "too low." More than a third of those who felt that they were not saving enough said they had already made plans to raise their 401(k) contributions in the next few months. But only 14% of them actually followed through on those good intentions.

So the problem is not that we don't know what's good for us. It's just that tomorrow seems like a better time to do it than today. And when tomorrow comes, the day after tomorrow will suddenly look even better.

That's because the benefits you reap from many decisions come mainly in the future, so they are vague and delayed. The costs, however, hit home right now—vivid, emotional, immediate. When you exercise, for example, you have to carve an hour out of your busy day and become a gasping, sweaty mess. Down the road, of course, you will probably lose weight and reap the rewards of a longer and healthier life. But the upside only comes later; the downside is now. Likewise, if you put more into your 401(k) today, you will have a richer retirement a few decades in the future—but first you have to fight your way through a thicket of paperwork, pick your funds from a stupefyingly complex menu, and live with the crimped spending that results from a slightly smaller paycheck. Again, the costs come now—"I want to buy those shoes instead!"—while the benefits do not accrue until much later. The emotional circuits in your reflexive brain fixate on the immediate costs, deterring your reflective brain from analyzing the future benefits.

Thus, by putting off until tomorrow what we really should do today, we can push the costs into the future. Going to all that trouble *then* is bound to be much easier than doing it *now.* And by playing this little trick on ourselves, we can pretend that we have put the two sides of our mental ledger—the costs on one hand, and the benefits on the other—back in balance. Now both the benefits and the costs are delayed. You can easily muddle through your whole life this way, renewing your promise to yourself every day that you will eat right, quit smoking, exercise more, save more, spend less—starting tomorrow.

These Are the Good Old Days

Imagine that you suddenly become old. If you are now in your twenties, thirties, or forties, you probably picture a ghost of your present self—agitated by the least annoyances, forgetting everything but your own name, unable to hear anything quieter than a chainsaw, falling asleep in front of the TV every day, taking your teeth out at night, and aching all over, inside and out, all the time. In his song "My Generation," Pete Townshend of The Who tapped into an almost universal fear when he wrote, "I hope I die before I get old."

Since old age seems like such a rotten deal to the young, most people believe that their ability to enjoy their money is bound to fade as the years go by. So they live for today: spending rather than saving, borrowing in great binges on their credit cards, trying to get rich quick on risky speculations instead of getting rich slowly with more reliable investments. "Many people dread getting older," says psychologist Heather Pond Lacey of Bryant University in Smithfield, Rhode Island. "That helps explain why young people often engage in risky behaviors instead of preserving themselves, and their assets, for the future."

But the intuitions that most young folks have about aging are remarkably wrong. Recently, nearly 300 young people (average age: 31), along with almost 300 older Americans (average age: 68), took part in an online survey. Both groups estimated how happy the typical person is at age 30 and age 70. Not only did the 30-year-olds feel that they would be less happy when they were 70, but those in the older group believed that they were happier when they were 30. Then came a shocker: Asked to rate their own happiness, people around the age of 70 scored more than 10% higher than did those around the age of 30! The surprising truth, then, is that people get happier as they get older, even though they do not expect it in advance and may not even realize it as it is happening.

Learning from your accumulated experience as you age, you screen out the things that used to upset you, focusing instead on whatever is most likely to give you pleasure. As time goes by, your brain becomes a better emotional manager. Your good moods last longer, you recover from your bad moods faster, and you become better at forgetting about past disappointments. Stanford University psychologist Laura Carstensen has shown that as you grow older, you become more inclined to let casual relationships lapse so you can spend more time with the people you already know you like the best, and to drop your passing interests so you can devote more energy to the activities you already know you enjoy the most. As your time horizons grow shorter, you derive happiness not so much from pursuing pleasures for their own sake, but from investing in experiences and relationships that add meaning and fulfillment to your life.

▪ Neurologists precisely measured the momentary eye movements of people viewing photographs. When a group in their twenties and a group with an average age of sixty-four saw an upsetting image—for instance, a soldier aiming a gun at a fleeing child—the younger people's eyes lingered about 25% longer.

- Comparing six different cars using a chart that rated them on dimensions like price, gas mileage, safety, and comfort, older folks spent about 10% more time evaluating the positive features and nearly 20% less time reviewing the negative features than younger people did.

- Younger and older people looked at rapid exposures of images that conveyed emotions like happiness, sadness, or fear. Later, the old folks recalled nearly as many of the positive photos as the young people had, but fewer than half as many of the upsetting ones—almost as if their memory banks were no longer accepting negative deposits.

- When people looked at emotional photographs inside an MRI brain scanner, the scans showed a huge, hot response to negative images in the amygdala of the typical young person. Inside the brains of older folks, however, the amygdala actually cooled down slightly when an emotionally upsetting picture came up.

Neuroscientists believe that older people have developed an automatic ability to counteract the reflexive response of the amygdala with the reflective powers of the prefrontal cortex. In effect, as you grow older, the more analytical parts of your brain step up to tamp down the responses that generated negative feelings when you were younger and more excitable. That leaves more mental space for experiences you expect to be positive—the very things you want to focus your attention on as time grows shorter.

By stifling the amygdala's reaction to negative events as you grow older, your reflective brain also changes the nature of your memories. As we saw in Chapter Seven, when your amygdala is aroused by fear, it helps burn the event into your memory like a branding iron. But when the amygdala becomes less active, fewer traces of the event may linger in future memory. Growing old not only gives us an impulse to dwell on good things; it gives us a kind of amnesia for the negative. The aging brain seems to hum along with the advice in the old Johnny Mercer song to "accentuate the positive" and "eliminate the negative."

Once you learn the truth about aging, it no longer seems surprising that depression is more common among people in their twenties than among those over the age of sixty-five. Only in the "oldest old" (those in their late eighties and beyond) does depression become as prevalent as it is among the young. Assuming your general health remains good, as you move past the age of sixty-five you should feel more satisfied with life than you were when you were younger.

That inner calm can make older investors more patient. In the terrible bear market of the 1970s, the only investors who steadily kept buying stocks as prices dropped were "old fogies" over the age of sixty-five. By 1979, as *BusinessWeek*'s cover declared "The Death of Equities," many younger investors had given up on the stock market in disgust. But older shareholders got the last laugh, as those falling stock prices of the 1970s set the stage for the bull market of the 1980s and 1990s. Those who had bought all the way through the downdraft got the biggest bang out of the upswing.

All in all, our fears of growing old are ungrounded. Saving makes sense: You will get more enjoyment out of your wealth as you become older and wiser. You will know then, as you never can when you are young, how best to use your money to add meaning and fulfillment to your life. Instead of being afraid of old age, you should look forward to reaping, with each passing year, a richer harvest from the money you save and invest when you are young. If you use the lessons in this book to get the best from both your reflexive and reflective brain, you should have nothing to fear as you face—and enter—the future. Rabbi Ben Ezra, in the poem by Robert Browning, was right when he exclaimed:

> *Grow old along with me!*
> *The best is yet to be,*
> *The last of life, for which the first was made.*

Get Happy

The most powerful and reassuring lesson from the new research into happiness is that you don't have to be rich to be happy. When it comes to increasing your sense of well-being, managing your emotions and expectations is at least as important as managing your money. There are many small steps you can take, and a few big ones, to get the maximum happiness out of your money with a minimum of effort. Let's take the little things first.

➢ TAKE A DEEP BREATH. As Richard Davidson's studies of Buddhist monks have shown, people who cultivate a sense of inner calm have greater activity in the left prefrontal cortex, one of the brain's centers for generating happiness. Every day, make a point of creating a few minutes of quiet time for yourself. Unplug all the wires: Turn off your cell phone, put away your BlackBerry, shut down your e-mail. Close your eyes, breathe deeply, and take a few moments to meditate, to pray, or to dwell in detail on a memory that makes you feel happy. As you come out of this exercise, think of one thing you could accomplish today that would make you feel good—the simpler the better. This technique will be especially important when your investments are not doing well.

➢ TURN OFF THE TUBE. Since envy and the "comparison complex" can make you feel shabby no matter how much you earn, you should compare yourself to others as seldom as possible. Prime-time TV shows and commercials bombard viewers with a stream of images that could make anyone feel like a loser for not being richer. Studies in the U.S., China, and Australia have shown that watching more television makes people more inclined to believe their happiness depends on things they cannot afford to buy—and to become less satisfied with their lives. If you want to feel better about the money you do have, turn off the TV and fill the time with a hobby, a night class, or a get-together with family or friends.

➢ THRIVE WHILE YOU DRIVE. Since commuting to work is at or near the bottom of nearly everyone's list of favorite activities, while spending time with friends is near the top, kill two birds with one stone: Set up a car pool with two or three good friends. You will turn a negative experience into a positive—and save money on gasoline to boot.

➢ GO WITH THE FLOW. If you find yourself struggling with your work, you probably won't find the solution by fighting even harder. An hour a day filled with the flow of exercise, art, or music will monopolize your attention, refocus your energy, make time seem to stop in its tracks, and take your mind off your troubles. When you arrive back at your desk, the solution may suddenly seem obvious. By the same token, it's amazing how a good vacation can suddenly make a falling stock market seem to go away.

➢ **THROW A PARTY.** Good managers realize that during difficult times it becomes more important than ever to give morale a boost. Mark Goldfine, a longtime options trader, reports that after an especially bad day on the trading floor, he has often treated his co-workers to drinks or dinner. "Anybody can celebrate after a good day," says Goldfine. "But I think it's better to celebrate after a bad day instead. That way, you keep the market from dictating your mood, and you can stop sulking and hurting. Happier people have more winning days on the trading floor."

➢ **END ON A HIGH NOTE.** Since emotional memories depend largely on how an experience concludes, you may be able to manipulate the future in the here and now. On a two-week vacation, for example, resist the urge to kick off the trip with an immediate splurge. Instead, save an extra-special occasion for the end: a romantic dinner, a boat ride beneath the night sky, a surprise reunion with family or friends—whatever seems most likely to create positive memories for you and your traveling companion(s). Ending your vacation on a high note will help make any prior letdowns disappear when you look back later. By the same token, brokers and financial planners can probably improve their clients' satisfaction by saving the best news for the end of a face-to-face meeting.

➢ **SURPRISE SOMEONE.** As we saw in Chapter Four, unexpected rewards trigger the brain to release a burst of dopamine. Just as you can't tickle yourself, you can't give yourself a present you weren't expecting. But it's easy to surprise someone else. Every dollar you spend on an unanticipated gift for another person will produce a lot more pleasure—for them and for you, too—than a dollar you spend on yourself. It doesn't have to be expensive jewelry; it really *is* the thought that counts.

➢ **GO BACK TO SCHOOL.** When people look back on their lives, they consistently say that one of their biggest regrets is not having gotten enough education, regardless of how many years of schooling they had. Whether you are curious about medieval history, the Civil War, cooking, computer maintenance, or the physics of baseball, a nearby college almost certainly offers a class that will give you an enormous bang for the buck. Get off your couch and into the classroom; you will learn something interesting and make new friends. As a by-product, you will pick up insights that might even help you earn more money.

➢ **DON'T LET THE OLD BECOME TOO BOLD.** Because aging makes us accentuate the positive and eliminate the negative, con men and shysters have long preyed on the elderly. Get-rich-quick schemes, which always emphasize the upside, have an almost irresistible appeal for older investors. (There are few places in America with a higher concentration of sleazy brokerage firms—as well as reputable ones—than Boca Raton, Florida, a traditional enclave for the elderly.) As people age, they simply don't like thinking about the downside—so it's vital to give them some protection. If you have aging parents or are over sixty-five yourself, install a spam filter to screen out unsolicited e-mails, sign up for caller ID to screen for telemarketers, always investigate anyone who offers to manage money for you or your parents, and never invest without using a checklist and an investment policy statement (see Appendices 2 and 3).

➢ **ACCENTUATE THE POSITIVE.** How often have you seen Grandma or Grandpa get really upset? The calmer and more positive outlook of older people means that fear is usually a poor way to motivate them. If you emphasize how much they stand to lose from a bad decision, they will tune you out. Instead, emphasize the positive outcomes of a better choice—especially the greater freedom it could give them to spend time with the people they value the most.

➢ **GO FOR THE GOAL.** The grasshopper instinct of your reflexive brain makes it hard to save for the future, since far-off rewards lack the emotional wallop you can get from spending your money today. The trick is to make your future goals as specific and vivid as possible. Give your savings or retirement account a nickname— "Emily's College Nest Egg" or "The Tuscan Villa Fund"—and a target date, like "Emily's 18th Birthday, Apr. 17, 2024" or "Christmas Eve, 2029." Cut a photo out of a magazine or download some clip art; display those pictures on your account records. The more imagery you can summon up about what the money can do for you in the future, the closer and more concrete the goal will seem and the more easily you will be able to set aside money for it today.

➢ **GET TIME ON YOUR SIDE.** Smart marketers know to put the rewards up front and the risks at the back end; that way, the upside is maximally exciting and the downside is minimally disturbing. Force yourself to focus not only on the current price of buying but also on the total cost of owning. If, for example, you are taking out a mortgage, don't pick a thirty-year loan just because the monthly payment is lower than a fifteen-year; ask to see the difference in the total amount of debt you will incur. And don't be fooled into assuming that a mutual fund with no "front-end" sales charge is cheaper than one with a lower annual fee; check the fee table in the prospectus first. Whenever you have a choice between different ways of paying for any major purchase, ask for a written analysis of what each method will cost you not just today, but five years from now. You may suddenly realize that the money you thought you were saving up front comes out of your hide later.

➢ **BOTTLE UP THE URGE TO SPLURGE.** If you find yourself going on spending binges, only to feel miserable about them later, segregate your money. Pay your regular living expenses from your normal checking account. Then create a separate checking account exclusively for splurges, and link a debit card to it. Cancel your credit cards and destroy them with a paper shredder. That way, you keep the convenience of paying with plastic—but, unlike with credit cards, you cannot spend more than you have on hand.

➢ **MAKE YOUR OWN LUCK.** Psychologist Richard Wiseman has spent years studying people who are unusually lucky (or unlucky). One of his most memorable examples is a woman who, before she goes to a party, thinks of a color. Then she systematically talks to everyone in the room wearing that color. By forcing herself to be friendly in a systematic way, she meets people she might never talk to otherwise—and gets asked out on more dates.

By breaking your routine and embracing new experiences, you can unleash your curiosity and open yourself to new ideas. The more often you can throw yourself in serendipity's path, the better the odds that you will get a lucky break in your investing or business life.

Try a new place for lunch once a week; go out for coffee with a person from another department; strike up a conversation with a stranger; if you usually take a bus to work, walk part of the way. Train yourself to notice something different each day about the world around you: what cars people are driving, what brand of shoes they are wearing, which make of mobile phones they are using. Seek out new sources of information—websites, magazines, trade publications—that are far afield from your usual interests and make you say, "Hmm." In 1998, out of idle curiosity, I bought a copy of *Scientific American* in an airport bookstore and read an article on neuroscience, purely because it opened with a pretty picture. To my amazement, I learned that people whose brains have been snipped in half calculate probabilities in a radically different way from the rest of us. That ultimately led to insights about investing I could never have found in my usual sources. If I had not ventured beyond my usual reading list that day, this book would never have come into being.

Wiseman also recommends keeping a list of "lucky goals." Be as specific and realistic as you can ("I want to get rich" doesn't count, but "I want ten new customers next month" does). Then monitor your progress toward those objectives. Set a few goals, so a lack of luck in one arena can be overcome by a lucky strike in another.

➢ **DO IT NOW.** Procrastination is one of the worst enemies of wealth and the biggest sources of unhappiness. We all know we should save more, and we are convinced we would—if we just could summon up a little more willpower. But willpower is rarely enough, as psychologist Roy Baumeister proved with a fiendishly clever experiment. Seating people in front of an oven in which fresh chocolate-chip cookies were baking, he told them they could eat as many radishes as they wanted from a bowl on the table—but the cookies were off-limits. Then he left them alone for a few minutes. Meanwhile, a second group of people could eat as many cookies as they wanted. Finally, Baumeister asked all his guinea pigs to solve a geometric puzzle. Those who had needed to focus their willpower on resisting the cookies quit trying to solve the puzzle more than twice as fast as those who had been free to eat a cookie whenever they wanted.

The obvious lesson: At least in the short run, willpower is not a renewable resource. Use it, and you lose it. Your self-control can so easily be depleted that you would be foolish to count on having it when you need it.

Fortunately, points out Harvard economist Laibson, "by committing today to lock in behavior tomorrow, you build a world in which you don't have to have self-control." Here are a few simple forms of precommitment:

Bond with a buddy. If someone you know is struggling to live up to the same resolution, work together. Agree on the date by which you will achieve your shared goal and the reward you will give each other if you both hit the target. (It could be a meal at your favorite restaurant, a night at the movies, whatever floats your boat.) If only one of you achieves the goal by the deadline, neither of you gets a prize.

Dare your mate. If there's some task you've been procrastinating on, tell your spouse or significant other that you are finally going to finish it by this Friday; if not, then next Saturday you will be obliged to do the household chore of his or her choice.

Get a tech check. Let's say you know you should raise the amount of money you put into your 401(k), but you feel yourself wavering. Laibson suggests that you tell a friend, "If I haven't done it by 5 p.m. next Friday, I owe you (a gift of your friend's choice)." Shake hands on it: You're not allowed to weasel out later. Then use your Blackberry, another PDA, or a shared online calendar like www.backpackit.com to create a message that will go out to your friend automatically on Friday morning. Then write an advance message to yourself, suggests Laibson, saying "He's gonna talk to me at 5!" If all this doesn't get you moving well before Friday, you're sure to kick into high gear around 4:45 that day.

Break it up. By taking baby steps, rather than trying to do everything at once, you can make the cost of action seem less painful. Let's say you've been meaning to sign up for online payment of all your bills. Don't try signing up for all of them at once; that's too daunting. Instead, pick one—long-distance phone service, for example—and register to pay that online. You'll be amazed how easy it is to sign up, and then you'll find out how good it feels not to have to write that check and stuff, stamp, lick, and mail that envelope. By getting your feet wet, instead of plunging in all at once, you'll understand that the costs are dwarfed by the benefits. Before long, you'll be paying all your bills online.

➢ **GIVE YOURSELF A HAPPINESS BOOST.** Psychologist Martin Seligman has shown that keeping a happiness diary once a day for a week—simply writing down three good things that happened, along with why you think they occurred—can make you measurably happier for months into the future. It's as if the act of counting your blessings somehow makes them appear to multiply.

You can also make what Seligman calls a gratitude visit: Think of someone who has had a great positive influence on your life. Write a 300-word letter of gratitude spelling out, in concrete detail, what the person did, what it has meant to the course of your life, and why you feel grateful. Then find where the person is today and ask to come for a visit. If the person asks why you want to visit, say it's a surprise. After you arrive, read your testimonial aloud; if you find that too emotional, just sit with the person while he or she reads it. That will give you both an enduring warm glow.

Another exercise is to visualize your "best possible self." Imagine what your life will be like in the future if all your goals are met, all your dreams come true, and you fulfill all your potential. Visualize looking back, from this future vantage point, on what you then will be proudest of. Describe your future self in as much detail as you can; carry the description with you, on a piece of paper in your wallet or a file in one of your portable electronic devices. Whenever you think of a new detail, add it to the list.

Finally, at the turn of every year—perhaps in January, when you are settling on New Year's resolutions and updating a spreadsheet of the year-end account balances of your investments—fill out a simple progress report on your happiness. Divide your life into a handful of major categories: for example, love, friendships, health, work, play, money, learning, giving, overall happiness. Rate yourself on each from 1 to 10. Adding these entries to your statement of net worth and tracking them year by year may be the simplest way to see what you need to work on to meet your goals for a richer life. By boosting your level of emotional wealth, you should end up raising your material wealth as well.

➢ **DOING AND BEING ARE BETTER THAN HAVING.** Finally, trying to get happy by making more money makes you overlook something simple but profound. There are three basic paths to happiness: having, doing, and being. *Having* centers around purchases and possessions—what Martin Seligman calls "the pleasant life." But, as a new SUV often shows, you can't buy your way to happiness for very long. *Doing* is about experiences and activities—what Seligman calls "the good life." Organizing a special occasion with family or friends, taking a college course just for learning's own sake, or taking up a hobby can all generate memories, create new skills, and broaden your horizons—adding up to a more durable kind of happiness. *Being* is about something larger than yourself: devoting a portion of your time and energy to an idea, a cause, or a community you want to be part of. Volunteering in a soup kitchen, becoming a leader at your house of worship, or fundraising for your favorite charity are the kinds of actions that can make you feel you are making a difference in other people's lives, enriching your own in the process. This kind of belonging and commitment create what Seligman calls "the meaningful life."

The bad news is that, as we have already seen, the money you spend on *having* creates diminishing returns; the more accustomed you become to whatever you buy, the less happiness you get from it. The good news is that the money you spend on *doing* and *being* gives a much longer afterglow, as experiences burnish themselves in your memory and belonging brightens your self-esteem. In the end, living a rich life depends less on how much you own than on how much you do, what you stand for, and how fully you reach your own potential. As Warren Buffett's business partner, Charles Munger, likes to say, "The best way to get what you want is to deserve what you want." To be a truly intelligent investor, you must remember that the money you can earn applying the lessons and techniques in this book is only a means, not an end in itself. Increasing your net worth is less important than maximizing your self-worth. The best "value investment" of all is channeling money into goals that will make your life more valuable: drawing out your innate gifts to make yourself matter to other people and to make the world around you a better place. Given the way the brain works, your happiness ultimately depends not on finding out how much you can buy but on learning how much you can be.

Appendix 1

Take the global view.

Keep calm by using a spreadsheet that emphasizes your total net worth—not the changes in each holding. Before you buy a stock or mutual fund, check whether it overlaps what you already own by using the Instant XRay tool at www.morningstar.com.

Hope for the best, but expect the worst.

Being braced for disaster—by diversifying and by learning market history—can help keep you from panicking. Every good investment performs badly some of the time. Intelligent investors stick around until the bad turns back to good.

Investigate, then invest.

A stock is not just a price; it's a piece of a living corporate organism. Study the company's financial statements. Read a mutual fund's prospectus before you buy. If you want to hire a broker or financial planner, do a background check before you write a check.

Never say always.

No matter how sure you are that an investment is a winner, don't put more than 10 percent of your portfolio in it. If you turn out to be right, you'll still make plenty of money; but if you turn out to be wrong, you'll be glad you kept most of your powder dry.

Know what you don't know.

Don't believe you are already an expert. Compare stock and fund returns against the overall market and across different time periods. Ask what might make this investment go down; find out if the people pushing it have their own money in it.

The past is not prologue.

On Wall Street, what goes up must come down, and what goes way up usually comes down with a sickening crunch. Never buy a stock or mutual fund just because it has been going up. Intelligent investors buy low and sell high, not the other way around.

Weigh what they say.

The easiest way to silence a market forecaster is to ask for the complete track record of all his predictions. If you can't get a complete list, don't listen. Before trying any strategy, gather objective evidence on the performance of others who have used it in the past.

If it sounds too good to be true, it probably is.

Not exactly: If it sounds too good to be true, it absolutely is. Anyone who offers high return at low risk in a short time is probably a fraud. Anyone who listens is definitely a fool.

Costs are killers.

Trading costs can eat up 1 percent of your money per year, while taxes and mutual fund fees can each take another 1 or 2 percent. If middlemen take 3 to 5 percent of your money per year, they will get rich. If you want to get rich, comparison-shop and trade at a snail's pace.

Eggs go splat.

So never put all your eggs in one basket. Spread your bets across U.S. and foreign stocks, bonds, and cash. No matter how much you like your job, don't put all your 401(k) into your own company's stock; the employees at Enron and WorldCom liked their company, too.

Appendix 2

Your Investing Checklist

Before buying a stock:

Do:

- Diversify, spreading some of your money across a wide range of U.S. stocks, some in foreign stocks, and some in bonds.
- Be sure you can afford to lose 100 percent of your money if you are wrong about this stock.
- Appraise your own ability to understand the business the company is in.
- Ask who this company's main competitors are and whether they are becoming weaker or stronger.
- Think about whether this company's customers would take their business elsewhere if it raised the prices of its goods or services.
- Look back at the company's annual report from its most profitable year and read the chairman's letter to shareholders. Did the CEO brag about the company's brilliant decisions and its infinite potential for growth, or did he warn not to count on such ideal conditions in the future?
- Imagine that the stock market closed down for five years. If you had no way of selling this stock to someone else, would you still be willing to own it?
- Go to www.sec.gov/edgar/searchedgar/webusers.htm and download at least three years' worth of the company's Form 10-Ks (annual reports), along with the last four 10-Qs (quarterly reports). Read them from back to front, paying special attention to the footnotes to the financial statements, where companies usually bury their dirty secrets.
- At the same site, download the latest Form 14-A proxy report to learn about the people who run the company. Are options awarded merely if

the stock goes up, or must managers exceed sensible performance targets? Look for "transactions with affiliated parties" which can warn you of unfair perks and conflicts of interest.

- Remember that this stock must go up more than 3 percent just for you to break even after paying all your brokerage costs and capital gains taxes. On short-term trades, you need a 4 percent gain before costs to break even after all your trading costs and taxes.
- Write down three reasons—having nothing to do with the stock price—why you want to be an owner of this business.
- Remember that you cannot buy a stock unless someone else is willing to sell it. What, exactly, do you know that this other person may have overlooked?

Don't:

- Buy a stock just because its price has been going up.
- Rationalize your investment with any reason that begins with the words "Everybody knows that . . ." or "It's obvious that . . ."
- Invest on a "tip" from a friend, a recommendation on TV, "technical analysis," or rumors about mergers or takeovers.
- Put more than 10 percent of your money in one company (including the company you work for, if its stock is a choice in your 401[k] plan).

Before buying a mutual fund:

Do:

- Read the prospectus, or owner's manual—from back to front, so that you don't miss the fine print that can alert you to potential problems.
- See how much the fund will charge you each year to own it (the "annual expense ratio" is included in the fund's prospectus and annual report).
- Learn how frequently the fund manager trades. Take the percentage given under "portfolio turnover rate" in the "financial highlights" of the prospectus and divide it into 1,200. The resulting number tells you how many months the fund typically hangs on to a stock. If it's less than twelve months, look elsewhere.
- Check the front of the prospectus to see how the fund did in its worst calendar quarter. Be sure you're comfortable losing at least that much money in a three-month period.

- Find out how fast the fund's total net assets have been growing. Funds tend to perform better when they have a few hundred million dollars to manage than when they have tens of billions of dollars.
- Ask whether the fund has ever "closed to new investors" to prevent a rapid inflow of new shareholders from making the fund grow too big too fast. You should favor funds that have closed in the past.
- Read the letter to shareholders from the manager. Good signs: The manager admits he has made a mistake, warns that future performance may not be as good, and urges investors to be patient. Bad signs: He brags about how fast the fund is growing, how high the recent returns have been, or how great the future looks.
- Realize that "average annual total return," even over a ten-year period, can be driven by a short stroke of good luck. Check the return over each calendar year as well, to see how consistently the fund has performed.
- Remember that you always can—and probably should—buy an index fund, a portfolio that owns every stock in a market benchmark like the Standard & Poor's 500-stock index. Because index funds operate at a fraction of the cost of other funds, they should be your default choice.

Don't:

- Buy a fund just because it's been hot lately.
- Invest just because you saw the manager on TV and he sounded smart.
- Consider any fund unless its annual expenses are lower than these thresholds:
  - government bond fund, 0.75%
  - blue-chip U.S. stock fund, 1.00%
  - small-stock or high-yield bond fund, 1.25%
  - foreign stock fund, 1.50%.
- Think you need a "specialty" or "sector" fund that specializes in stocks of the industry you work in. Your career is already riding on that industry; your money should be elsewhere, to spread your risks.
- Purchase a fund unless you're willing to own it for at least five years.

Appendix 3

Investment Policy Statement for John and Jane Doe

PURPOSE OF PORTFOLIO

To provide steady growth of capital and at least $________ in annual income after inflation and taxes, enabling us to provide for our needs now and after retirement.

EXPECTATIONS

The average annual return on stocks is roughly 7 percent after inflation. After tax, it is about 5 percent; after trading and management fees, it is less than 4 percent. Long-term returns much higher than that are not sustainable. If we want to have more, we will have to save more.

TIME HORIZON

We understand that short-term drops in the price of individual investments can be sudden, severe, and upsetting. But we are committed to focusing on the permanent growth of our entire portfolio, rather than fixating on temporary declines of isolated investments. Since we intend this portfolio to last the rest of our lives, and we then intend to bequeath it to our children, our investment horizon is 50 to 100 years. How our investments behave in the days to come matters much less than how we behave as investors in the decades to come.

DIVERSIFICATION STRATEGY

Our portfolio will consist of cash, individual stocks or bonds, and mutual funds. We will diversify broadly across the following asset classes: cash (bank accounts and money-market funds), bonds, U.S. stocks, foreign stocks, real estate securities (such as REITs), and hedges against inflation (such as Treasury Inflation-Protected Securities or TIPS).

REBALANCING

We will set "target allocations" for each of these asset classes (for example, 10% in cash, 10% in bonds, 40% in U.S. stocks, 30% in foreign stocks, 5% in real estate and 5% in TIPS). Every six months—on January 1 and July 1—we will rebalance back to those targets by selling whatever has gone up and buying whatever has gone down.

EVALUATION OF PERFORMANCE

We will compare the parts of our portfolio against appropriate performance benchmarks (for example, the Wilshire 5000 index for U.S. stocks, the Lehman Bros. U.S. bond aggregate index for bonds). We will look at the performance of individual investments only after calculating the total return of our entire portfolio and the return of each broad asset class.

FREQUENCY OF EVALUATION

We will evaluate investment performance every three months, comparing the total value of our portfolio against its value three months, one year, three years, five years, and ten years earlier. At least once a year, we will seek to cut our investing costs by preparing an itemized statement of our total expenses, including all brokerage charges, management fees, and taxes on dividends and capital gains. Short-term trading raises both our brokerage expenses and our tax bill; the less we trade, the more we will keep.

ADDING AND SUBTRACTING

We will add more money to our accounts, in proportion to our target allocations, whenever we have spare cash to invest. When we withdraw money, we will take it first from the cash account. If we have to take money out of stock or bond accounts, we will first make sure we have minimized the tax consequences of any withdrawal. In retirement, when this portfolio may be our primary means of support, we will not withdraw more than ________% of the account value annually.

WE WILL NEVER . . .

. . . put more than 10 percent of our total assets into a single stock, buy any investment solely because its price has gone up, sell any investment solely because its price has gone down, trade futures or options, trade with borrowed money ("margin"), trade on "tips" or "hunches," or respond to unsolicited investing e-mails.

This IPS is a simplified sample for illustration purposes only. Each investor should customize an IPS based on his or her goals and needs.

Notes

Abbreviations

AER	*American Economic Review*
ANYAS	*Annals of the New York Academy of Sciences*
AP	*American Psychologist*
APMR	*Archives of Physical Medicine and Rehabilitation*
BHJDM	Derek J. Koehler and Nigel Harvey, *Blackwell Handbook of Judgment and Decision Making* (Oxford, U.K.: Blackwell Publishing, 2004)
BP	*Biological Psychiatry*
CB	*Current Biology*
CC	*Cerebral Cortex*
CDPS	*Current Directions in Psychological Science*
C&E	*Cognition and Emotion*
CNE	Richard D. Lane and Lynn Nadel, eds., *Cognitive Neuroscience of Emotion* (Oxford: Oxford University Press, 2000)
COIN	*Current Opinion in Neurobiology*
EJN	*European Journal of Neuroscience*
FAJ	*Financial Analysts Journal*
HAB	Thomas Gilovich et al., *Heuristics and Biases* (Cambridge, U.K.: Cambridge University Press, 2002)
IJF	*International Journal of Forecasting*
JAP	*Journal of Applied Psychology*
JASP	*Journal of Applied Social Psychology*
JB	*Journal of Business*

JBDM	*Journal of Behavioral Decision Making*
JBF	*Journal of Behavioral Finance* (formerly *Journal of Psychology and Financial Markets*)
JCN	*Journal of Cognitive Neuroscience*
JCR	*Journal of Consumer Research*
JEAB	*Journal of the Experimental Analysis of Behavior*
JEBO	*Journal of Economic Behavior and Organization*
JESP	*Journal of Experimental Social Psychology*
JEP	*Journal of Experimental Psychology (General)*
JEPHLM	*Journal of Experimental Psychology: Human Learning and Memory*
JEPHPP	*Journal of Experimental Psychology: Human Perception & Performance*
JF	*Journal of Finance*
JFE	*Journal of Financial Economics*
JN	*Journal of Neuroscience*
JOHS	*Journal of Happiness Studies*
JPE	*Journal of Political Economy*
JPM	*Journal of Portfolio Management*
JPSP	*Journal of Personality and Social Psychology*
JRU	*Journal of Risk and Uncertainty*
JUU	Daniel Kahneman et al., eds., *Judgment under Uncertainty* (Cambridge, U.K.: Cambridge University Press, 1982)
JZ	Jason Zweig
MM	*Money* (articles by Jason Zweig are archived and freely available at www.jasonzweig.com)
MS	*Management Science*
NBER	National Bureau of Economic Research
NNS	*Nature Neuroscience*
NRN	*Nature Reviews Neuroscience*
NYT	*New York Times*
OBHDP	*Organizational Behavior and Human Decision Processes* (formerly *Organizational Behavior and Human Performance*)

PB	*Psychological Bulletin*
PNAS	*Proceedings of the National Academy of the Sciences*
PR	*Psychological Review*
PRSLB	*Proceedings of the Royal Society of London B: Biological Sciences*
PS	*Psychological Science*
PSPB	*Personality and Social Psychology Bulletin*
PTRSLB	*Philosophical Transactions of the Royal Society of London B: Biological Sciences*
QJE	*Quarterly Journal of Economics*
RA	*Risk Analysis*
RFS	*Review of Financial Studies*
SA	*Scientific American*
SIR	*Social Indicators Research*
TAD	George Loewenstein et al., eds., *Time and Decision* (New York: Russell Sage Foundation, 2003)
TICS	*Trends in Cognitive Sciences*
TII	Benjamin Graham with Jason Zweig, *The Intelligent Investor* (New York: HarperBusiness, 2003)
W-B	Daniel Kahneman et al., eds., *Well-Being* (New York: Russell Sage Foundation, 1999)
WSJ	*Wall Street Journal*

1. NEUROECONOMICS

1 *BRAIN, n.:* Ambrose Bierce, *The Devil's Dictionary* (New York: Hill & Wang, 1957), p. 19.

3 *An investor I'll call:* JZ interview with "Ed," Nov. 13, 2004.

4 *In the 1950s:* JZ, "How the Big Brains Invest at TIAA-CREF," *MM,* Jan. 1998, p. 118.

4 *Jack and Anna Hurst:* JZ interview with Hurst, Nov. 12, 2004; JZ, "The Soul of an Investor," *MM,* March 2005, pp. 66–71.

5 *"Financial decision-making":* Remarks by Daniel Kahneman, panel discussion moderated by JZ, Oxford Programme on Investment Decision-Making, Saïd Business School, Oxford University, U.K. Oct. 22, 2004.

6 *Most of all:* "Adam Smith," *The Money Game* (New York: Random House, 1968), p. 41.

2. "THINKING" AND "FEELING"

8 *"It is necessary to know":* Benedict de Spinoza, *On the Improvement of the Understanding, The Ethics, and Correspondence* (New York: Dover, 1955), p. 200 (*The Ethics,* Part IV, Proposition 17).

8 *Not long ago:* To protect his identity, I have changed the doctor's name; but the story is true. In related research, psychologists have shown that people are much more likely to buy consumer products whose brands contain the same letters as their own names. Even more remarkably, people are more inclined to marry someone whose initials are the same as their own. See John T. Jones et al., "How Do I Love Thee? Let Me Count the Js . . . ," *JPSP,* vol. 87, no. 5 (Nov. 2004), pp. 665–83; Michael J. Cooper, Orlin Dimitrov, and P. Raghavendra Rau, "A Rose.com by Any Other Name," *JF,* vol. 56, no. 6 (Dec. 2001), pp. 2371–88; Gregory W. Brown and Jay C. Hartzell, "Market Reaction to Public Information," *JFE,* vol. 60, nos. 2–3 (May/June 2001), pp. 333–70; JZ e-mail interview with Gregory Brown, Nov. 15, 2001; Yahoo!Finance message board for KKD, Oct. 2, 2002 (message number 59863), and July 15, 2003 (message number 69986).

9 *Nor is this kind:* Robert A. Olsen, "Professional Investors as Naturalistic Decision Makers: Evidence and Market Implications," *JBF,* vol. 3, no. 3 (2002), pp. 161–67; Malcolm Gladwell, "Blowing Up," *The New Yorker,* April 22–29, 2002, p. 162; Malcolm Gladwell, *Blink* (New York: Little, Brown, 2005), p. 14.

11 *Psychologists call this process:* For more on anchoring, see Gary Belsky and Thomas Gilovich, *Why Smart People Make Big Money Mistakes and How to Correct Them* (New York: Fireside, 1999), pp. 129–49; Gretchen B. Chapman and Eric J. Johnson, "Incorporating the Irrelevant," in *HAB,* pp. 120–38; Nicholas Epley and Thomas Gilovich, "Putting Adjustment Back in the Anchoring and Adjustment Heuristic," in *HAB,* pp. 139–49.

11 *But your intuition:* Amos Tversky and Daniel Kahneman, "Judgment Under Uncertainty," in *JUU,* pp. 3–20; J. Edward Russo and Paul J. H. Schoemaker, *Decision Traps* (New York: Simon & Schuster, 1989), pp. 90–91. Tversky died in 1996 and would have shared Kahneman's 2002 Nobel Prize in Economics. At the time of the original experiment, approximately 30% of U.N. members were African countries.

12 *A candy bar:* Adapted from Daniel Kahneman and Shane Frederick, "Representativeness Revisited," in *HAB,* pp. 49–81.

12 *You can correctly solve:* Lieberman credits Jean-Paul Sartre with coining the terms "reflexive" and "reflective" in the existentialist essay *The Transcendence of the Ego* (1936–37). Matthew Lieberman, "Reflective and Reflexive Judgment Processes," in Joseph P. Forgas et al., *Social Judgments* (New York: Cambridge University Press, 2003), pp. 44–67; Matthew Lieberman et al., "Reflection and Reflexion," *Advances in Experimental Social Psychology,* vol. 34 (2002), pp. 199–249; remarks by Daniel Kahneman, panel dis-

cussion moderated by JZ, Oxford Programme on Investment Decision-Making, Saïd Business School, Oxford University, U.K., Oct. 22, 2004. Not all neuroscientists agree that the brain uses discrete systems to process reason and emotion; for a compelling statement of the dissenting view, see Paul W. Glimcher et al., "Physiological Utility Theory and the Neuroeconomics of Choice," *Games and Economic Behavior,* vol. 52 (2005), pp. 213–56.

14 *All mammals:* The term *limbic* comes from the Latin *limbus,* which means "edge" or "border" or "transitional state" (as in limbo, the no-man's-land between heaven and hell in Catholic theology); Raymond J. Dolan, "Emotion, Cognition, and Behavior," *Science,* vol. 298 (Nov. 8, 2002), pp. 1191–94.

15 *That enables us:* JZ e-mail interview with Dukas, March 24, 2005; Reuven Dukas, "Behavioural and Ecological Consequences of Limited Attention," *PTRSLB,* vol. 357 (2002), pp. 1539–47; Reuven Dukas, "Causes and Consequences of Limited Attention," *Brain, Behavior and Evolution,* vol. 63 (2004), pp. 197–210; Reuven Dukas and Alan C. Kamil, "Limited Attention," *Behavioral Ecology,* vol. 12, no. 2 (2001), pp. 192–99; Marcus E. Raichle and Debra A. Gusnard, "Appraising the Brain's Energy Budget," *PNAS,* vol. 99, no. 16 (Aug. 6, 2002), pp. 10237–39; JZ interview with Matthew Lieberman, March 29, 2005.

16 *But the moment anything:* Arne Öhman et al., "Unconscious Emotion," in *CNE,* pp. 296–327; JZ telephone interview with Paul Slovic, Feb. 3, 2005.

16 *Economist Colin Camerer:* JZ telephone interview with Colin Camerer, April 6, 2005.

17 *But there's more:* JZ, "Are You Wired for Wealth?" *MM,* Oct. 2002, p. 79; JZ interview with Matthew Lieberman, March 29, 2005; Stanislas Dehaene, "Arithmetic and the Brain," *COIN,* vol. 14 (2004), pp. 218–24; Mark Jung-Beeman et al., "Neural Activity When People Solve Verbal Problems with Insight," *PLoS Biology,* vol. 2, no. 4 (April 2004), pp. 500–10.

17 *Jordan Grafman has shown:* Marian Gomez-Beldarrain et al., "Patients with Right Frontal Lesions Are Unable to Assess and Use Advice to Make Predictive Judgments," *JCN,* vol. 16, no. 1 (Jan. 2004), pp. 74–89; JZ e-mail interview with Grafman, March 24, 2005; Baba Shiv and Alexander Fedorikhin, "Heart and Mind in Conflict," *JCR,* vol. 26, no. 3 (Dec. 1999), pp. 278–92.

18 *But the reflective brain:* Robin M. Hogarth, "Deciding Analytically or Trusting Your Intuition?" (Oct. 2002), www.econ.upf.edu/eng/research/onepaper.php?id=654.

19 *Computational neuroscientists:* JZ e-mail interview with Nathaniel Daw, March 25, 2005.

20 *When a problem is hard:* Hillel J. Einhorn and Robin M. Hogarth, "Confidence in Judgment," *PR,* vol. 85 (1978), pp. 395–416.

21 *And the human mind:* Susan T. Fiske and Shelley E. Taylor, *Social Cognition* (Reading, Mass.: Addison-Wesley, 1984).

21 *The conflict between:* Veronika Denes-Raj and Seymour Epstein, "Conflict Between Intuitive and Rational Processing," *JPSP,* vol. 66, no. 5 (1994), pp. 819–29.

23 *In the late 1980s:* Paul B. Andreassen, "On the Social Psychology of the Stock Market," *JPSP,* vol. 53, no. 3 (1987), pp. 490–96; "Explaining the Price-Volume Relationship," *OBHDP,* vol. 41 (1988), pp. 371–89; and "Judgmental Extrapolation and Market Overreaction," *JBDM,* vol. 3 (1990), pp. 153–74.

23 *Investment professionals:* Noel Capon et al., "An Individual Level Analysis of the Mutual Fund Investment Decision," *Journal of Financial Services Research,* vol. 10 (1996), pp. 59–82; Investment Company Institute, *Understanding Shareholders' Use of Information and Advisers* (Washington, D.C. 1997), p. 21; Ronald T. Wilcox, "Bargain Hunting or Star Gazing?" *JB,* vol. 76, no. 4 (Oct. 2003), pp. 645–63; Michael A. Jones, Vance P. Lesseig, and Thomas I. Smythe, "Financial Advisors and Mutual Fund Selection," *JFP* (March 2005), www.fpanet.org/journal/articles/2005_Issues/jfp0305-art8.cfm.

24 *When it comes to trust:* JZ interview with Fred Kobrick, portfolio manager, MetLife State Street Capital Appreciation Fund, July 1, 1993; JZ telephone interview with Robin Hogarth, March 17, 2005.

25 *Know when reflex will rule:* JZ telephone interview with Paul Slovic, Feb. 3, 2005; JZ, "What Fund Investors Really Need to Know," *MM,* June 2002, pp. 110–15.

26 *Ask another question:* Daniel Kahneman, "Maps of Bounded Rationality," http://nobelprize.org/economics/laureates/2002/kahnemann-lecture.pdf; JZ telephone interview with Christopher K. Hsee, Oct. 22, 2001.

27 *Motion imparts a power:* J.Y. Lettvin et al., "What the Frog's Eye Tells the Frog's Brain," *Proceedings of the Institute for Radio Engineers,* vol. 47 (1959), pp. 1940–51, http://jerome.lettvin.info/lettvin/Jerome/WhatTheFrogsEyeTellsThe FrogsBrain.pdf; Michael W. Morris et al., "Metaphors and the Market," *OBHDP,* vol. 102, no. 2 (2007), pp. 174–92.

28 *Only fools:* Patricia Dreyfus, "Investment Analysis in Two Easy Lessons," *MM,* July 1976, p. 37.

28 *Count to ten:* Piotr Winkielman et al., "Unconscious Affective Reactions to Masked Happy Versus Angry Faces Influence Consumption Behavior and Judgments of Value," *PSPB,* vol. 31, no. 1 (Jan. 2005), pp. 121–35; JZ telephone interview with Winkielman, April 19, 2005; JZ telephone interview with Schwarz, April 20, 2005.

29 *Nearly all of us:* David Hirshleifer and Tyler Shumway, "Good Day Sunshine," *JF,* vol. 58, no. 3 (June 2003), pp. 1009–32; Ralf Runde, "Lunar Cycles and Capital Markets: An Empirical Analysis of the Moon and Ger-

man Stock Returns," working paper, University of Dortmund, 2000; Ilia D. Dichev and Troy D. Janes, "Lunar Cycle Effects in Stock Returns," http://ssrn.com/abstract=281665; Mark Kamstra et al., "Winter Blues: A SAD Stock Market Cycle," www.frbatlanta.org/filelegacydocs/wp0213.pdf; Alex Edmans, Diego Garcia, and Oyvind Norli, "Sports Sentiment and Stock Returns" (Nov. 2005), http://ssrn.com/abstract=677103.

29 *Companies can exploit:* Adam L. Alter and Daniel Oppenheimer, "Predicting Short-Term Stock Fluctuations by Using Processing Fluency," *PNAS,* vol. 103, no. 24 (June 13, 2006), pp. 8907–8; Alex Head et al., "Would a Stock by Any Other Ticker Smell as Sweet?" www.economics.pomona.edu/GarySmith/frames/GaryFrameset.html.

29 *So unless you guard:* Ronald S. Friedman and Jens Förster, "The Effects of Approach and Avoidance Motor Actions on the Elements of Creative Insight," *JPSP,* vol. 79 (2000), pp. 477–92; John T. Cacioppo et al., "Rudimentary Determinants of Attitudes II," *JPSP,* vol. 65, no. 1 (1993), pp. 5–17; JZ telephone interview with Meir Statman, April 12, 2005.

30 *When the market "blinks":* Adrienne Carter, "Investing with Style—Any Style," *BusinessWeek,* Feb. 7, 2005; Warren Buffett, remarks at Berkshire Hathaway Inc. annual meeting, April 30, 2005.

31 *Stocks have prices:* Michael S. Rashes, "Massively Confused Investors Making Conspicuously Ignorant Choices (MCI-MCIC)," *JF,* vol. 56, no. 5 (Oct. 2001), pp. 1911–27; "In Mannatech IPO, Tech Craze Formally Surpasses Tulips," *Investment Dealers' Digest,* Feb. 22, 1999, pp. 6–7; Warren Buffett, speech to University of Tennessee College of Business Administration (2003; videotape kindly provided to JZ by Al Auxier); Claudia Goldin and Cecilia Rouse, "Orchestrating Impartiality," (Jan. 1997), www.nber.org/papers/w5903; Malcolm Gladwell, *Blink* (New York: Little, Brown, 2005), pp. 245–48; JZ telephone interview with Buffett, May 5, 2005.

32 *Take account:* Remarks by Daniel Kahneman, panel discussion moderated by JZ, Oxford Programme on Investment Decision-Making, Saïd Business School, Oxford University, U.K., Oct. 22, 2004.

3. GREED

34 *He that loveth:* Ecclesiastes 5: 10.

34 *Laurie Zink:* JZ interviews with Zink, Jan. 19 and Mar. 30, 2005; the odds of winning the grand prize in California's SuperLotto Plus game are posted at www.calottery.com/Games/SuperLottoPlus/HowtoPlay/. Data on Ohio winners: Charles T. Clotfelter and Philip J. Cook, *Selling Hope* (Cambridge, Mass.: Harvard University Press, 1991), p. 122.

35 *Of course, you don't have:* Hans C. Breiter et al., "Imaging the Neural Systems for Motivated Behavior and Their Dysfunction in Neuropsychiatric Illness," in Thomas S. Deisboeck and J. Yasha Kresh, eds., *Complex Sys-*

tems Science in Biomedicine (New York: Springer, 2006), pp. 763–810; JZ telephone interview with P. Read Montague, June 1, 2005; Brooks King-Casas et al., "Getting to Know You," *Science,* vol. 308 (April 1, 2005), pp. 78–83; JZ e-mail interview with Peter Kirsch, Feb. 4, 2005; Peter Kirsch et al., "Anticipation of Reward in a Nonaversive Differential Conditioning Paradigm and the Brain Reward System," *NeuroImage,* vol. 20 (2003), pp. 1086–95.

36 *Long before neuroeconomists:* Mark Twain, *Roughing It* (Berkeley, Calif.: University of California Press, 1993), pp. 258–70; Twain's "The $30,000 Bequest" can be read in a handsome online edition at www2.hn.psu.edu/faculty/jmanis/twain/bequest.pdf. His financial fumbles are detailed in Justin Kaplan, *Mr. Clemens and Mark Twain* (New York: Touchstone, 1983).

37 *I lived through the rush:* JZ participated in Knutson's experiment in the cognitive neuroscience laboratory at Stanford University, May 26, 2004. See also Brian Knutson et al., "Distributed Neural Representation of Expected Value," *JN,* vol. 25, no. 19 (May 11, 2005), pp. 4806–12; Patricio O'Donnell et al., "Modulation of Cell Firing in the Nucleus Accumbens," *ANYAS,* vol. 877 (1999), pp. 157–75; Hans C. Breiter and Bruce R. Rosen, "Functional Magnetic Resonance Imaging of Brain Reward Circuitry in the Human," *ANYAS,* vol. 877 (1999), pp. 523–47; Hugo D. Critchley et al., "Neural Activity in the Human Brain Relating to Uncertainty and Arousal During Anticipation," *Neuron,* vol. 29 (2001), pp. 537–45; Scott C. Matthews et al., "Selective Activation of the Nucleus Accumbens During Risk-Taking Decision Making," *NeuroReport,* vol. 15, no. 13 (Sept. 15, 2004), pp. 2123–27.

39 *"Reward is experienced":* JZ telephone interview with Breiter, June 22, 2005; John P. O'Doherty et al., "Neural Responses During Anticipation of a Primary Taste Reward," *Neuron,* vol. 33 (Feb. 28, 2002), pp. 815–26.

39 *Why does the reflexive:* Satoshi Ikemoto and Jaak Panksepp, "The Role of Nucleus Accumbens Dopamine in Motivated Behavior," *Brain Research Reviews,* vol. 31 (1999), pp. 6–41; JZ telephone interview with Paul Slovic, Feb. 3, 2005.

40 *As the French essayist:* Michel de Montaigne, "How Our Mind Hinders Itself," in *The Complete Essays of Montaigne* (Stanford, Calif.: Stanford University Press, 1965), p. 462 (this is an ancient literary device, used by Dante at the beginning of the fourth canto of *Il Paradiso* [ca. 1320] and by Aristotle in Book II, chapter 13 of his *On the Heavens* [ca. 360 B.C.]). John Barth, *The Floating Opera* and *The End of the Road* (New York: Doubleday, 1988), pp. 331–34.

40 *We can learn more:* JZ e-mail interview with Taketoshi Ono, Feb. 15, 2005; Yutaka Komura et al., "Retrospective and Prospective Coding for Predicted Reward in the Sensory Thalamus," *Nature,* vol. 412 (Aug. 2, 2001), pp. 546–49; Emily Dickinson, poem 995 ("This was in the White of the Year"),

The Complete Poems of Emily Dickinson (New York: Little, Brown, 1960), pp. 462–463.

41 *Lately, psychologists:* JZ e-mail interviews with Rudolf Cardinal, Jan. 22, 2005, and Feb. 16, 2005; Rudolf N. Cardinal et al., "Impulsive Choice Induced in Rats by Lesions of the Nucleus Accumbens Core," *Science,* vol. 292 (June 29, 2001), pp. 2499–2501.

42 *One of Wall Street's:* Historical data on Celera's stock price from Yahoo! Finance; www.celera.com/celera/pr_1056647999; www.ornl.gov/sci/techresources/Human_Genome/project/clinton2.shtml; http://archives.cnn.com/2000/HEALTH/06/26/human.genome.04/.

43 *In a remarkable recent experiment:* JZ e-mail interview with Emrah Düzel, Feb. 8, 2005; Bianca C. Wittmann et al., "Reward-Related fMRI Activation of Dopaminergic Midbrain . . . ," *Neuron,* vol. 45 (Feb. 3, 2005), pp. 459–67; Brian Knutson and R. Alison Adcock, "Remembrance of Rewards Past," *Neuron,* vol. 45 (Feb. 3, 2005), pp. 331–32; JZ telephone interview with Peter Shizgal, June 27, 2005.

44 *Experiments in rats:* Paul E. M. Phillips et al., "Subsecond Dopamine Release Promotes Cocaine Seeking," *Nature,* vol. 422 (April 10, 2003), pp. 614–18; John N. J. Reynolds et al., "A Cellular Mechanism of Reward-Related Learning," *Nature,* vol. 413 (Sept. 6, 2001), pp. 67–70; Fyodor Dostoyevsky, *The Gambler* (1866), chapter 17 (Constance Garnett translation).

45 *When rewards are near:* JZ e-mail interviews with Hiroyuki Nakahara, June 27 and 29, 2005; Michael L. Platt, "Caudate Clues to Rewarding Cues," *Neuron,* vol. 33 (Jan. 31, 2002), pp. 316–18; Johan Lauwereyns et al., "Feature-Based Anticipation of Cues that Predict Reward in Monkey Caudate Nucleus," *Neuron,* vol. 33 (Jan. 31, 2002), pp. 463–73; Johan Lauwereyns et al., "A Neural Correlate of Response Bias in Monkey Caudate Neurons," *Nature,* vol. 418 (July 25, 2002), pp. 413–17. Cisco stock performance: Bloomberg L.P. and Time Inc. Business Information Research Center.

46 *Anticipation has another:* Brian Knutson et al., "Distributed Neural Representation of Expected Value," *JN,* vol. 25, no. 19 (May 11, 2005), pp. 4806–12; George F. Loewenstein et al., "Risk as Feelings," *PB,* vol. 127, no. 2 (2001), pp. 267–86; JZ telephone interview with Loewenstein, Jan. 26, 2005.

47 *There's another thing:* JZ telephone interview with Mellers, Jan. 27, 2005; see also Barbara A. Mellers, "Choice and the Relative Pleasure of Consequences," *PB,* vol. 126, no. 6 (2000), pp. 910–24; Hans C. Breiter et al., "Functional Imaging of Neural Responses to Expectancy and Experience of Monetary Gains and Losses," *Neuron,* vol. 30 (May 2001), pp. 619–39.

48 *Among stock promoters:* Richard Dale, *The First Crash* (Princeton, N.J.: Princeton University Press, 2004), p. 101; data on stock returns for 2000 through 2002 kindly provided to JZ by Aronson + Johnson + Ortiz, L.P.

49 *There's only one sure thing:* For a fuller list of jackpot jargon, see *TII*, p. 275.

50 *Control your cues:* Howard Rachlin, *The Science of Self-Control* (Cambridge, Mass.: Harvard University Press, 2000), pp. 126–27; JZ telephone interview with Knutson, Feb. 2, 2005; www.berkshirehathaway.com/letters/letters.html.

4. PREDICTION

53 *Pecuniary motives:* Coleridge, letter to William Sotheby, Nov. 9, 1828, in *The Portable Coleridge* (New York: Viking Penguin, 1950), p. 302. Coleridge, of course, knew a narcotic when he saw one, since he was an opium addict for much of his life.

53 *In the Mesopotamian galleries:* British Museum, ANE 92668. You can view an image of this artifact online by visiting www.thebritishmuseum.ac.uk/compass/ and entering "liver" in the search window.

53 *Just like an ancient* baru: *BusinessWeek*'s annual "Where to Invest" forecast issue, 1996 through 2005, inclusive; *WSJ,* Aug. 13, 1982, p. 33; *NYT,* Aug. 13, 1982, pp. D1, D6; *WSJ,* April 17, 2000, pp. A20, C1, C4; *USA Today,* April 17, 2000, p. 6B; Peter L. Bernstein, *The Power of Gold* (New York: John Wiley, 2000), pp. 357–58; data on accuracy of analysts' earnings forecasts kindly updated by David Dreman, e-mail to JZ, May 6, 2005; Peter L. Bernstein, "The King of Siam and the Gentle Art of Postcasting," *Economics & Portfolio Strategy,* Aug. 1, 2005.

56 *It took two psychologists:* Daniel Kahneman and Amos Tversky, "Subjective Probability," in *JUU,* pp. 32–47. Note: The probability of any given sequence of heads and tails is $1/2^n$, where *n* is the number of coin flips. Flip a coin six times and the odds of each possible sequence are $1/2^6$ or 1/64.

58 *The pursuit:* I am grateful to my colleague, the formidable classics scholar Michael Sivy, for helping me settle on the appropriate Latin for "man the pattern-seeker." *Pareidolia* is discussed in Sagan's brilliant essay "The Man in the Moon and the Face on Mars," in his *The Demon-Haunted World* (New York: Ballantine, 1996), pp. 41–59.

58 *Others sift through mountains:* Cheol-Ho Park and Scott H. Irwin, "The Profitability of Technical Analysis" (Nov. 2004), http://ssrn.com/abstract=603481; Wei Jiang, "A Nonparametric Test of Market (Mis-) Timing (Aug. 2001), http://papers.ssm.com/sol3/papers.cfm?abstract_id=287102; *WSJ,* Jan. 17, 2003, p. C4, and Jan. 30, 2004, p. C4; *Fortune,* Dec. 24, 2001, p. 156; Edward A. Dyl, "Did Joe Montana Save the Stock Market?" *FAJ,* Sept.–Oct., 1989, pp. 4–5. The "Super Bowl Predictor" was first presented in Leonard Koppett, "If the Bulls and the Bears Have You Buffaloed, Try Our Foxy Formulas," *Sports Illustrated,* April 23, 1979, p. 8. Koppett, who intended his article as a spoof on the confusion between correlation and causation, was amazed that anyone ever took the Super Bowl Predictor seriously. Late in his life he called it "an embarrassment to

rational thought" and "too stupid to believe" (JZ telephone interview with Koppett, Dec. 13, 2001).

59 *What drives this behavior?:* JZ telephone interviews with John Staddon, Duke University, Sept. 1, 2000, and George Wolford, Dartmouth College, Sept. 6, 2000; George Wolford et al., "The Left Hemisphere's Role in Hypothesis Formation," *JN,* vol. 20 (2000), RC64, pp. 1–4; Michael S. Gazzaniga, "The Split Brain Revisited," *SA,* July 1998, pp. 50–55; JZ, "The Trouble with Humans," *MM,* Nov. 2000, pp. 67–71; John I. Yellott Jr., "Probability Learning with Noncontingent Success," *Journal of Mathematical Psychology,* vol. 6 (1969), pp. 541–75; R. J. Herrnstein, "On the Law of Effect," *JEAB,* vol. 13, no. 2 (1970), pp. 243–66; Richard J. Herrnstein, *The Matching Law* (Cambridge, Mass.: Harvard University Press, 1997); David R. Shanks et al., "A Re-examination of Probability Matching and Rational Choice," *JBDM,* vol. 15, no. 3 (2002), pp. 233–50; Leo P. Sugrue et al., "Matching Behavior and the Representation of Value in the Parietal Cortex," *Science,* vol. 304 (June 18, 2004), pp. 1782–87.

60 *Meanwhile, the rest of us:* Matthew Rabin, "Inference by Believers in the Law of Small Numbers," *QJE,* vol. 117, no. 3 (Aug. 2002), pp. 775–816; http://emlab.berkeley.edu/users/rabin/.

61 *Why are we cursed:* JZ interview with Paul Glimcher, Feb. 19, 2002.

62 *"The main difference":* JZ telephone interview with Preuss, Dec. 29, 2004; Jean de Heinzelin et al., "Environment and Behavior of 2.5-Million-Year-Old Bouri Hominids," *Science,* vol. 284 (April 23, 1999), pp. 625–29; Tim D. White et al., "Pleistocene *Homo sapiens* from Middle Awash, Ethiopia," *Nature,* vol. 423 (June 12, 2003), pp. 742–47; William H. Calvin, *A Brief History of the Mind* (Oxford, U.K.: Oxford University Press, 2004). More recent, fossil remains of anatomically modern humans unearthed near Kibish, Ethiopia, have been dated to about 195,000 years ago, but no intact skull from this deposit has yet been found.

62 *It's easy to visualize:* Peter B. deMenocal, "African Climate Change and Faunal Evolution During the Pliocene-Pleistocene," *Earth and Planetary Science Letters,* vol. 220 (2004), pp. 3–24; Todd M. Preuss, "What Is It Like to Be a Human?" in Michael S. Gazzaniga (ed.), *The Cognitive Neurosciences III* (Cambridge, Mass.: MIT Press, 2004), pp. 5–22; Jay Quade et al., "Paleoenvironments of the Earliest Stone Toolmakers, Gona, Ethiopia," *Geological Society of America Bulletin,* vol. 116, no. 11/12 (Nov./Dec. 2004), pp. 1529–44; Simon M. Reader and Kevin N. Laland, "Social Intelligence, Innovation, and Enhanced Brain Size in Primates," *PNAS,* vol. 99, no. 7 (April 2, 2002), pp. 4436–41.

63 *We like to imagine:* Jared Diamond, "Evolution, Consequences and Future of Plant and Animal Domestication," *Nature,* vol. 418 (Aug. 8, 2002), pp. 700–7; Alice Louise Slotsky, *The Bourse of Babylon* (Bethesda, Md.: CDL Press, 1997); Karl Moore and David Lewis, *Birth of the Multinational*

(Copenhagen: Copenhagen Business School Press, 1999); Edward J. Swan, *Building the Global Market* (New York: Kluwer Law International, 2000); JZ interview with P. Read Montague, Feb. 28, 2002.

63 *Wolfram Schultz:* JZ interview with Schultz, March 13, 2002; Wolfram Schultz, "Getting Formal with Dopamine and Reward," *Neuron,* vol. 36 (Oct. 10, 2002), pp. 241–63.

64 *as neuroscientist Antoine Bechara:* JZ telephone interview with Bechara, June 20, 2005; JZ e-mail interview with Paul Glimcher, June 9, 2005; Wolfram Schultz, "Predictive Reward Signal of Dopamine Neurons," *Journal of Neurophysiology,* vol. 80 (1998), pp. 1–27; Irene A. Yun et al., "The Ventral Tegmental Area Is Required for the Behavioral and Nucleus Accumbens Neuronal Firing Responses to Incentive Cues," *JN,* vol. 24, no. 12 (March 24, 2004), pp. 2923–33; JZ telephone interview with Kent Berridge, May 12, 2005. Dopamine signals actually travel below the usual speed limits of the brain, but they spread quickly because these neurons are so massively interconnected to other areas.

64 *Researchers Montague and Schultz:* JZ interview with Montague, Feb. 28, 2002; JZ interview with Schultz, March 13, 2002; Wolfram Schultz et al., "A Neural Substrate of Prediction and Reward," *Science,* vol. 275 (March 14, 1997), pp. 1593–99.

66 *That's more than a metaphor:* M. Leann Dodd et al., "Pathological Gambling Caused by Drugs Used to Treat Parkinson Disease," *Archives of Neurology,* vol. 62 (Sept. 2005), pp. 1–5; John C. Morgan et al., "Impulse Control Disorders and Dopaminergic Drugs," *Archives of Neurology,* vol. 63 (Feb. 2006), pp. 298–99; Gaetano Di Chiara and Assunta Imperato, "Drugs Abused by Humans Preferentially Increase Synaptic Dopamine Concentrations in the Mesolimbic System of Freely Moving Rats," *PNAS,* vol. 85 (July 1988), pp. 5274–78; Amanda J. Roberts and George F. Koob, "The Neurobiology of Addiction," *Alcohol Health and Research World,* vol. 21, no. 2 (1997), pp. 101–6; Hans C. Breiter et al., "Acute Effects of Cocaine on Human Brain Activity and Emotion," *Neuron,* vol. 19 (Sept. 1997), pp. 591–611; Roy A. Wise, "Drug-Activation of Brain Reward Pathways," *Drug and Alcohol Dependence,* vol. 41 (1998), pp. 13–22; Garret D. Stuber et al., "Extinction of Cocaine Self-Administration Reveals Functionally and Temporally Distinct Dopaminergic Signals . . . ," *Neuron,* vol. 46 (May 19, 2005), pp. 661–69.

66 *If laboratory rats:* James Olds and Peter Milner, "Positive Reinforcement Produced by Electrical Stimulation of Septal Area and Other Regions of Rat Brain," *Journal of Comparative and Physiological Psychology,* vol. 47 (1954), pp. 419–27; M. E. Olds and J. L. Fobes, "The Central Basis of Motivation: Intracranial Self-Stimulation Studies," *Annual Review of Psychology,* vol. 32 (1981), pp. 523–74; Antonio P. Strafella et al., "Repetitive Transcranial Magnetic Stimulation of the Human Prefrontal Cortex Induces Dopamine

Release in the Caudate Nucleus," *JN,* vol. 21 (2001), RC 157, pp. 1–4; Bart J. Nuttin et al., "Long-Term Electrical Capsular Stimulation in Patients with Obsessive-Compulsive Disorder," *Neurosurgery,* vol. 52, no. 6 (June 2003), pp. 1263–74.

66 *Neuroscientists do not yet know:* It is not yet clear whether dopamine is directly involved in the sensation of pleasure, or whether it is just the precursor of a pleasure signal conducted by one or more of the brain's other chemical messengers, like glutamate, orexin, norepinephrine, or acetylcholine. Edwin C. Clayton et al., "Phasic Activation of Monkey Locus Ceruleus Neurons . . ." *JN,* vol. 24, no. 44 (Nov. 3, 2004), pp. 9914–20; Angela J. Yu and Peter Dayan, "Uncertainty, Neuromodulation, and Attention," *Neuron,* vol. 46 (May 19, 2005), pp. 681–92; Glenda C. Harris et al., "A Role for Lateral Hypothalamic Orexin Neurons in Reward Seeking," *Nature,* vol. 437 (Sept. 22, 2005), pp. 556–59; JZ telephone interviews with P. Read Montague, June 1, 2005, and Peter Shizgal, June 27, 2005; JZ e-mail interviews with Gregory Berns, Paul Glimcher, and Brian Knutson, June 3, 2005, and Wayne Drevets and Wolfram Schultz, July 18, 2005; Wayne C. Drevets et al., "Amphetamine-Induced Dopamine Release in Human Ventral Striatum Correlates with Euphoria," *BP,* vol. 49 (2001), pp. 81–96.

66 *At Harvard Medical School:* JZ telephone interview with Breiter, June 22, 2005.

67 *What is the underlying force:* Pavlov's classic experiments are summarized at http://nobelprize.org/medicine/laureates/1904/pavlov-bio.html and http://en.wikipedia.org/wiki/Ivan_Pavlov. For more detail, see "Experimental Psychology and Psychopathology in Animals" and "The Conditioned Reflex," in I. P. Pavlov, *Selected Works* (Moscow: Foreign Languages Publishing House, 1955), pp. 151–68, 245–70.

67 *And if the reward is big enough:* JZ telephone interviews with P. Read Montague, June 1, 2005, and Peter Shizgal, June 27, 2005; see also P. Read Montague et al., "Computational Roles for Dopamine in Behavioural Control," *Nature,* vol. 431 (Oct. 14, 2004), pp. 760–67; Wolfram Schultz, "Neural Coding of Basic Reward Terms . . . ," *COIN,* vol. 14 (2004), pp. 139–47; P. Read Montague and Gregory S. Berns, "Neural Economics and the Biological Substrates of Valuation," *Neuron,* vol. 36 (Oct. 10, 2002); Wolfram Schultz et al., "A Neural Substrate of Prediction and Reward," *Science,* vol. 275 (March 14, 1997), pp. 1593–99; Udi E. Ghitza et al., "Persistent Cue-Evoked Activity of Accumbens Neurons after Prolonged Abstinence from Self-Administered Cocaine," *JN,* vol. 23, no. 19, pp. 7239–45.

67 *I learned how automatically:* JZ participated in the Montague-Berns experiment, administered by Sam McClure, on July 1, 2002; JZ telephone interview with P. Read Montague, June 1, 2005; JZ interview with Gregory Berns, July 1, 2002; Gregory S. Berns et al., "Brain Regions Responsive to

Novelty in the Absence of Awareness," *Science,* vol. 276 (May 23, 1997), pp. 1272–75.

69 *It's remarkable:* Scott A. Huettel et al., "Perceiving Patterns in Random Series: Dynamic Processing of Sequence in Prefrontal Cortex," *NNS,* vol. 5, no. 5 (May 2002), pp. 485–90; Charles T. Clotfelter and Philip J. Cook, *Selling Hope* (Cambridge, Mass.: Harvard University Press, 1991), p. 54; *TII,* pp. 436–37 (Graham originally wrote this passage in 1971); JZ e-mail interviews with Huettel, May 20 and May 24, 2005.

70 *Huettel's findings:* Burton G. Malkiel, *A Random Walk Down Wall Street* (New York: W. W. Norton, 2003), p. 17 (50% odds in each year over three years amounts to 50% x 50% x 50% = 12.5% for the entire period); *Morningstar Mutual Funds;* Gregory Bresiger, "Pit Stop for a Pit Bull," *Financial Planning,* Sept. 2000; performance data for Grand Prix Fund Class A and price and volume data for Taser International are from http://quicktake.morningstar.com and http://finance.yahoo.com; Amit Goyal and Sunil Wahal, "The Selection and Termination of Investment Managers by Plan Sponsors" (Sept. 2005), http://ssrn.com/abstract=675970.

71 *There's more to investing:* JZ telephone interview with Glimcher, May 31, 2005; Hannah M. Bayer and Paul W. Glimcher, "Midbrain Dopamine Neurons Encode a Quantitative Reward Prediction Error Signal," *Neuron,* vol. 47 (July 7, 2005), pp. 129–41; Paul W. Glimcher, *Decisions, Uncertainty, and the Brain* (Cambridge, Mass.: MIT Press, 2003), pp. 330–34.

73 *You might wonder:* JZ telephone interview with Odean, July 12, 2005; Brad M. Barber and Terrance Odean, "All That Glitters" (March 2006), http://ssrn.com/abstract=460660; Terrance Odean, "Do Investors Trade Too Much?" *AER,* vol. 89 (Dec. 1999), pp. 1279–98, www.odean.us; Kenneth L. Fisher and Meir Statman, "Bubble Expectations," *Journal of Wealth Management* (Fall 2002), pp. 17–22; David Dreman et al., "A Report on the Mar. 2001 Investor Sentiment Survey," *JBF,* vol. 2, no. 3 (2001), pp. 126–34; Patricia Fraser, "How Do U.S. and Japanese Investors Process Information and How Do They Form Their Expectations of the Future?" (Sept. 2000), http://ssrn.com/abstract=257440. Stock returns: Ibbotson Associates, Chicago. Survey of individual investors: Werner De Bondt, "Betting on Trends," *IJF,* vol. 9 (1993), pp. 355–71.

73 *Similarly, investors chase:* Firsthand: JZ, "What Fund Investors Really Need to Know," *MM,* June 2002, pp. 110–15. Energy funds: data courtesy of Avi Nachmany, Strategic Insight, New York. Roger G. Clarke and Meir Statman, "Bullish or Bearish?" *FAJ,* May/June 1998, pp. 63–72; "Total net assets, cash position, sales and redemptions of own shares," 1954 to present, monthly data set provided to JZ by Investment Company Institute. See also Erik R. Sirri and Peter Tufano, "Costly Search and Mutual Fund Flows," *JF,* vol. 53, no. 5 (Oct. 1998), pp. 1589–1622; Jason Karceski, "Returns-Chasing Behavior, Mutual Funds, and Beta's Death," *Journal of*

Financial and Quantitative Analysis, vol. 37, no. 4 (Dec. 2002), pp. 559–94; and Anthony W. Lynch and David K. Musto, "How Investors Interpret Past Fund Returns," *JF,* vol. 58, no. 5 (Oct. 2003), pp. 2033–58.

74 *The first step:* JZ interview with Eric Johnson, July 29, 2005. For more on the prefrontal cortex in predicting reward, see Masataka Watanabe, "Reward Expectancy in Primate Prefrontal Neurons," *Nature,* vol. 382 (Aug. 15, 1996), pp. 629–32; Edmund T. Rolls, "The Orbitofrontal Cortex and Reward," *CC,* vol. 10 (March 2000), pp. 284–94; Jay A. Gottfried et al., "Encoding Predictive Reward Value in Human Amygdala and Orbitofrontal Cortex," *Science,* vol. 301 (Aug. 22, 2003), pp. 1104–7; Scott A. Huettel et al., "Decisions under Uncertainty," *JN,* vol. 25, no. 3 (March 30, 2005), pp. 3304–11; Hiroyuki Oya et al., "Electrophysiological Correlates of Reward Prediction Error Recorded in the Human Prefrontal Cortex," *PNAS,* vol. 102, no. 23 (June 7, 2005), pp. 8351–56.

76 *Stop predicting:* For more on dollar-cost averaging, see JZ, "Tie Me Down and Make Me Rich," *MM,* May 2004, p. 118.

76 *Ask for the evidence:* Michel de Montaigne, "Of Cannibals," in *The Complete Essays of Montaigne* (Stanford, Calif.: Stanford University Press, 1965), p. 155; Deuteronomy 18:10–12; J. Scott Armstrong, "Review of Ravi Batra, *The Great Depression of 1990,*" *IJF,* vol. 4 (1988), p. 493; JZ telephone and e-mail interviews with Bob Billett, March 7, 2005; JZ e-mail interviews with Sherwood Vine, July 7, Aug. 16, and Sept. 26, 2005; Einstein letter to Jost Winteler, July 8, 1901, in Alice Calaprice, *The New Quotable Einstein* (Princeton, N.J.: Princeton University Press, 2005), p. 253.

77 *Practice, practice, practice:* Remarks by Richard Zeckhauser, panel discussion moderated by JZ, Oxford Programme on Investment Decision-Making, Saïd Business School, Oxford University, U.K., Oct. 22, 2004; Patricia Dreyfus, "Investment Analysis in Two Easy Lessons," *MM,* July 1976, p. 37.

78 *Face up to base rates:* Howard Rachlin, *The Science of Self-Control* (Cambridge, Mass.: Harvard University Press, 2000), pp. 155–56; Michael Lewis, *Moneyball* (New York: W. W. Norton, 2003), p. 274; Kahneman: JZ, "Do You Sabotage Yourself?" *MM,* May 2001, pp. 75–78.

80 *Correlation is not causation:* Pat Regnier, "How High Is Up?" *MM,* Dec. 1999, pp. 108–14; David Leinweber and David Krider, "Stupid Data Miner Tricks," research monograph, First Quadrant Corp., Pasadena, Calif., 1997. Dent fund performance: e-mail to JZ from Annette Larson, senior research analyst, Morningstar Inc., Aug. 16, 2005. Foolish Four performance: *TII,* pp. 44–46.

81 *Take a break:* JZ e-mail interview with Wolford, Sept. 9, 2000. See also Richard Ivry and Robert T. Knight, "Making Order from Chaos," *NNS,* vol. 5, no. 5 (May 2002), pp. 394–96.

82 *One of our oddest mental quirks*: http://news.bbc.co.uk/1/hi/world/europe/

4256595.stm; www.mccombs.utexas.edu/faculty/jonathan.koehler/docs/sta309h/Gambler's_Fallacy_in_Italy.htm.

82 *There's a simple way:* Eric Gold and Gordon Hester, "The Gambler's Fallacy and the Coin's Memory," working paper, Carnegie Mellon University (Nov. 1987); copy kindly provided to JZ by Robyn Dawes.

83 *By using technology:* M. J. Koepp et al., "Evidence for Striatal Dopamine Release during a Video Game," *Nature,* vol. 393 (May 21, 1998), pp. 266–68; *TII,* p. 38.

83 *"If owning stocks":* JZ, "Do You Sabotage Yourself?" *MM,* May 2001, pp. 75–78; Richard H. Thaler et al., "The Effect of Myopia and Loss Aversion on Risk Taking," *QJE,* May 1997, pp. 647–61; Uri Gneezy and Jan Potters, "An Experiment on Risk Taking and Evaluation Periods," *QJE,* May 1997, pp. 631–45; JZ e-mail interview with Uri Gneezy, Nov. 2, 1997; Shlomo Benartzi and Richard H. Thaler, "Risk Aversion or Myopia?" *MS,* vol. 45, no. 3 (March 1999), pp. 364–81; Uri Gneezy et al., "Evaluation Periods and Asset Prices in a Market Experiment," *JF,* vol. 58, no. 2 (April 2003), pp. 821–37; *Seinfeld,* "The Stock Tip," originally broadcast June 21, 1990, www.seinfeld scripts.com/TheStockTip.htm; tracking study of *MM* subscribers, 2002, data courtesy of Douglas King, Time Inc. Consumer Marketing.

5. CONFIDENCE

85 *"Before I was Pope":* Cited in Paul A. Samuelson, "Is There Life After Nobel Coronation?," at http://nobelprize.org/economics/articles/samuelson/index.html.

85 *In 1965:* Caroline E. Preston and Stanley Harris, "Psychology of Drivers in Traffic Accidents," *JAP,* vol. 49, no. 4 (1965), pp. 284–88; Ola Svenson, "Are We All Less Risky and More Skillful than Our Fellow Drivers?" *Acta Psychologica,* vol. 47 (1981), pp. 143–48.

86 *When I give speeches:* JZ first ran this little test, inspired by Richard Thaler of the University of Chicago, around 1998. The lowest ratio between "my savings" and "average savings" seems to be around 1.5; people often think they will save at least twice as much as the typical person in the room.

86 *Of course, being overconfident:* JZ, "Did You Beat the Market?" *MM,* Jan. 2000, p. 56; Arnold C. Cooper et al., "Entrepreneurs' Perceived Chances for Success," *Journal of Business Venturing,* vol. 3 (1988), pp. 97–108; Daniel Kahneman, remarks at the Institute of Certified Financial Planners' Wealth Management Symposium, New York, April 30, 1999.

88 *Every month:* Kenneth L. Fisher and Meir Statman, "Bubble Expectations," *Journal of Wealth Management,* fall 2002, pp. 17–22; press release, Montgomery Asset Management, Jan. 12, 1998; survey conducted by Intersearch Corp., Nov. 17–Dec. 9, 1997; Robert F. Bruner, "Does M&A Pay?" http://faculty.darden.edu/brunerb/Bruner_PDF/Does%20M&A%20Pay.pdf;

JZ email interview with Bruner, Nov. 7, 2005; Neil D. Weinstein, "Unrealistic Optimism about Future Life Events, *JPSP,* vol. 39, no. 5 (1980), pp. 806–20 (this experiment was conducted with 120 students at Rutgers's Cook College, all of them women); Barna Research Group nationwide telephone survey of more than 1,000 adult Americans, released Oct. 21, 2003, www.barna.org.

89 *In short, to evaluate:* Justin Kruger and David Dunning, "Unskilled and Unaware of It," *JPSP,* vol. 77, no. 6 (1999), pp. 1121–34; Shelley E. Taylor and Jonathon D. Brown, "Illusion and Well-Being," *PB,* vol. 103, no. 2 (1988), pp. 193–210. The term "con man," of course, is short for "confidence man," someone who earns trust by projecting a suave sense of ability, power, and knowledge (see David W. Maurer, *The Big Con* [New York: Anchor, 1999].)

90 *That's surprisingly hard:* JZ, "Did You Beat the Market?" *MM,* Jan. 2000, pp. 55–57; Don A. Moore et al., "Positive Illusions and Forecasting Errors in Mutual Fund Investment Decisions," *OBHDP,* vol. 79, no. 2 (Aug. 1999), pp. 95–114; JZ telephone interviews with Max Bazerman and Don A. Moore, Nov. 16, 1999.

91 *In March 2002:* JZ spoke at the *Boston Globe* Personal Finance Conference, Mar. 23, 2002; Bethany McLean and Peter Elkind, *The Smartest Guys in the Room* (New York: Portfolio, 2003), pp. 242, 314, 401.

92 *Around the world:* Gur Huberman, "Familiarity Breeds Investment," *RFS,* vol. 14, no. 3 (Fall 2001), pp. 659–80; Joshua D. Coval and Tobias J. Moskowitz, "Home Bias at Home," *JF,* vol. 54, no. 6 (Dec. 1999), pp. 2045–73; Kalok Chan et al., "What Determines the Domestic Bias and Foreign Bias?," *JF,* vol. 60, no. 3 (June 2005), pp. 1495–1534; Kenneth R. French and James M. Poterba, "Investor Diversification and International Equity Markets" (Jan. 1991), www.nber.org/papers/w3609; Yesim Tokat, "International Equity Investing: Long-Term Expectations and Short-Term Departures," Vanguard Investment Counseling & Research, May 2004, p. 11; Morgan Stanley Capital International *Blue Book,* Dec. 2005, p. 4; Michael Kilka and Martin Weber, "Home Bias in International Stock Return Expectations," *JBF,* vol. 1, no. 3–4 (2000), pp. 176–92; Shlomo Benartzi, "Excessive Extrapolation and the Allocation of 401(k) Accounts to Company Stock," *JF,* vol. 56, no. 5 (Oct. 2001), pp. 1747–64.

93 *The frontier between:* I am indebted to Professor Zeev Mankowitz of Hebrew University for this metaphor.

94 *Almost forty years ago:* R. B. Zajonc, "Attitudinal Effects of Mere Exposure," *JPSP, Monograph Supplement,* vol. 9 (1968), pp. 1–27; JZ telephone interview with Robert Zajonc, Oct. 12, 2005; William Raft Kunst-Wilson and R. B. Zajonc, "Affective Discrimination of Stimuli That Cannot Be Recognized," *Science,* vol. 207 (Feb. 1, 1980), pp. 557–58; Jennifer L. Monahan et al., "Subliminal Mere Exposure: Specific, General, and Diffuse Effects,"

PS, vol. 11, no. 6 (Nov. 2000), pp. 462–66; R. B. Zajonc, "Mere Exposure: A Gateway to the Subliminal," *CDPS,* vol. 10, no. 6 (Dec. 2001), pp. 228. Although the duration of a "blink of an eye" is highly variable, it averages roughly one-third of a second (Frans VanderWerf et al., "Eyelid Movements: Behavioral Studies of Blinking in Humans under Different Stimulus Conditions," *Journal of Neurophysiology,* vol. 89 (2003), pp. 2784–96).

96 *What's going on:* Rebecca Elliott and Raymond J. Dolan, "Neural Response During Preference and Memory Judgments for Subliminally Presented Stimuli: A Functional Neuroimaging Study," *JN,* vol. 18, no. 12 (June 15, 1998), pp. 4697–4704; Samuel M. McClure et al., "Neural Correlates of Behavioral Preference for Culturally Familiar Drinks," *Neuron,* vol. 44 (Oct. 14, 2004), pp. 379–87.

96 *There are tantalizing hints:* The theory of place cells was first proposed in John O'Keefe and Lynn Nadel's masterpiece, *The Hippocampus as a Cognitive Map* (Oxford, U.K.: Oxford University Press, 1978), www.cognitivemap.net. See also Eleanor A. Maguire et al., "Recalling Routes Around London: Activation of the Right Hippocampus in Taxi Drivers," *JN,* vol. 17, no. 18 (Sept. 15, 1997), pp. 7103–10; Eleanor A. Maguire et al., "Navigation-Related Structural Change in the Hippocampi of Taxi Drivers," *PNAS,* vol. 97, no. 8 (April 11, 2000), pp. 4398–4403; Gabriel Kreiman et al., "Category-Specific Visual Responses of Single Neurons in the Human Medial Temporal Lobe," *NNS,* vol. 3, no. 9 (Sept. 2000), pp. 946–53; Arne D. Ekstrom et al., "Cellular Networks Underlying Human Spatial Navigation," *Nature,* vol. 425 (Sept. 11, 2003), pp. 184–87; R. Quian Quiroga et al., "Invariant Visual Representation by Single Neurons in the Human Brain," *Nature,* vol. 435 (June 23, 2005), pp. 1102–7.

98 *Brain scans in Peter Kenning's:* JZ interview with Kenning, Society for Neuroeconomics annual meeting, Sept. 10, 2006; Peter Kenning et al., "The Role of Fear in Home-Biased Decision Making," University of Münster, working paper (January 2007).

98 *Once you understand:* Nicholas Epley et al., "What Every Skeptic Should Know about Subliminal Persuasion," *Skeptical Inquirer,* Sept./Oct. 1999, pp. 40–45; JZ interview with Piotr Winkielman, April 19, 2005; JZ interview with Zajonc, Oct. 12, 2005; Shlomo Benartzi et al., "Company Stock, Market Rationality, and Legal Reform," University of Chicago Law School working paper (July 2004), pp. 2, 8, 28; www.ici.org/stats/res/per12-01_appendix.pdf, p. 26; Merrill Lynch & Co. Inc., Form 11-K, June 26, 2006, www.sec.gov/Archives/edgar/data/65100/000095012306008132/y22548e11vk.htm.

99 *The mere-exposure effect underlies:* Yoav Ganzach, "Judging Risk and Return of Financial Assets," *OBHDP,* vol. 83, no. 2 (Nov. 2000), pp. 353–70; Michael J. Brennan et al., "Alternative Factor Specifications, Security Characteristics, and the Cross-Section of Expected Stock Returns," *JFE,* vol.

49, no. 3 (Sept. 1998), pp. 345–73; Charles M. C. Lee and Bhaskaran Swaminathan, "Price Momentum and Trading Volume," *JF,* vol. 55, no. 5 (Oct. 2000), pp. 2017–69; Brad M. Barber and Terrance Odean, "All That Glitters" (March 2006), http://ssrn.com/abstract=460660; JZ interview with Zajonc, Oct. 12, 2005.

99 *Unfortunately, history shows:* David Hirshleifer et al., "Fear of the Unknown" (May 2004), http://public.kenan-flagler.unc.edu/faculty/zhangha/familiarity-5-30-2004-v2.pdf.

100 *Juanita Edwards:* This is not her real name, but the story is true.

101 *Long ago, the psychologist:* B. F. Skinner, " 'Superstition' in the Pigeon," *JEP,* vol. 38 (1948), pp. 168–72, http://psychclassics.yorku.ca/Skinner/Pigeon/; John Staddon, *The New Behaviorism* (Philadelphia: Psychology Press, 2001), pp. 54–68. While later research refuted some of Skinner's observations, the basic finding still holds: that the pigeons behave "as if" their behavior causes them to be fed.

101 *To see how sensitive:* The bet on the stock tables is adapted from Chip Heath and Amos Tversky, "Preference and Belief," *JRU,* vol. 4 (1991), pp. 5–28. The dice experiment is Lloyd H. Strickland et al., "Temporal Orientation and Perceived Control as Determinants of Risk-Taking," *JESP,* vol. 2, no. 2 (1966), pp. 143–51; copy kindly provided to JZ by Roy Lewicki.

102 *Psychologist Ellen Langer:* Ellen J. Langer, "The Illusion of Control," *JPSP,* vol. 32, no. 2 (Aug. 1975), pp. 311–28; Rosa Bersabé and Rosario Martínez Arias, "Superstition in Gambling," *Psychology in Spain,* vol. 4, no. 1 (2000), www.psychologyinspain.com/content/reprints/2000/3.pdf; see also Deborah Davis et al., "Illusory Personal Control as a Determinant of Bet Size and Type in Casino Craps Games," *Journal of Applied Social Psychology,* vol. 30, no. 6 (2000), pp. 1224–42. Retirement investors: William N. Goetzmann and Nadav Peles, "Cognitive Dissonance and Mutual Fund Investors," *Journal of Financial Research,* vol. 20, no. 2 (1997), pp. 145–58.

102 *Study after study:* Robert E. Knox and James A. Inkster, "Postdecision Dissonance at Post Time," *JPSP,* vol. 8, no. 4 (April 1968), pp. 319–23; Jonathan C. Younger et al., "Postdecision Dissonance at the Fair," *PSPB,* vol. 3, no. 2 (1977), pp. 284–87; Robert B. Cialdini, *Influence* (New York: William Morrow, 1993), pp. 57–59; Paul Rosenfeld et al., "Decision Making," *Journal of Social Psychology,* vol. 126, no. 5 (2001), pp. 663–65.

103 *The illusion of control is stronger:* Ellen J. Langer, "The Illusion of Control," *JPSP,* vol. 32, no. 2 (Aug. 1975), pp. 311–28; Ellen J. Langer, "The Psychology of Chance," *Journal for the Theory of Social Behaviour,* vol. 7, no. 2 (Oct. 1977), pp. 185–207.

103 *From small-timers:* Meghan Collins, "Traders Ward off Evil Spirits," cnn.com, Oct. 29, 2003, http://money.cnn.com/2003/10/28/markets/trader_superstition /index.htm; William M. O'Barr and John M. Conley, *Fortune and Folly* (Homewood, Ill.: Business One Irwin, 1992), p. 155; Mark

Fenton-O'Creevy et al., *Traders* (Oxford, U.K.: Oxford University Press, 2005), p. 87; *BusinessWeek,* Sept. 3, 2001, p. 70; *NYT,* Oct. 15, 2001, p. C1; www.globalcrossing.com/xml/news/2002/january/28.xml.

104 *Neuroeconomists are now exploring:* Elizabeth M. Tricomi et al., "Modulation of Caudate Activity by Action Contingency," *Neuron,* vol. 41 (Jan. 22, 2004), pp. 281–92; JZ e-mail interviews with Delgado, Oct. 31 and Dec. 12, 2005; Arthur Aron et al. "Reward, Motivation, and Emotion Systems Associated with Early-Stage Intense Romantic Love," *Journal of Neurophysiology,* vol. 94 (2005), pp. 327–37; Caroline F. Zink et al., "Human Striatal Responses to Monetary Reward Depend on Saliency," *Neuron,* vol. 42 (May 13, 2004), pp. 509–17; JZ telephone interview with Caroline Zink, Jan. 18, 2005. Caroline is the sister of Laurie Zink (see "I Know How Good It Would Feel," Chapter Three, p. 34).

105 *Researchers at the University of Wisconsin:* Tim V. Salomons et al., "Perceived Controllability Modulates the Neural Response to Pain," *JN,* vol. 24, no. 32 (Aug. 11, 2004), pp. 7199–7203.

105 *That seems to be a basic part:* Michael T. Rogan et al., "Distinct Neural Signatures for Safety and Danger in the Amygdala and Striatum of the Mouse," *Neuron,* vol. 46 (Apr. 21, 2005), pp. 309–20; JZ e-mail interview with Kandel and Rogan, Dec. 14, 2005.

106 *In late 1999 and early 2000:* JZ e-mail interviews with Brad Russell, Jan. 10, 2005; Richard H. Thaler and Eric J. Johnson, "Gambling with the House Money and Trying to Break Even," *MS,* vol. 36, no. 6 (June 1990), pp. 643–60.

107 *Secondly, a streak:* Amanda Bischoff-Grethe et al., "The Context of Uncertainty Modulates the Subcortical Response to Predictability," *JCN,* vol. 13, no. 7 (2001), pp. 986–93.

107 *Next, a hot streak:* Ellen J. Langer and Jane Roth, "Heads I Win, Tails It's Chance," *JPSP,* vol. 32, no. 6 (Dec. 1975), pp. 951–55; Willem A. Wagenaar and Gideon B. Keren, "Chance and Luck Are Not the Same," *JBDM,* vol. 1, no. 2 (1988), pp. 65–75; Nehemia Friedland, "On Luck and Chance," *JBDM,* vol. 5, no. 4 (1992), pp. 267–82.

107 *It's not just small investors:* www.ge.com/annual94/iba3a18.htm; http://pages.stern.nyu.edu/~lcabral/teaching/ge.pdf; Robert L. Conn et al., "Why Must All Good Things Come to an End?" (Feb. 2004), http://ssrn.com/abstract=499310; Matthew T. Billett and Yiming Qian, "Are Overconfident Managers Born or Made?" (May 2006), http://ssrn.com/abstract=696301; Richard Roll, "The Hubris Hypothesis of Corporate Takeovers," *JB,* vol. 59, no. 2 (Apr. 1986), pp. 197–216; Gilles Hilary and Lior Menzly, "Does Past Success Lead Analysts to Become Overconfident?" *MS,* vol. 52, no. 4 (April 2006), pp. 489–500; Vernon L. Smith et al., "Bubbles, Crashes, and Endogenous Expectations in Experimental Spot Asset Markets," *Econometrica,* vol. 56, no. 5 (Sept. 1988), pp. 1119–51; David P. Porter and Vernon

L. Smith, "Stock Market Bubbles in the Laboratory," *Applied Mathematical Finance*, vol. 1 (1994), pp. 111–27.

108 *People may also be:* JZ telephone interview with John Allman, Dec. 28, 2004; Stephanie Kovalchik and John Allman, "Measuring Reversal Learning," *C&E*, vol. 20, no. 5 (2005), pp. 714–28.

109 *People who have suffered:* J. Hornak et al., "Reward-Related Reversal Learning after Surgical Excisions in Orbito-frontal or Dorsolateral Prefrontal Cortex in Humans," *JCN*, vol. 16, no. 3 (2004), pp. 463–78. Most of these patients had suffered minor brain damage as a result of corrective surgery for stroke, epilepsy, brain tumors, or head trauma.

109 *Once you understand:* Alanna Nash, *The Colonel* (New York: Simon & Schuster, 2003), p. 322; Carolyn Abraham, "Diapers Keep Gamblers at Slots," *Ottawa Citizen*, May 3, 1997, p. A1.

109 *What does a financial hot streak:* Wayne C. Drevets et al., "Subgenual Prefrontal Cortex Abnormalities in Mood Disorders," *Nature*, vol. 386 (Apr. 24, 1997), pp. 824–27; Rita Carter, *Mapping the Mind* (Berkeley: University of California Press, 1999), p. 197; Rebecca Elliott et al., "Dissociable Neural Responses in Human Reward Systems," *JN*, vol. 20, no. 16 (Aug. 15, 2000), pp. 6159–65; Edmund T. Rolls et al., "Activity of Primate Subgenual Cingulate Cortex Neurons Is Related to Sleep," *Journal of Neurophysiology*, vol. 90 (2003), pp. 134–42.

110 *In the waning days:* Ravi Dhar and William N. Goetzmann, "Bubble Investors" (Aug. 2006), http://ssrn.com/abstract=683366; JZ e-mail interview with Will Goetzmann, Nov. 9, 2005.

111 *This quirk of human behavior:* Baruch Fischhoff and Ruth Beyth, " 'I Knew It Would Happen,' " *OBHDP*, vol. 13 (1975), pp. 1–16; Baruch Fischhoff, "Hindsight Does Not Equal Foresight . . ." *JEPHPP*, vol. 1, no. 3 (1975), pp. 288–99; JZ telephone interview with Baruch Fischhoff, Sept. 21, 2005; remarks by Daniel Kahneman, panel discussion moderated by JZ, Oxford Programme on Investment Decision-Making, Saïd Business School, Oxford University, U.K. Oct. 22, 2004.

112 *Pundits have the same problem:* Gary Rivlin, "The Madness of King George," *Wired* (July 2002), www.wired.com/wired/archive/10.07/gilder.html; Jeanne Lee, "Crash Test," *MM*, June 2000, p. 39; Pablo Galarza and Jeanne Lee, "The Sensible Internet Portfolio," *MM*, Dec. 1999, p. 129. (Disclosure: In May 1999, JZ wrote a column for *Money* entitled "Baloney.com," warning investors not to buy Internet stocks. But he did not stop his friend Pablo Galarza from recommending Ariba.)

112 *Hindsight bias also skews:* This example is inspired by Meir Statman and Jonathan Scheid, "Buffett in Foresight and Hindsight," *FAJ*, vol. 58, no. 4 (July/Aug. 2002), pp. 11–18. The number of funds in existence at year-end 1996 is from www.icifactbook.org/pdf/06_fb_table06.pdf. The odds of flip-

ping heads ten times in a row are $\frac{1}{2}^{10}$, or $\frac{1}{1024}$, or 0.001. Thus the probability against a given manager getting ten heads in a row is 99.9%. The probability that none of the 1,325 managers will flip ten heads in a row is $.999^{1325}$, or 27.4%. Thus, the odds that at least one manager will get ten straight heads are 1 - .274, or 72.6%.

114 *Here are three:* www.cea.fn and *Science,* vol. 309 (Aug. 19, 2005), pp. 1170–71. The percentages of people's certainty about the answers to these three questions are based on audience responses at speeches JZ has given over the past ten years. "The Inner Tube of Life," *Science,* vol. 307 (Mar. 25, 2005), p. 1914.

115 *Years ago:* Baruch Fischhoff et al., "Knowing with Certainty," *JEPHPP,* vol. 3, no. 4 (1977), pp. 552–64. A $1 bet in 1977, adjusted for inflation, is equivalent to roughly $3 today.

115 *The same ignorance:* 2006 EBRI Retirement Confidence Survey, www.ebri.org/surveys/rcs/2006/.

115 *That's why the old proverb:* This message applies to all the people who confidently attribute this proverb to Will Rogers. According to Steven Gragert, archival historian at the Will Rogers Memorial Museums in Claymore, Okla., there is no evidence that Rogers ever said it. *Bartlett's Familiar Quotations* suggests that it is probably based on an aphorism by the 19th-century American humorist Josh Billings: "It is better to know nothing than to know what ain't so."

115 *That's as true:* Carl-Axel S. Staël von Holstein, "Probabilistic Forecasting," *OBHDP,* vol. 8 (1972), pp. 139–58; J. Frank Yates et al., "Probabilistic Forecasts of Stock Prices and Earnings," *OBHDP,* vol. 49 (1991), pp. 60–79; Gustaf Törngren and Henry Montgomery, "Worse than Chance?" *JBF,* vol. 5, no. 3 (2004), pp. 148–53; Brad M. Barber and Terrance Odean, "The Courage of Misguided Convictions," *FAJ,* vol. 55, no. 6 (Nov./Dec. 1999); "Trading Is Hazardous to Your Wealth," *JF,* vol. 55, no. 2 (Apr., 2000); "Individual Investors," in Richard H. Thaler, ed., *Advances in Behavioral Finance,* vol. 2 (Princeton, N.J.: Princeton University Press, 2005), pp. 543–69, www.odean.us; Richard Deaves et al., "The Dynamics of Overconfidence" (Nov. 2005), http://ssrn.com/abstract=868970.

117 *Our confidence has another quirk:* Sarah Lichtenstein and Baruch Fischhoff, "Do Those Who Know More Also Know More about How Much They Know?" *OBHDP,* vol. 20, no. 2 (Dec. 1977), pp. 159–83; Dale Griffin and Amos Tversky, "The Weighing of Evidence and the Determinants of Confidence," in *HAB,* pp. 230–49.

117 *Experiments with monkeys:* Christopher D. Fiorillo et al., "Discrete Coding of Reward Probability and Uncertainty by Dopamine Neurons," *Science,* vol. 299 (March 21, 2003), pp. 1898–1902; Peter Shizgal and Andreas Arvanitogiannis, "Gambling on Dopamine," *Science,* vol. 299 (March 21,

2003), pp. 1856–58; Hugo D. Critchley et al., "Neural Activity in the Human Brain Relating to Uncertainty and Arousal during Anticipation," *Neuron,* vol. 29 (Feb. 2001), pp. 537–45.

117 *Some psychologists contend:* Gerd Gigerenzer, *Adaptive Thinking* (Oxford, U.K.: Oxford University Press, 2000), and "Fast and Frugal Heuristics," in *BHJDM,* pp. 62–88; *WSJ,* Jan. 18, 1995, p. A1; U.S. Securities and Exchange Commission, Office of Municipal Securities, *Cases and Materials* (1999), www.sec.gov/pdf/mbondcs.pdf, pp. 59–100; *Pensions & Investments,* March 6, 2006, p. 28; *WSJ,* Sept. 20, 2006, p. C1; www.sdcera.org/pdf/SDCERAInvestmentPerformanceAward.pdf.

118 *Probably the main reason:* Hart Blanton et al., "Overconfidence as Dissonance Reduction," *JESP,* vol. 37 (2001), pp. 373–85; Robert H. Thouless, "The Tendency to Certainty in Religious Belief," *British Journal of Psychology,* vol. 26 (1935), pp. 16–31, http://psychclassics.yorku.ca/Thouless/certainty.pdf.

118 *Warren Buffett:* "Chairman's Letter," 1992 Berkshire Hathaway Inc. annual report, www.berkshirehathaway.com/letters/1992.html.

119 *Many investors act:* Nicholas Dawidoff, *The Fly Swatter* (New York: Pantheon, 2002), pp. 212–13.

119 *"I don't know, and I don't care":* JZ, "I Don't Know, I Don't Care," cnn.com, Aug. 29, 2001, http://money.cnn.com/2001/08/29/investing/Zweig/.

120 *Create a "too hard" pile:* JZ telephone interview with Buffett, May 5, 2005; see also Peter D. Kaufman, ed., *Poor Charlie's Almanack* (Virginia Beach, Va.: Donning Publishers, 2005), pp. 43, 95.

120 *Measure twice, cut once:* Gary Belsky and Thomas Gilovich, *Why Smart People Make Big Money Mistakes and How to Correct Them* (New York: Fireside, 1999), pp. 172–73.

121 *Write it down, right away:* JZ telephone interview with Baruch Fischhoff, Sept. 21, 2005; Baruch Fischhoff, "For Those Condemned to Study the Past: Heuristics and Biases in Hindsight," *JUU,* pp. 335–51; Elizabeth Loftus, "Our Changeable Memories: Legal and Practical Implications," *NNS,* vol. 4 (March 2003), pp. 231–34, www.seweb.uci.edu/faculty/loftus/.

121 *Learn what works:* JZ telephone interview with Robin Hogarth, March 17, 2005; Robin Hogarth, *Educating Intuition* (Chicago: University of Chicago Press, 2001), pp. 81–91.

122 *Handcuff your inner con man:* S&P 500 return from *Stocks, Bonds, Bills, and Inflation 2005 Yearbook* (Chicago: Ibbotson Associates, 2005). Return adjusted for cash added and withdrawn is from Ilia D. Dichev, "What Are Stock Investors' Actual Historical Returns?" (Jan. 2006), http://ssrn.com/abstract=544142. These three questions are inspired by Daniel Kahneman and Amos Tversky, "Intuitive Prediction," in *JUU,* pp. 414–21.

123 *Embrace the mistake:* JZ interview with Christopher Davis, Dec. 8, 2005.

123 *Don't just "buy what you know":* Peter Lynch with John Rothchild, *One up*

on Wall Street (New York: Penguin, 1990), pp. 18–19; Peter Lynch with John Rothchild, *Beating the Street* (New York: Simon & Schuster, 1993), pp. 152–59.

124 *Don't get stuck:* Merck & Co. press release, Sept. 30, 2004, at www.merck.com; Merck & Co. Form 11-K, July 8, 2005, www.sec/gov/Archives/edgar/data/64978/000095014405007240/g96187e11vk.htm; N.Y. State Attorney General press release, Oct. 14, 2004, www.oag.state.ny.us/press/2004/oct/oct14a_04.html; Marsh & McLennan press releases, Nov. 9, 2004, and March 1, 2005, at www.marsh.com; Marsh and McLennan Cos., Inc., Form 11-K, June 29, 2005, www.sec.gov/Archives/edgar/data/62709/000006270905000183/000006 2709-05-000183-index.htm; Lisa Meulbroek, "Company Stock in Pension Plans: How Costly Is It?" *Journal of Law and Economics,* vol. 48, no. 2 (Oct. 2005), pp. 443–74; JZ e-mail interview with Meulbroek, Nov. 1, 2005; JZ e-mail interview with David Laibson, Oct. 8, 2005. See also JZ, "Don't Try This with Your 401(k)," cnn.com, Nov. 4, 1999, http://money.cnn.com/1999/11/04/mailbag/mailbag/. Meulbroek's calculation assumes that you have no savings outside of your 401(k). But even if your 401(k) is just half of your total net worth, a 25% allocation of your 401(k) to company stock for ten years would have an effective value of only 84 cents on the dollar.

125 *Diversification is the best:* Rodger W. Bridwell, *High-Tech Investing* (New York: New York Times Book Co., 1983), pp. 12–13; Robert Metz, *Future Stocks* (New York: Harper & Row, 1982), pp. 31–32.

6. RISK

127 *"If you burn your mouth":* As quoted by Turkish Prime Minister Turgut Ozal, *WSJ,* July 25, 1984.

127 *If ever there was:* JZ telephone interview with Bobbi Bensman, Dec. 13, 2005; Meir Statman, "Lottery Players/Stock Traders," *FAJ,* Jan.–Feb. 2002, pp. 14–21, and "What Do Investors Want?" *JPM,* 30th Anniversary Issue (2004), pp. 153–61.

128 *For investors and their financial advisors:* www.ifa.com/SurveyNET/index.aspx (accessed March 28, 2006); www.firstambank.com/464.html (accessed March 28, 2006); www.myscudder.com, Planning and Retirement Worksheet (accessed Nov. 8, 2004); Ken C. Yook and Robert Everett, "Assessing Risk Tolerance," *JFP,* vol. 16, no. 8 (Aug. 2003), pp. 48–55.

130 *To an astonishing degree:* Margo Wilson and Martin Daly, "Do Pretty Women Inspire Men to Discount the Future?" *PRSLB* (Suppl.), vol. 271 (2004), pp. S177–S179; Roy F. Baumeister, "The Psychology of Irrationality," in Isabelle Brocas and Juan D. Carrillo, *The Psychology of Economic Decisions,* vol. 1 (Oxford, U.K.: Oxford University Press, 2003), pp. 3–16; Karen Pezza Leith and Roy F. Baumeister, "Why Do Bad Moods Increase

Self-Defeating Behavior?" *JPSP,* vol. 71, no. 6 (1996), pp. 1250–67; Rajagopal Raghunathan and Michel Tuan Pham, "All Negative Moods Are Not Equal," *OBHDP,* vol. 79, no. 1 (1999), pp. 56–77; Michel Tuan Pham, "The Logic of Feeling," *Journal of Consumer Psychology,* vol. 14, no. 4 (2004), pp. 360–69; Jennifer S. Lerner et al., "Heart Strings and Purse Strings," *PS,* vol. 15, no. 5 (2004), pp. 337–41; Denise Chen et al., "Chemosignals of Fear Enhance Cognitive Performance in Humans," *Chemical Senses,* vol. 31, no. 4 (2006), pp. 415–23; JZ e-mail interview with Chen, April 8, 2006; Norbert Schwarz, "Metacognitive Experiences in Consumer Judgment and Decision Making," *Journal of Consumer Psychology,* vol. 14, no. 4 (2004), pp. 332–48; Alexander J. Rothman and Norbert Schwarz, "Constructing Perceptions of Vulnerability," *PSPB,* vol. 24, no. 10 (Oct. 1998), pp. 1053–64; Norbert Schwarz, "Situated Cognition and the Wisdom of Feelings," in Lisa Feldman Barrett and Peter Salovey, *The Wisdom in Feeling* (New York: Guilford Press, 2002), pp. 144–66; JZ telephone interview with Schwarz, April 20, 2005; Matthew C. Keller et al., "A Warm Heart and a Clear Head," *PS,* vol. 16, no. 9 (2005), pp. 724–31.

131 *Imagine yourself transported:* Amos Tversky, "The Psychology of Decision Making," in Arnold S. Wood, ed., *Behavioral Finance and Decision Theory in Investment Management* (Charlottesville, Va.: AIMR Publications, 1995), p. 6; Alex Kacelnik and Melissa Bateson, "Risky Theories," *American Zoologist,* vol. 36, no. 4 (Sept., 1996), pp. 402–34; Sharoni Shafir, "Risk-Sensitive Foraging," *Oikos,* vol. 88, no. 3 (2000), pp. 663–69.

132 *Ecologist Leslie Real:* Leslie A. Real, "Animal Choice Behavior and the Evolution of Cognitive Architecture," *Science,* vol. 253 (Aug. 30, 1991), pp. 980–86; JZ interview with Real, Jan. 26, 1999; Joseph M. Wunderle, Jr. and Zoraida Cotto-Navarro, "Constant vs. Variable Risk-Aversion in Foraging Bananaquits," *Ecology,* vol. 69, no. 5 (1988), pp. 1434–38; letter to JZ from Wunderle, Oct. 17, 1997.

133 *It's not just the birds:* Elke U. Weber et al., "Predicting Risk Sensitivity in Humans and Lower Animals," *PR,* vol. 111, no. 2 (2004), pp. 430–45; Ralph Hertwig et al., "Decisions from Experience and the Effect of Rare Events in Risky Choice," *PS,* vol. 15, no. 8 (2004), pp. 534–39; JZ telephone and e-mail interviews with Elke Weber, March 6 and June 10, 2006; Sarah Holden and Jack VanDerhei, "401(k) Plan Asset Allocation, Account Balances, and Loan Activity in 2005," www.ici.org.

134 *Are there other reasons:* Craig R. M. McKenzie and Jonathan D. Nelson, "What a Speaker's Choice of Frame Reveals," *Psychonomic Bulletin & Review,* vol. 10, no. 3 (2003), pp. 596–602; JZ interview with Slovic, June 29, 2005; Irwin P. Levin and Gary J. Gaeth, "How Consumers Are Affected by the Framing of Attribute Information before and after Consuming the Product," *JCR,* vol. 15 (Dec. 1988), pp. 374–78; Philip Sedgwick and Angela Hall, "Teaching Medical Students and Doctors How to Communicate

Risk," *British Medical Journal,* vol. 327 (Sept. 27, 2003), pp. 694–95; Barbara J. McNeil et al., "On the Elicitation of Preferences for Alternative Therapies," *New England Journal of Medicine,* vol. 306, no. 21 (May 27, 1982), pp. 1259–62.

135 *The classic example of framing:* Amos Tversky and Daniel Kahneman, "The Framing of Decisions and the Psychology of Choice," *Science,* vol. 211 (Jan. 30, 1981), pp. 453–58; see also Kahneman and Tversky, "Choices, Values, and Frames," in *CVF,* pp. 1–16.

137 *Framing can lead:* This example is adapted from Kahneman and Tversky's "Prospect Theory," *Econometrica,* vol. 47, no. 2 (1979), pp. 263–91; Richard H. Thaler and Eric J. Johnson, "Gambling with the House Money and Trying to Break Even," *MS,* vol. 36, no. 6 (June 1990), pp. 643–60; John S. Hammond et al., "The Hidden Traps in Decision Making," *Harvard Business Review,* Jan. 2006, pp. 118–26.

137 *In the financial world:* JZ interview with Nicholas Epley, Sept. 12, 2003; Nicholas Epley et al., "Bonus or Rebate?" *JBDM,* vol. 19, no. 3 (2006), pp. 213–27; Yahoo! Inc. press release, April 7, 2004; stock performance from http://finance.yahoo.com/; JZ, "Splitsville," *MM,* March 2001, pp. 55–56; JZ interview with Daniel Kahneman, Sept. 18, 1996; Daniel Kahneman and Dan Lovallo, "Timid Choices and Bold Forecasts," in *CVF,* pp. 393–413; JZ interview with Paul Slovic, June 29, 2005; Paul Slovic et al., "Rational Actors or Rational Fools?" www.decisionresearch.org/pdf/dr498v2.pdf; Eldar Shafir et al., "Money Illusion," in *CVF,* pp. 335–55.

138 *What creates the frames:* JZ e-mail interview with Cleotilde Gonzalez and Jason Dana, Jan. 25, 2006; Cleotilde Gonzalez et al., "The Framing Effect and Risky Decisions," *Journal of Economic Psychology,* vol. 26 (2005), pp. 1–20; see also Kip Smith et al., "Neuronal Substrates for Choice under Ambiguity, Risk, Gains, and Losses," *MS,* vol. 48, no. 6 (June 2002), pp. 711–18; Alumit Ishai et al., "Distributed Neural Systems for the Generation of Visual Images," *Neuron,* vol. 28 (Dec. 2000), pp. 979–90; Elia Formisano et al., "Tracking the Mind's Image in the Brain," *Neuron,* vol. 35 (July 3, 2002), pp. 185–204; Scott A. Huettel et al., "Decisions under Uncertainty," *JN,* vol. 25, no. 13 (March 30, 2005), pp. 3304–11; Antonia F. de C. Hamilton and Scott T. Grafton, "Goal Representation in Human Anterior Intraparietal Sulcus," *JN,* vol. 26, no. 4 (Jan. 25, 2006), pp. 1133–37.

139 *Now imagine two scenarios:* Benedetto De Martino et al., "Frames, Biases, and Rational Decision-Making in the Human Brain," *Science,* vol. 313 (Aug. 4, 2006), pp. 684–87.

140 *One of the cleverest forms:* Walter Updegrave, "The Surprise Inside the Perfect Investment," *MM,* Oct. 2005, pp. 48–50; *Investment News,* July 17, 2006, pp. 1, 29.

140 *Besides half-full:* Paul Slovic et al., "Violence Risk Assessment and Risk Communication," *Law and Human Behavior,* vol. 24, no. 3 (2000), pp.

271–96; Paul Slovic et al., "The Affect Heuristic," in *HAB,* pp. 397–420; Kimihiko Yamagishi, "When a 12.86% Mortality Is More Dangerous than 24.14%," *Applied Cognitive Psychology,* vol. 11 (1997), pp. 495–506; JZ interview with Slovic, June 29, 2005.

142 *Do investors really act:* Ivo Welch, "Herding Among Security Analysts," *JFE,* vol. 58, no. 3 (2000), pp. 369–96; JZ e-mail interview with Welch, March 8, 2006; Zoran Ivkovic and Scott Weisbenner, "Information Diffusion Effects in Individual Investors' Common Stock Purchases" (April 2004), www.nber.org/papers/w10436; Bing Liang, "Alternative Investments," *Journal of Investment Management,* vol. 2, no. 4 (2004), pp. 76–93; Nicole M. Boyson, "Is There Hedge Fund Contagion?" (March 2006), www.nber.org/papers/w12090; Esther Duflo and Emmanuel Saez, "Participation and Investment Decisions in a Retirement Plan," *Journal of Public Economics,* vol. 85 (2002), pp. 121–48; Karen W. Arenson, "Embarrassing the Rich," *NYT,* May 21, 1995, p. E4; Steve Wulf, "Too Good to Be True," *Time,* May 29, 1995, p. 34; Michael Lewis, "Separating Rich People from Their Money," *NYT Magazine,* June 18, 1995, p. 18; Josef Lakonishok et al., "What Do Money Managers Do?," working paper, University of Illinois, 1997; Stanley G. Eakins et al., "Institutional Portfolio Composition" (1998), http://ssrn.com/abstract=45754; Richard W. Sias, "Institutional Herding," *RFS,* vol. 17, no. 1 (2004), pp. 165–206; JZ e-mail interview with Sias, Mar. 9, 2006; Vivek Sharma et al., "Institutional Herding and the Internet Bubble" (May 2006) http://ssrn.com/abstract=501423; Robert J. Shiller and John Pound, "Survey Evidence on Diffusion of Interest among Institutional Investors" (May 1986), http://cowles.econ.yale.edu/P/cd/dy1986.htm.

143 *Ideas, like yawns:* Sushil Bikhchandani et al., "A Theory of Fads, Fashion, Custom, and Cultural Change as Informational Cascades," *JPE,* vol. 100, no. 5 (Oct. 1992), pp. 992–1026; "Informational Cascades and Rational Herding," http://welch.econ.brown.edu/cascades/; David Hirshleifer and Siew Hong Teoh, "Herd Behavior and Cascading in Capital Markets," *European Financial Management,* vol. 9, no. 1 (March 2003), pp. 25–66; Jean-Marc Amé et al., "Collegial Decision Making Based on Social Amplification Leads to Optimal Group Formation," *PNAS,* vol. 103, no. 15 (April 11, 2006), pp. 5385–40; Ian T. Baldwin et al., "Volatile Signaling in Plant-Plant Interactions," *Science,* vol. 311 (Feb. 10, 2006), pp. 812–15; Lars Chittka and Ellouise Leadbeater, "Social Learning," *CB,* vol. 15, no. 21 (2005), R869–R871; Isabelle Coolen et al., "Social Learning in Noncolonial Insects," *CB,* vol. 15, no. 21 (Nov. 8, 2005), pp. 1931–35; Etienne Danchin et al., "Public Information," *Science,* vol. 305 (July 23, 2004), pp. 487–91; Julie W. Smith et al., "The Use and Misuse of Public Information by Foraging Red Crossbills," *Behavioral Ecology,* vol. 10, no. 1 (1999), pp. 54–62. The virtues of collective intelligence among humans are discussed in James Surowiecki, *The Wisdom of Crowds* (New York: Doubleday, 2004).

144 *Humans are animals:* Matthew J. Salganik et al., "Experimental Study of Inequality and Unpredictability in an Artificial Cultural Market," *Science,* vol. 311 (Feb. 10, 2006), pp. 854–66.

145 *How much money you made:* Melissa Bateson, "Recent Advances in Our Understanding of Risk-Sensitive Foraging Preferences," *Proceedings of the Nutrition Society,* vol. 61 (2002), pp. 1–8; Thomas Caraco et al., "An Empirical Demonstration of Risk-Sensitive Foraging Preferences," *Animal Behaviour,* vol. 28, no. 3 (Aug. 1980), pp. 820–30; D. W. Stephens and J. R. Krebs, *Foraging Theory* (Princeton, N.J.: Princeton University Press, 1986), pp. 134–50; John M. McNamara and Alasdair I. Houston, "Risk-Sensitive Foraging," *Bulletin of Mathematical Biology,* vol. 54, no. 2/3 (1992), pp. 355–78.

145 *If birds run out of seeds:* www.ops.gov.ph/news/archives2006/feb05.htm; http://news.bbc.co.uk/2/hi/asia-pacific/4680040.stm; www.manilatimes.net/national/2006/feb/05/yehey/images/front.pdf.

145 *Even when no one dies:* Charles T. Clotfelter, "Do Lotteries Hurt the Poor?" summary of congressional testimony, April 28, 2000, www.pubpol.duke.edu/people/faculty/clotfelter/lottsum.pdf; Consumer Federation of America press release, "How Americans View Personal Wealth," Jan. 9, 2006, www.consumerfed.org; Rui Yao et al., "The Financial Risk Tolerance of Blacks, Hispanics and Whites," *Financial Counseling and Planning,* vol. 16, no. 1 (2005), pp. 51–62; Keith C. Brown et al., "Of Tournaments and Temptations," *JF,* vol. 51, no. 1 (March 1996), pp. 85–110; Joshua D. Coval and Tyler Shumway, "Do Behavioral Biases Affect Prices?" *JF,* vol. 60, no. 1 (Feb. 2005), pp. 1–34; Alok Kumar, "Who Gambles in the Stock Market?" (May 2005), http://ssrn.com/abstract=686022.

147 *Step outside yourself:* Richard S. Tedlow, "The Education of Andy Grove," *Fortune,* Dec. 12, 2005, p. 122; JZ, "What Can We Learn from History?" cnn.com, Sept. 21, 2001, http://money.cnn.com/2001/09/21/investing/zweig/index.htm.

148 *When the price drops:* JZ e-mail interview with Nygren, April 24, 2006; *BusinessWeek,* Aug. 13, 1979, p. 56. For detailed advice on how to compare stock price and business value, see *TII,* especially Chapters 14 and 15.

150 *Write yourself a policy:* JZ thanks Stephen Barnes, Thomas J. Connelly, Roger Gibson, and Robert N. Veres for their help in preparing a streamlined IPS.

150 *Get reframed:* Peter D. Kaufman, ed., *Poor Charlie's Almanack* (Virginia Beach, Va.: Donning Publishers, 2005), p. 145.

151 *Try to prove yourself wrong:* JZ interview with Bernstein, July 28, 2004; JZ, "Peter's Uncertainty Principle," *MM,* Nov. 2004, pp. 142–49.

151 *Economists used to say:* Daniel Kahneman and Amos Tversky, "Prospect Theory," *Econometrica,* vol. 47, no. 2 (1979), pp. 263–91; JZ, "Do You Sabotage Yourself?," *MM,* May 2001, pp. 74–78.

152 *Since no one really has:* JZ interview with Slovic, June 29, 2005.

152 *When you're on fire:* JZ interview with Bernstein, July 28, 2004; JZ, "Peter's Uncertainty Principle," *MM,* Nov. 2004, pp. 142–49.

7. FEAR

154 *Neither a man:* Bertrand Russell, "An Outline of Intellectual Rubbish" (1943), www.solstice.us/russell/intellectual_rubbish.html.

154 *Here are a few questions:* Mark Peplow, "Counting the Dead," *Nature,* vol. 440 (April 20, 2006), pp. 982–83; Dillwyn Williams and Keith Baverstock, "Too Soon for a Final Diagnosis," *Nature,* vol. 440 (April 20, 2006), pp. 993–94; www.nature.com/news/2005/050905/full/437181b.html; www.who.int/mediacentre/news/releases/2005/pr38/en/index.html; seer.cancer.gov/statfacts/html/melan.html. Roughly 4,000 cases of thyroid cancer resulted from the Chernobyl accident; so far, 15 have been fatal. While the U.N.'s worst-case estimate is that more than 9,000 people may eventually die as a result of Chernobyl, nearly all the potential victims remain alive two decades after the accident.

155 *None of this means: NYT,* Nov. 12, 2002, p. F4; www.flmnh.ufl.edu/fish/sharks/attacks/relariskanimal.htm; Ricky L. Langley, "Alligator Attacks on Humans in the United States," *Wilderness and Environmental Medicine,* vol. 16 (2005), pp. 119–24; www.natural-resources.wsu.edu/research/bear-center/bear-people.htm; www.cdc.gov/nasd; *World Report on Violence and Health* (U.N. World Health Organization, 2002), www.who.int/violence_injury_prevention/violence/world_report/en/, p. 10. United Nations Development Programme, "The Human Consequences of the Chernobyl Nuclear Accident" (UNDP and UNICEF, Jan. 25, 2002), www.undp.org; Douglas Chapin et al., "Nuclear Power Plants and Their Fuel as Terrorist Targets," *Science,* vol. 297 (Sept. 20, 2002), pp. 1997–99. These data are for calendar year 2000, but even the war in Iraq has not changed the numbers enough to alter the order they are listed in.

155 *We're no different:* John Ameriks et al., "Expectations for Retirement," Vanguard Center for Retirement Research (Nov. 2004), pp. 12–14.

156 *If we were strictly logical:* JZ, "When the Stock Market Plunges, Will You Be Brave or Will You Cave?" *MM,* Jan. 1997, p. 106.

156 *On the other hand:* Paul Slovic, "Informing and Educating the Public about Risk," *RA,* vol. 6, no. 4 (1986), pp. 403–15; www.planecrashinfo.com/cause.htm; Kukoc: *Sports Illustrated,* Feb. 24, 1997, p. 46; *National Transportation Statistics 2005* (U.S. Department of Transportation), table 2-1, www.bts.gov/publications/national_transportation_statistics/2005/index.html; Michael Sivak and Michael J. Flannagan, "Flying and Driving after the September 11th Attacks," *American Scientist,* Jan.–Feb. 2003, http://american

scientist.org/template/AssetDetail/assetid/16237; Gerd Gigerenzer, "Out of the Frying Pan into the Fire," *RA*, vol. 26, no. 2 (2006), pp. 347–51.

157 *The more vivid and easily imaginable:* Eric J. Johnson et al., "Framing, Probability Distortions, and Insurance Decisions," in *CVF*, pp. 224–40.

157 *When an intangible feeling:* Consumer confidence data courtesy of the Conference Board's Carol Courter, e-mail to JZ, March 14, 2006. Eric J. Johnson and Amos Tversky, "Affect, Generalization, and the Perception of Risk," *JPSP*, vol. 45, no. 1 (1983), pp. 20–31; JZ e-mail interview with Eric Johnson, Feb. 14, 2006.

157 *Our intuitive sense of risk:* Sarah Lichtenstein et al., "Judged Frequency of Lethal Events," *JEPHLM*, vol. 4, no. 6 (Nov. 1978), pp. 551–78; Paul Slovic, "Perception of Risk," *Science*, vol. 236 (April 17, 1987), pp. 280–85; JZ interview with Paul Slovic and Ellen Peters, June 29, 2005. Besides dread and knowability, there is a third factor—how many people are exposed to the risk—but it appears to play a less significant role.

158 *That's why so many people:* www.floodsmart.gov; Mark J. Browne and Robert E. Hoyt, "The Demand for Flood Insurance," *JRU*, vol. 20, no. 3 (2000), pp. 291–300; Howard Kunreuther, "Has the Time Come for Comprehensive Natural Disaster Insurance?" in Ronald J. Daniels et al., *On Risk and Disaster* (Philadelphia: University of Pennsylvania Press, 2006), pp. 175–202.

159 *In the stock market:* Press release, Wendy's International Inc., July 7, 2005; Wendy's 10-Q report, Aug. 11, 2005; www.cnn.com/2005/LAW/09/09/wendys.finger.ap; stock data from http://finance.yahoo.com; Patricia Sellers, "eBay's Secret," *Fortune*, Oct. 18, 2004, p. 160; www.forbes.com/forbes/1999/0726/6402238a.html.

160 *Remarkably, the amygdala:* Antonio Damasio, *Descartes' Error* (New York: Penguin, 1994); Joseph LeDoux, *The Emotional Brain* (New York: Simon & Schuster, 1996); Andrew J. Calder et al., "Neuropsychology of Fear and Loathing," *NRN*, vol. 2 (May, 2001), pp. 352–63; K. Luan Phan et al., "Functional Neuroanatomy of Emotion," *NeuroImage*, vol. 16 (2002), pp. 331–48; M. Davis and P. J. Whalen, "The Amygdala," *Molecular Psychiatry*, vol. 6 (2001), pp. 13–34; Nathan J. Emery and David G. Amaral, "The Role of the Amygdala in Primate Social Cognition," in *CNE*, pp. 156–91; JZ interview with Antoine Bechara, April 2, 2002; D. Caroline Blanchard and Robert J. Blanchard, "Innate and Conditioned Reactions to Threat in Rats with Amygdaloid Lesions," *Journal of Comparative and Physiological Psychology*, vol. 81, no. 2 (1972), pp. 281–90.

161 *Social signals can set off:* Paul J. Whalen et al., "Masked Presentation of Emotional Facial Expressions Modulate Amygdala Activity without Explicit Knowledge," *JN*, vol. 18, no. 1 (Jan. 1, 1998), pp. 411–18; Beatrice de Gelder, "Towards the Neurobiology of Emotional Body Language," *NRN*, vol. 7

(March, 2006), pp. 242–49; Beatrice de Gelder et al., "Fear Fosters Flight," *PNAS*, vol. 101, no. 47 (Nov. 23, 2004), pp. 16701–706.

161 *Finally, the amygdala is sensitive:* N. Isenberg et al., "Linguistic Threat Activates the Human Amygdala," *PNAS,* vol. 96 (Aug., 1999), pp. 10456–459; Laetitia Silvert et al., "Autonomic Responding to Aversive Words without Conscious Valence Discrimination," *International Journal of Psychophysiology,* vol. 53 (2004), pp. 135–45; Elizabeth L. Loftus and John C. Palmer, "Reconstruction of Automobile Destruction," *Journal of Verbal Learning and Verbal Behavior,* vol. 13 (1974), pp. 585–9. A normal eyeblink lasts about 320 milliseconds (e-mail to JZ from SUNY Stony Brook neurobiologist Craig Evinger, Mar. 23, 2006).

162 *Both actual and imagined losses:* Tiziana Zalla et al., "Differential Amygdala Responses to Winning and Losing," *EJN,* vol. 12 (2000), pp. 1764–70; JZ interview with Grafman, March 6, 2002; Hans C. Breiter et al., "Functional Imaging of Neural Responses to Expectancy and Experience of Monetary Gains and Losses," *Neuron,* vol. 30 (May 2001), pp. 619–39; Gleb P. Shumyatsky et al., "Stathmin, a Gene Enriched in the Amygdala, Controls Both Learned and Innate Fear," *Cell,* vol. 123 (Nov. 18, 2005), pp. 697–709; R. Douglas Fields, "Making Memories Stick," *SA,* Feb. 2005, pp. 75–81; Karim Nader et al., "Fear Memories Require Protein Synthesis in the Amygdala for Reconsolidation after Retrieval," *Nature,* vol. 406 (Aug. 17, 2000), pp. 722–26; James L. McGaugh, "Memory—a Century of Consolidation," *Science,* vol. 287 (Jan. 14, 2000), pp. 248–51; B. A. Strange and R. J. Dolan, "ß-Adrenergic Modulation of Emotional Memory-Evoked Human Amygdala and Hippocampal Responses," *PNAS,* vol. 101, no. 31 (Aug. 3, 2004), pp. 11454–58; James L. McGaugh et al., "Modulation of Memory Storage by Stress Hormones and the Amygdaloid Complex," in Michael S. Gazzaniga, ed., *The New Cognitive Neurosciences* (Cambridge, Mass.: MIT Press, 2000), pp. 1081–98; Joe Guillaume Pelletier et al., "Lasting Increases in Basolateral Amygdala Activity After Emotional Arousal," *Learning & Memory,* vol. 12 (2005), pp. 96–102; Rebecca Elliott et al., "Dissociable Neural Responses in Human Reward Systems," *JN,* vol. 20, no. 16 (Aug. 15, 2000), pp. 6159–65. "Adrenaline" is the common term for epinephrine.

163 *What's so bad: 2005 Investment Company Fact Book* (Washington, D.C.: Investment Company Institute, 2005), p. 77; *1996 Mutual Fund Fact Book* (Washington, D.C.: Investment Company Institute, 1996), p. 125; Donald B. Keim and Ananth Madhavan, "The Relation Between Stock Market Movements and NYSE Seat Prices," *JF,* vol. 55, no. 6 (Dec. 2000), pp. 2817–40; William James, *The Principles of Psychology* (New York: Henry Holt, 1890, reprinted Dover Press, 1950), vol. 1, p. 670. (Italics in original.)

163 *I learned how my own:* JZ participated in the Iowa Gambling Task (and interviewed Antoine Bechara and Antonio Damasio) at the University of Iowa,

April 2, 2002. The experiment is also described in Antonio Damasio, *Descartes' Error* (New York: Avon, 1994), pp. 212–22; Antonio R. Damasio, "The Somatic Marker Hypothesis and the Possible Functions of the Prefrontal Cortex," *PTRSLB,* vol. 351 (1996), pp. 1413–20; Antoine Bechara et al., "Different Contributions of the Human Amygdala and Ventromedial Prefrontal Cortex to Decision-Making," *JN,* vol. 19, no. 13 (July 1, 1999), pp. 5473–81; Antoine Bechara et al., "The Somatic Marker Hypothesis and Decision-Making," in Jordan Grafman, ed., *Handbook of Neuropsychology* (London: Elsevier, 2002), pp. 117–43. For a divergent view, see Alan G. Sanfey and Jonathan D. Cohen, "Is Knowing Always Feeling?" *PNAS,* vol. 101, no. 48 (Nov. 30, 2004), pp. 16709–10, and Tiago V. Maia and James L. McClelland, "A Reexamination of the Evidence for the Somatic Marker Hypothesis," *PNAS,* vol. 101, no. 45 (Nov. 9, 2004), pp. 16075–80.

165 *A team of researchers:* Baba Shiv et al., "Investment Behavior and the Negative Side of Emotion," *PS,* vol. 16, no. 6 (2005), pp. 435–39.

166 *Nowadays, investment herds:* Luc-Alain Giraldeau, "The Ecology of Information Use," in John R. Krebs and Nicholas B. Davies, *Behavioural Ecology* (Oxford: Blackwell, 1997), pp. 42–68; Isabelle Coolen et al., "Species Difference in Adaptive Use of Public Information in Sticklebacks," *PRSLB,* vol. 270, no. 1531 (Nov. 22, 2003), pp. 2413–19; Theodore Stankowich and Daniel T. Blumstein, "Fear in Animals," *PRSLB,* vol. 272, no. 1581 (Dec. 22, 2005), pp. 2627–34; JZ e-mail interview with Blumstein, March 6, 2006. The science and mathematical laws of herding are explored in depth in Luc-Alain Giraldeau and Thomas Caraco, *Social Foraging Theory* (Princeton, N.J.: Princeton University Press, 2000).

167 *Of course, anyone who has ever:* Gregory S. Berns et al., "Neurobiological Correlates of Social Conformity and Independence during Mental Rotation," *BP,* vol. 58 (2005), pp. 245–53; Jaak Panksepp, "Feeling the Pain of Social Loss," *Science,* vol. 302 (Oct. 10, 2003), pp. 237–39; Naomi I. Eisenberger et al., "Does Rejection Hurt?" *Science,* vol. 302 (Oct. 10, 2003), pp. 290–92.

167 *Once you join the crowd:* Uri Hasson et al., "Intersubject Synchronization of Cortical Activity During Natural Vision," *Science,* vol. 303 (March 12, 2004), pp. 1634–40; Luiz Pessoa, "Seeing the World in the Same Way," *Science,* vol. 303 (March 12, 2004), pp. 1617–18.

168 *Military-intelligence scholar:* Ellsberg's biography and his classic article "Risk, Ambiguity, and the Savage Axioms" (*QJE,* vol. 75, no. 4 [Nov. 1961], pp. 643–69) are available at www.ellsberg.net. The Ellsberg Paradox has been replicated in many subsequent experiments; see Colin Camerer and Martin Weber, "Recent Developments in Modeling Preferences," *JRU,* vol. 5 (1992), pp. 325–70, and Catrin Rode et al., "When and Why Do People Avoid Unknown Probabilities in Decisions Under Uncertainty?," *Cognition,* vol. 72 (1999), pp. 269–304. Rumsfeld remarks: Department of Defense

news briefing, Feb. 12, 2002, www.defenselink.mil/transcripts/2002/t02122002_t212sdv2.html.

169 *That's no surprise:* Aldo Rustichini, "Emotion and Reason in Making Decisions," *Science,* vol. 310 (Dec. 9, 2005), pp. 1624–25; Ming Hsu et al., "Neural Systems Responding to Degrees of Uncertainty in Human Decision-Making," *Science,* vol. 310 (Dec. 9, 2005), pp. 1680–83. (The frontal lobe is also involved: Scott A. Huettel et al., "Neural Signatures of Economic Preferences for Risk and Ambiguity," *Neuron,* vol. 49 [March 2, 2006], pp. 765–75.) Not knowing what the odds are is very different from knowing that the odds are low; as we saw in Chapter Three, nothing is quite as thrilling as a long-shot gamble on a big jackpot. When the probabilities of winning are remote, many people prefer an ambiguous over a certain gamble; see Hillel J. Einhorn and Robin M. Hogarth, "Decision Making Under Ambiguity," *JB,* vol. 59, no. 4, pt. 2 (1986, pp. S225–S250).

170 *Ellsberg's Paradox often shows up:* Michael J. Brennan, "The Individual Investor," *Journal of Financial Research,* vol. 18, no. 1 (1995), pp. 59–74; Robert A. Olsen and George H. Troughton, "Are Risk Premium Anomalies Caused by Ambiguity?" *FAJ,* vol. 56, no. 2 (March–April 2000), pp. 24–31; Thomas K. Philips, "The Source of Value," *JPM,* vol. 28, no. 4 (2002), pp. 36–44; Brad Barber et al., "Reassessing the Returns to Analysts' Stock Recommendations," *FAJ,* vol. 59, no. 2 (March–April 2003), pp. 88–96; John A. Doukas et al., "Divergent Opinions and the Performance of Value Stocks," *FAJ,* vol. 60, no. 6 (Nov.–Dec. 2004), pp. 55–64; Eugene F. Fama and Kenneth R. French, "The Anatomy of Value and Growth Stock Returns" (Sept. 2005), http://ssrn.com/abstract=806664.

171 *Get it off your mind:* Paul Zimmerman, "The Ultimate Winner," *Sports Illustrated,* Aug. 13, 1990, pp. 72–83; Larry Schwartz, "No Ordinary Joe," http://espn.go.com/classic/biography/s/Montana_Joe.html.

171 *Use your words:* James J. Gross, "Antecedent- and Response-Focused Emotion Regulation," *JPSP,* vol. 74, no. 1 (1998), pp. 224–37.

172 *As we've learned, if you view:* JZ e-mail interview with Ahmad Hariri, April 14, 2005; Ahmad R. Hariri et al., "Modulating Emotional Responses," *NeuroReport,* vol. 11, no. 1 (Jan. 2000), pp. 43–48; Ahmad R. Hariri et al., "Neocortical Modulation of the Amygdala Response to Fearful Stimuli," *BP,* vol. 53 (2003), pp. 494–501; Kezia Lange et al., "Task Instructions Modulate Neural Responses to Fearful Facial Expressions," *BP,* vol. 53 (2003), pp. 226–32; Florin Dolcos and Gregory McCarthy, "Brain Systems Mediating Cognitive Interference by Emotional Distraction," *JN,* vol. 26, no. 7 (Feb. 15, 2006), pp. 2072–79.

173 *Track your feelings:* JZ interview with Antoine Bechara, April 2, 2002; JZ, "What's Eating You," *MM,* Dec. 2001, pp. 63–64; Beverly Goodman, "Family Tradition," *SmartMoney,* March 2006, pp. 64–67; JZ e-mail interview with Davis, June 27, 2006.

173 *Get away from the herd:* Stanley Milgram, *Obedience to Authority* (New York: Harper & Row, 1974), pp. 4, 23, 107, 117–19, 123.

175 *Alfred P. Sloan Jr.:* Irving L. Janis, *Groupthink* (Boston: Houghton Mifflin, 1982), p. 271; Tacitus, *Germania,* www.fordham.edu/HALSALL/basis/tacitus-germanygord.html.

8. SURPRISE

176 *What can no longer be imagined:* Harry Zohn, ed., *In These Great Times: A Karl Kraus Reader* (Manchester, U.K.: Carcanet, 1984), p. 70.

176 *What's striking about this experience:* According to leading manufacturers Kohler and American Standard, the typical distance from the bathroom floor to the rim of a toilet bowl is 16.5 to 18 inches with the seat up and 18 to 19.5 inches with the seat down.

176 *It takes a while:* Kalanit Grill-Spector et al., "Repetition and the Brain," *TICS,* vol. 10, no. 1 (Jan. 2006), pp. 14–23; JZ e-mail interview with Douglas Fields, July 13, 2005; remarks by Michael Gazzaniga, Institute of Behavioral Finance conference, New York, Oct. 11, 2001.

177 *On January 31, 2006:* Google Inc. press release, Jan. 31, 2006; www.thestreet.com/tech/internet/10265459.html; Yahoo! Finance message board for Google stock, http://finance.yahoo.com/q/mb?s=GOOG, Jan. 31, 2006, messages 550355, 550407, 550455, 551097.

178 *Every investor should know:* David N. Dreman and Michael A. Berry, "Analyst Forecasting Errors and Their Implications for Security Analysis," *FAJ* (May–June 1995), pp. 30–41; David Dreman, *Contrarian Investment Strategies* (New York: Simon & Schuster, 1998), pp. 91–93; Eli Bartov et al., "The Rewards to Meeting or Beating Earnings Expectations" (Oct. 2000), http://ssrn.com/abstract=247435; Thomas J. Lopez and Lynn Rees, "The Effect of Meeting Analysts' Forecasts and Systematic Positive Forecast Errors," (Feb. 2001), http://ssrn.com/abstract=181929. I am grateful to Langdon Wheeler, Shanta Puchtler, and Yucong Huang of Numeric Investors L.P. for providing extensive data on 2005 results (e-mail to JZ from Huang, May 31, 2006).

179 *Humans and great apes:* Esther A. Nimchinsky et al., "A Neuronal Morphologic Type Unique to Humans and Great Apes," *PNAS,* vol. 96 (April 1999), pp. 5268–73; John Allman et al., "Two Phylogenetic Specializations in the Human Brain," *The Neuroscientist,* vol. 8, no. 4 (2002), pp. 335–46; John M. Allman et al., "The Anterior Cingulate Cortex," *ANYAS,* vol. 935 (May 2001), pp. 107–17; JZ telephone interview with John Allman, Dec. 28, 2004; Hugo D. Critchley et al., "Human Cingulate Cortex and Autonomic Control," *Brain,* vol. 126 (2003), pp. 2139–52; JZ e-mail interview with Scott Huettel, May 24, 2006; Phan Luu and Michael I. Posner, "Anterior Cingulate Cortex Regulation of Sympathetic Activity," *Brain,* vol. 126 (2003), pp. 2119–20; Matthew M.

Botvinick et al., "Conflict Monitoring and Anterior Cingulate Cortex," *TICS*, vol. 8, no. 12 (Dec. 2004), pp. 539–46; Nick Yeung et al., "The Neural Basis of Error Detection," *PR*, vol. 111, no. 4 (2004), pp. 931–59; Shigehiko Ito et al., "Performance Monitoring by the Anterior Cingulate Cortex during Saccade Countermanding," *Science*, vol. 302 (Oct. 3, 2003), pp. 120–22; Chunmao Wang et al., "Responses of Human Anterior Cingulate Cortex Microdomains," *JN*, vol. 25, no. 3 (Jan. 19, 2005), pp. 604–13; JZ e-mail interview with George Bush, June 2, 2006; Tomas Paus, "Primate Anterior Cingulate Cortex," *NRN*, vol. 2 (June 2001), pp. 417–24. Spindle neurons have recently been found in the brains of some species of whales.

180 *You can sense your own ACC:* J. Ridley Stroop, "Studies of Interference in Serial Verbal Reactions," *JEP*, vol. 18 (1935), pp. 643–62, http://psychclassics.yorku.ca/Stroop/; http://faculty.washington.edu/chudler/java/ready.html; Kenji Matsumoto and Keiji Tanaka, "Conflict and Cognitive Control," *Science*, vol. 303 (Feb. 13, 2004), pp. 969–70; John G. Kerns et al., "Anterior Cingulate Conflict Monitoring and Adjustments in Control," *Science*, vol. 303 (Feb. 13, 2004), pp. 1023–26; Matthew M. Botvinick et al., "Conflict Monitoring and Cognitive Control," *PR*, vol. 108, no. 3 (2001), pp. 624–52; Jonathan D. Cohen et al., "Anterior Cingulate and Prefrontal Cortex," *NNS*, vol. 3, no. 5 (May 2000), pp. 421–23; Joshua W. Brown and Todd S. Braver, "Learned Predictions of Error Likelihood in the Anterior Cingulate Cortex," *Science*, vol. 307 (Feb. 18, 2005), pp. 1118–21.

180 *Even in monkeys:* Keisetsu Shima and Jun Tanji, "Role for Cingulate Motor Area Cells in Voluntary Movement Selection Based on Reward," *Science*, vol. 282 (Nov. 13, 1998), pp. 1335–38; George Bush et al., "Dorsal Anterior Cingulate Cortex," *PNAS*, vol. 99, no. 1 (Jan. 8, 2002), pp. 523–28; Ziv M. Williams et al., "Human Anterior Cingulate Neurons and the Integration of Monetary Reward with Motor Responses," *NNS*, vol. 7, no. 12 (Dec. 2004), pp. 1370–75; William J. Gehring and Stephan F. Taylor, "When the Going Gets Tough, the Cingulate Gets Going," *NNS*, vol. 7, no. 12 (Dec. 2004), pp. 1285–87; William J. Gehring and Adrian R. Willoughby, "The Medial Frontal Cortex and the Rapid Processing of Monetary Gains and Losses," *Science*, vol. 295 (March 22, 2002), pp. 2279–82; Stephan F. Taylor et al., "Medial Frontal Cortex Activity and Loss-Related Responses to Errors," *JN*, vol. 26, no. 15 (April 12, 2006), pp. 4063–70; JZ e-mail interview with Gehring, April 24, 2006.

181 *Using tiny electrodes:* Ziv M. Williams et al., "Human Anterior Cingulate Neurons and the Integration of Monetary Reward with Motor Responses," *NNS*, vol. 7, no. 12 (Dec. 2004), pp. 1370–75; JZ e-mail interview with George Bush, June 2, 2006; Thomas J. Lopez and Lynn Rees, "The Effect of Meeting Analysts' Forecasts and Systematic Positive Forecast Errors" (Feb. 2001), http://ssrn.com/abstract=181929.

181 *The intensity of your surprise:* Scott A. Huettel et al., "Perceiving Patterns

in Random Series," *NNS,* vol. 5, no. 2 (May 2002), pp. 485–90; JZ e-mail interview with Huettel, May 29, 2006. After eight repetitions of a stimulus, the amplitude of the fMRI signal was three times greater than after three repetitions.

181 *The stock market provides:* Irene Y. Kim, "An Analysis of the Market Reward and Torpedo Effect of Firms That Consistently Meet Expectations" (Jan. 2002), http://ssrn.com/abstract=314381; Edward Keon Jr. et al., "A Requiem for SUE," *Journal of Investing* (Winter 2002), pp. 11–14; Jennifer Conrad et al., "When Is Bad News Really Bad News?" *JF,* vol. 57, no. 6 (Dec. 2002), pp. 2507–32; Jennifer Conrad et al., "How Do Analyst Recommendations Respond to Major News?" (Aug. 2004), http://ssrn.com/abstract=305167; James N. Myers et al., "Earnings Momentum and Earnings Management" (May 2005), http://ssrn.com/abstract=741244.

182 *So there's a lot more at stake:* Douglas J. Skinner and Richard G. Sloan, "Earnings Surprises, Growth Expectations, and Stock Returns" (July 1999), http://ssrn.com/abstract=172060; Lawrence D. Brown, "Small Negative Surprises," *IJF,* vol. 19 (2003), pp. 149–59; Dave Jackson and Jeff Madura, "Profit Warnings and Timing," *Financial Review,* vol. 38 (2003), pp. 497–513; Tim Loughran and Jennifer Marietta-Westberg, "Divergence of Opinion Surrounding Extreme Events" (Jan. 2005), http://ssrn.com/abstract=298647.

182 *How big and bad:* Apple press release, Sept. 28, 2000, www.apple.com/investor; "Apple Bruises Tech Sector," Sept. 29, 2000, http://money.cnn.com/2000/09/29/markets/techwrap/index.htm. Historical prices for AAPL stock from http://finance.yahoo.com/q/hp?s=AAPL.

183 *Of course, the corporate world:* John R. Graham et al., "The Economic Implications of Corporate Financial Reporting" (June 2004), http://papers.nber.org/papers/W10550 and http://ssrn.com/abstract=491627; Mei Cheng et al., "Earnings Guidance and Managerial Myopia" (Nov. 2005), http://ssrn.com/abstract=851545; François Degeorge et al., "Earnings Management to Exceed Thresholds," *JB,* vol. 72, no. 1 (1999), pp. 1–33; Sanjeev Bhojraj et al., "Making Sense of Cents" (Feb. 2006), http://ssrn.com/abstract=418100.

184 *A few years ago:* Qwest Communications International Inc., 2002 Form 10-K filing, www.qwest.com/about/investor/financial/reports/2002/Final_10K _AR_10_28_03.pdf; U.S. Securities and Exchange Commission civil complaint against Joseph P. Nacchio et al., March 15, 2005, www.sec.gov/litigation/litreleases/lr19136.htm and www.sec.gov/litigation/complaints/comp_nacchio19136.pdf. The SEC alleged that Nacchio and his senior associates fabricated roughly $3 billion in fraudulent revenues in a desperate attempt to meet outside expectations; the U.S. Department of Justice later indicted Nacchio for securities fraud. (He insisted he is innocent.)

184 *After writing about Wall Street:* Dupin, the detective in Poe's "The Mystery

of Marie Roget" (1850), says that "modern science has resolved to calculate upon the unforeseen." In Chesterton's story "The Blue Cross" (1911), the policeman Valentin muses that "in the paradox of Poe, wisdom should reckon on the unforeseen."

185 *High hopes:* David Dreman, *Contrarian Investment Strategies* (New York: Simon & Schuster, 1998), pp. 117–36; Lawrence D. Brown, "Small Negative Surprises," *IJF,* vol. 19 (2003), pp. 149–59; Scott A. Huettel et al., "Perceiving Patterns in Random Series," *NNS,* vol. 5, no. 2 (May 2002), pp. 485–90; JZ e-mail interview with Huettel, May 29, 2006; Laura Frieder, "Evidence on Behavioral Biases in Trading Activity" (Dec. 2003), http://ssrn.com/abstract=479983.

186 *Track the "whys":* Jerome S. Bruner and Leo Postman, "On the Perception of Incongruity," *Journal of Personality,* vol. 18, no. 2 (Dec. 1949), pp. 206–23.

186 *Stay away:* Joseph Fuller and Michael C. Jensen, "Just Say No to Wall Street" (Winter 2002), http://ssrn.com/abstract=297156; Amy P. Hutton, "Determinants of Managerial Earnings Guidance . . ." (July 2004), http://ssrn.com/abstract=567441.

187 *Look under the statistical hood:* Craig H. Wisen, "The Bias Associated with New Mutual Fund Returns" (Jan. 2002), http://ssrn.com/abstract=290463; letter to JZ from Wisen, Jan. 15, 2002; Richard B. Evans, "Does Alpha Really Matter?" (Jan. 2004), http://lcb.uoregon.edu/departments/finl/conference/papers/evans.pdf; "The Morningstar Rating for Funds" (July 2002), Morningstar Inc.; e-mail to JZ from Annette Larson, Morningstar, June 6, 2006; "Understanding Biases in Performance Benchmarking," *Windows into the Mutual Funds Industry,* March 2006 (Strategic Insight, New York), p. 10; JZ, "New Cause for Caution on Stocks," *Time,* May 6, 2002, p. 71, www.time.com/time/magazine/article/0,9171,1101020506-234145,00.html.

9. REGRET

190 *Clear, unscaleable:* "Autumn Song" (1936), in *W. H. Auden: Collected Poems* (New York: Random House, 1976), p. 118.

190 *To this day:* JZ telephone interviews with Robertson, Feb. 15 and 17, 2005.

191 *Imagine that I show you:* Colin Camerer, "Individual Decision Making," in John H. Kagel and Alvin E. Roth, *The Handbook of Experimental Economics* (Princeton, N.J.: Princeton University Press, 1995), pp. 587–703; Jack L. Knetsch and J. A. Sinden, "Willingness to Pay and Compensation Demanded," *QJE,* vol. 99, no. 3 (Aug. 1984), pp. 507–21; John K. Horowitz and Kenneth E. McConnell, "A Review of WTA/WTP Studies," *Journal of Environmental Economics and Management,* vol. 44 (2002), pp. 426–47; George Loewenstein and Daniel Adler, "A Bias in the Prediction of Tastes," *Economic Journal,* vol. 105, no. 431 (July 1995), pp. 929–37; Daniel Kahneman et al., "Experimental Tests of the Endowment Effect and the Coase

Theorem," *JPE*, vol. 98, no. 6 (1990), pp. 1325–47; Ziv Carmon and Dan Ariely, "Focusing on the Foregone," *JCR*, vol. 27 (Dec. 2000), pp. 360–70; Ying Zhang and Ayelet Fischbach, "The Role of Anticipated Emotions in the Endowment Effect," *Journal of Consumer Psychology*, vol. 15, no. 4 (2005), pp. 316–24; Eric J. Johnson et al., "Aspects of Endowment," Columbia University working paper (Oct. 2004); Daniel Kahneman et al., "The Endowment Effect, Loss Aversion, and Status Quo Bias," in Richard H. Thaler, *The Winner's Curse* (Princeton, N.J.: Princeton University Press, 1992), pp. 63–78. For a technical criticism of the endowment effect, see Charles R. Plott and Kathryn Zeiler, "The Willingness to Pay–Willingness to Accept Gap, the 'Endowment' Effect, Subject Misconceptions, and Experimental Procedures for Eliciting Valuations," *AER*, vol. 95, no. 3 (June 2005), pp. 530–45.

192 *How different are stocks:* Brigitte C. Madrian and Dennis F. Shea, "The Power of Suggestion" (May 2000), www.nber.org/papers/w7682; Jeffrey R. Brown et al., "410(k) Matching Contributions in Company Stock" (April 2004), www.nber.org/papers/w10419; Henrik Cronqvist and Richard H. Thaler, "Design Choices in Privatized Social-Security Systems," *AER*, vol. 94, no. 2 (2004), pp. 424–28; Ted Martin Hedesstrom et al., "Identifying Heuristic Choice Rules in the Swedish Premium Pension Scheme," *JBF*, vol. 5, no. 1 (2004), pp. 32–42.

193 *In a psychology lab:* Maya Bar Hillel and Efrat Neter, "Why Are People Reluctant to Exchange Lottery Tickets?," *JPSP*, vol. 70, no. 1 (1996), pp. 17–27.

193 *Sometimes, investing inertia:* Olivia S. Mitchell et al., "The Inattentive Participant" (April 2006), http://ssrn.com/abstract=881854; William Samuelson and Richard Zeckhauser, "Status Quo Bias in Decision Making," *JRU*, vol. 1, no. 1 (1988), pp. 7–59; John Ameriks and Stephen P. Zeldes, "How Do Household Portfolio Shares Vary with Age?" (Sept. 2004), www2.gsb.columbia.edu/faculty/szeldes/Research/; JZ e-mail interview with John Ameriks, June 20, 2006.

194 *Imagine that you can choose:* Daniel Kahneman and Amos Tversky, "Prospect Theory," *Econometrica*, vol. 47, no. 2 (March 1979), pp. 263–92; David Romer, "Do Firms Maximize?" *JPE*, vol. 114, no. 2 (April 2006), pp. 340–65; Cass Sunstein and Richard H. Thaler, "Market Efficiency and Rationality," *Michigan Law Review*, vol. 102, no. 6 (May 2004), pp. 1390–1403; M. Keith Chen et al., "How Basic Are Behavioral Biases?" *JPE*, vol. 114, no. 3 (June 2006), pp. 517–37; www.som.yale.edu/faculty/keith.chen.

196 *What happens to investors:* This example (adjusted for inflation) is taken from Daniel Kahneman and Amos Tversky, "The Psychology of Preferences," *SA*, vol. 246 (1982), pp. 160–73, and Daniel Kahneman and Dale T. Miller, "Norm Theory," in *HAB*, pp. 348–366.

196 *Who feels worse:* JZ telephone interview with Thomas Gilovich, July 17, 2006; Daniel Kahneman, remarks at Oxford Programme on Investment Decision-Making, Saïd School of Business, Oxford, U.K. Oct. 22, 2004.

197 *Holding losers too long:* Hersh Shefrin and Meir Statman, "The Disposition to Sell Winners Too Early and Ride Losers Too Long," *JF,* vol. 40, no. 3 (July 1985), pp. 777–90; Erik R. Sirri and Peter Tufano, "Costly Search and Mutual Fund Flows," *JF,* vol. 53, no. 5 (Oct. 1998), pp. 1589–1622; David W. Harless and Steven P. Peterson, "Investor Behavior and the Persistence of Poorly-Performing Mutual Funds," *JEBO,* vol. 37 (1998), pp. 257–76; William N. Goetzmann and Nadav Peles, "Cognitive Dissonance and Mutual Fund Investors," *Journal of Financial Research,* vol. 20, no. 2 (Summer 1997), pp. 145–58; Martin Weber and Colin F. Camerer, "The Disposition Effect in Securities Trading," *JEBO,* vol. 33 (1998), pp. 167–84; Stephen P. Ferris et al., "Predicting Contemporary Volume with Historic Volume at Differential Price Levels," *JF,* vol. 43, no. 3 (July 1998), pp. 677–97; Mark Grinblatt and Matti Keloharju, "What Makes Investors Trade?" *JF,* vol. 56, no. 2 (April 2001), pp. 589–615; Terrance Odean, "Are Investors Reluctant to Realize Their Losses?" *JF,* vol. 53, no. 5 (Oct. 1998), pp. 1775–97; Ravi Dhar and Ning Zhu, "Up Close and Personal" (Aug. 2002), http://ssrn.com/abstract=302245; Li Jin and Anna Scherbina, "Inheriting Losers" (Feb. 2006), http://ssrn.com/abstract=895765; Gjergji Cici, "The Relation of the Disposition Effect to Mutual Fund Trades and Performance" (July 2005), http://ssrn.com/abstract=645841; Zur Shapira and Itzhak Venezia, "Patterns of Behavior of Professionally Managed and Independent Investors," *Journal of Banking & Finance,* vol. 25, no. 8 (Aug. 2001), pp. 1573–87; Karl E. Case and Robert J. Shiller, "The Behavior of Home Buyers in Boom and Post-Boom Markets" (Oct. 1988), www.nber.org/papers/w2748; David Genesove and Christopher Mayer, "Loss Aversion and Seller Behavior," *QJE,* vol. 116, no. 4 (Nov. 2001), pp. 1233–60.

198 *As Hamlet said:* William Shakespeare, *Hamlet,* 3.1.88–89; Justin Kruger et al., "Counterfactual Thinking and the First Instinct Fallacy," *JPSP,* vol. 88, no. 5 (2005), pp. 725–35; Dale T. Miller and Brian R. Taylor, "Counterfactual Thought, Regret, and Superstition," in Neal J. Roese and James M. Olson, eds., *What Might Have Been* (Mahwah, N.J.: Erlbaum, 1995), pp. 305–32; Orit Tykocinski et al., "Inaction Inertia in the Stock Market," *Journal of Applied Social Psychology,* vol. 34, no. 6 (2004), pp. 1166–75.

198 *You've just received:* Carrie M. Heilman et al., "Pleasant Surprises," *Journal of Marketing Research,* vol. 39, no. 2 (May 2002), pp. 242–52; Hal R. Arkes et al., "The Psychology of Windfall Gains," *OBHDP,* vol. 59 (1994), pp. 331–47; Pamela W. Henderson and Robert A. Peterson, "Mental Accounting and Categorization," *OBHDP,* vol. 51 (1992), pp. 92–117; Nicholas Epley and Ayelet Gneezy, "The Framing of Financial Windfalls and Implications for Public Policy," *Journal of Socio-Economics,* vol. 36, no. 1 (2007), pp. 36–47.

200 *For years:* Franklin mentions his asbestos purse in Chapter Five of his autobiography. It is described in James E. Alleman and Brooke T. Mossman, "Asbestos Revisited," *SA,* July 1997, pp. 70–75, and can be viewed at http://piclib.nhm.ac.uk/piclib/www/image.php?img=46575. On the sadness of lottery winners, see Philip Brickman et al., "Lottery Winners and Accident Victims," *JPSP,* vol. 36, no. 8 (1978), pp. 917–27.

200 *A windfall that you believe:* I was unable to reach Mr. X directly, so this story is based on interviews with two of his former colleagues, who requested anonymity.

201 *We all believe it:* A. Charles Catania and Terje Sagvolden, "Preference for Free Choice over Forced Choice in Pigeons," *JEAB,* vol. 34, no. 1 (1980), pp. 77–86; A. Charles Catania, "Freedom and Knowledge," *JEAB,* vol. 24, no. 1 (1975), pp. 89–106.

201 *In a classic experiment:* Sheena S. Iyengar and Mark R. Lepper, "When Choice Is Demotivating," *JPSP,* vol. 79, no. 6 (2000), pp. 995–1006; Barry Schwartz et al., "Maximizing versus Satisficing," *JPSP,* vol. 83, no. 5 (2002), pp. 1178–97; Sheena Sethi Iyengar et al., "How Much Choice Is Too Much?" in Olivia S. Mitchell and Stephen P. Utkus, *Pension Design and Structure* (Oxford, U.K.: Oxford University Press, 2004), pp. 83–95; Olivia S. Mitchell et al., "Turning Workers into Savers?" (Oct. 2005), www.nber.org/papers/w11726; Henrik Cronqvist and Richard H. Thaler, "Design Choices in Privatized Social-Security Systems," *AER,* vol. 94, no. 2 (2004), pp. 424–28; Jiwoong Sin and Dan Ariely, "Keeping Doors Open," *MS,* vol. 50, no. 5 (May 2004), pp. 575–86.

202 *True or false:* This example is adapted from Richard Thaler, "Toward a Positive Theory of Consumer Choice," *JEBO,* vol. 1 (1980), pp. 39–60.

202 *In Holland:* Marcel Zeelenberg and Rik Pieters, "Consequences of Regret Aversion in Real Life," *OBHDP,* vol. 93 (2004), pp. 155–68.

203 *In the 2006 Winter Olympics:* NBC television broadcast of the women's snowboardcross final, Feb. 17, 2006; Victoria Husted Medvec et al., "When Less Is More," *JPSP,* vol. 69, no. 4 (1995), pp. 603–10.

203 *An athlete:* Karl Halvor Teigen, "When the Unreal Is More Likely than the Real," *Thinking and Reasoning,* vol. 4, no. 2 (1998), pp. 147–77; Jonathan Parke and Mark Griffiths, "Gambling Addiction and the Evolution of the 'Near Miss,'" *Addiction Research and Theory,* vol. 12, no. 5 (2004), pp. 407–11; R. L. Reid, "The Psychology of the Near Miss," *Journal of Gambling Behavior,* vol. 2, no. 1 (1986), pp. 32–39; Daniel Kahneman and Carol A. Varey, "Propensities and Counterfactuals," *JPSP,* vol. 59, no. 6 (1990), pp. 1101–10; Michael J. A. Wohl and Michael E. Enzle, "The Effects of Near Wins and Near Losses on Self-Perceived Personal Luck and Subsequent Gambling Behavior," *JESP,* vol. 39 (2003), pp. 184–91.

204 *Students in Scranton:* Keith D. Markman and Philip E. Tetlock, "Accountability and Close-Call Counterfactuals," *PSPB,* vol. 26, no. 10 (Oct. 2000),

pp. 1213–24; Brad M. Barber et al., "Once Burned, Twice Shy" (Sept. 2004), http://ssrn.com/abstract=611267; JZ telephone interview with Terrance Odean, July 12, 2005.

204 *The human brain:* John Allman et al., "Two Phylogenetic Specializations in the Human Brain," *The Neuroscientist,* vol. 8, no. 4 (2002), pp. 335–46; Edmund T. Rolls, "The Orbitofrontal Cortex and Reward," *CC,* vol. 10 (March 2000), pp. 284–94; D. Ongur and J. L. Price, "The Organization of Networks Within the Orbital and Medial Prefrontal Cortex of Rats, Monkeys and Humans," *CC,* vol. 10 (March 2000), pp. 206–19; Antoine Bechara et al., "Emotion, Decision Making and the Orbitofrontal Cortex," *CC,* vol. 10 (March 2000), pp. 295–307; Jacqueline N. Wood and Jordan Grafman, "Human Prefrontal Cortex," *NRN,* vol. 4 (Feb. 2003), pp. 139–47; Daeyeol Lee, "Best to Go with What You Know?" *Nature,* vol. 441 (June 15, 2006), pp. 822–23; Nathaniel D. Daw et al., "Cortical Substrates for Exploratory Decisions in Humans," *Nature,* vol. 441 (June 15, 2006), pp. 876–79. The orbitofrontal cortex lies on the underside of the VMPFC.

205 *At least one region:* Katerina Semendeferi et al., "Prefrontal Cortex in Humans and Apes," *American Journal of Physical Anthropology,* vol. 114 (2001), pp. 224–41; Katerina Semendeferi et al., "Humans and Great Apes Share a Large Frontal Cortex," *NNS,* vol. 5, no. 3 (March 2002), pp. 272–76; Morten L. Kringelbach and Edmund T. Rolls, "The Functional Neuroanatomy of the Human Orbitofrontal Cortex," *Progress in Neurobiology,* vol. 72 (2004), pp. 341–72; Narender Ramnani and Adrian M. Owen, "Anterior Prefrontal Cortex," *NRN,* vol. 5 (March 2004), pp. 184–94.

205 *People with injuries:* Vinod Goel et al., "A Study of the Performance of Patients with Frontal Lobe Lesions in a Financial Planning Task," *Brain,* vol. 120 (1997), pp. 1805–22; JZ interview with Jordan Grafman, March 6, 2002; E. T. Rolls et al., "Emotion-Related Learning in Patients with Social and Emotional Changes Associated with Frontal Lobe Damage," *Journal of Neurology, Neurosurgery, and Psychiatry,* vol. 57 (1994), pp. 1518–24; Antoine Bechara, "Disturbances of Emotion Regulation after Focal Brain Lesions," *International Review of Neurobiology,* vol. 62 (2004), pp. 159–93; Antoine Bechara et al., "Deciding Advantageously Before Knowing the Advantageous Strategy," *Science,* vol. 275 (Feb. 28, 1997), pp. 1293–95; Antonio Damasio's *Descartes' Error* (New York: Avon, 1995) describes "Elliot," who became incapable of long-term planning after a tumor damaged his VMPFC, and the 19th-century railroad worker Phineas Gage, who became impetuous and impulsive after an iron rod shot through his head in a blasting accident.

206 *Neuroeconomics is now explaining:* Jean-Claude Dreher et al., "Neural Coding of Distinct Statistical Properties of Reward Information in Humans," *CC,* vol. 16, no. 4 (2006), pp. 561–73; Hackjin Kim et al., "Is Avoiding an Aversive Outcome Rewarding?" *PloS Biology,* vol. 4, no. 8 (Aug. 2006); JZ e-mail interview with John O'Doherty, July 7, 2006.

207 *In Bron:* Nathalie Camille et al., "The Involvement of the Orbitofrontal Cortex in the Experience of Regret," *Science,* vol. 304 (May 21, 2004), pp. 1167–70; David M. Eagleman, "Comment on 'The Involvement of the Orbitofrontal Cortex in the Experience of Regret'" and Giorgio Coricelli et al., "Response to Comment on 'The Involvement of the Orbitofrontal Cortex in the Experience of Regret,'" *Science,* vol. 308 (May 27, 2005), pp. 1260b-c; Giorgio Coricelli et al., "Regret and Its Avoidance," *NNS,* vol. 8, no. 9 (Sept. 2005), pp. 1255–62.

208 *Thus these areas:* Lesley K. Fellows, "Deciding How to Decide," *Brain,* vol. 129, no. 4 (2006), pp. 944–52.

208 *On June 13:* e-mail to JZ from sender who requested anonymity, June 13, 2006; JZ, "Murphy Was an Investor," *MM,* July 2002, pp. 61–62; JZ interview with Matthews, March 12, 2002; Angus Maddison, *The World Economy* (Paris: OECD, 2001), p. 262; William J. Bernstein and Robert D. Arnott, "Earnings Growth: The 2% Dilution," *FAJ* (Sept.–Oct. 2003), pp. 47–55; William J. Bernstein, *The Birth of Plenty* (New York: McGraw-Hill, 2004), pp. 17–27; Warren Buffett, chairman's letter, Berkshire Hathaway annual report, 1997, www.berkshirehathaway.com/letters/1997.html; Peter L. Bernstein, *Against the Gods* (New York: John Wiley, 1996), pp. 167–86.

212 *The regret you feel:* Donald A. Redelmeier and Robert J. Tibshirani, "Why Cars in the Next Lane Seem to Go Faster," *Nature,* vol. 401 (Sept. 2, 1999), pp. 35–36; Donald A. Redelmeier and Robert J. Tibshirani, "Are Those Other Drivers Really Going Faster?" *Chance,* vol. 13, no. 3 (2000), pp. 8–14; Dale T. Miller and Brian R. Taylor, "Counterfactual Thought, Regret, and Superstition," in Neal J. Roese and James M. Olson, eds., *What Might Have Been* (Mahwah, N.J.: Erlbaum, 1995), pp. 305–32.

213 *Why is it so hard:* John M. Allman et al., "Intuition and Autism," *TICS,* vol. 9, no. 8 (Aug. 2005), pp. 367–73; R. A. Borman et al., "5-HT2B Receptors Play a Key Role in Mediating the Excitatory Effects of 5-HT in Human Colon in Vitro," *British Journal of Pharmacology,* vol. 135 (2002), pp. 1144–51; Eduardo P. M. Vianna et al., "Increased Feelings with Increased Body Signals," *Social, Cognitive, and Affective Neuroscience,* vol. 1, no. 1 (2006), pp. 37–48. The special molecule is a serotonin 2b receptor.

213 *The abundance:* Andrew J. Calder et al., "Neuropsychology of Fear and Loathing," *NRN,* vol. 2 (May 2001), pp. 352–63; K. Luan Phan et al., "Functional Neuroanatomy of Emotion," *NeuroImage,* vol. 16 (2002), pp. 331–48; A. D. Craig, "How Do You Feel?," *NRN,* vol. 3 (Aug. 2002), pp. 655–66; Antoine Bechara and Nasir Naqvi, "Listening to Your Heart," *NNS,* vol. 7, no. 2 (Feb. 2004), pp. 102–3; Hugo D. Critchley et al., "Neural Systems Supporting Interoceptive Awareness," *NNS,* vol. 7, no. 2 (Feb. 2004), pp. 189–95.

213 *Despite its name:* S. Dupont et al., "Functional Anatomy of the Insula," *Surgical and Radiologic Anatomy,* vol. 25 (2003), pp. 113–19; Paul W. Glimcher

and Brian Lau, "Rethinking the Thalamus," *NNS,* vol. 8, no. 8 (Aug. 2005), pp. 983–84; Takafumi Minamimoto et al., "Complementary Process to Response Bias in the Centromedian Nucleus of the Thalamus," *Science,* vol. 308 (June 17, 2005), pp. 1798–1801.

214 *You don't need:* M. L. Phillips et al., "A Specific Neural Substrate for Perceiving Facial Expressions of Disgust," *Nature,* vol. 389 (Oct. 2, 1987), pp. 495–98; Andrew J. Calder et al., "Neuropsychology of Fear and Loathing," *NRN,* vol. 2 (May 2001), pp. 352–63; Andrew J. Calder et al., "Impaired Recognition and Experience of Disgust Following Brain Injury," *NNS,* vol. 3, no. 11 (Nov. 2000), pp. 1077–78; Bruno Wicker et al., "Both of Us Disgusted in My Insula," *Neuron,* vol. 40 (Oct. 30, 2003), pp. 655–64; Pierre Krolak-Salmon et al., "An Attention Modulated Response to Disgust in Human Ventral Anterior Insula," *Annals of Neurology,* vol. 53, no. 4 (April 2003), pp. 446–53; www.open2.net/humanmind/article_faces_2.htm.

214 *There's something else:* Martin P. Paulus et al., "Increased Activation in the Right Insula during Risk-Taking Decision Making Is Related to Harm Avoidance and Neuroticism," *NeuroImage,* vol. 19 (2003), pp. 1439–48.

214 New research also shows: Brian Knutson et al., "Neural Predictors of Purchases," *Neuron,* vol. 53, no. 1 (January 4, 2007), pp. 1–10.

214 *I felt my own:* JZ participated in Huettel's experiment at the Brain Imaging and Analysis Center, Duke University, June 22, 2004. (There was also a large, but less intense, area of activation in my dorsolateral prefrontal cortex.)

215 *In another experiment:* Baba Shiv et al., "Investment Behavior and the Negative Side of Emotion," *PS,* vol. 16, no. 6 (2005), pp. 435–39; Tetsuo Koyama et al., "The Subjective Experience of Pain," *PNAS,* vol. 102, no. 36 (Sept. 6, 2005), pp. 12950–55; George Loewenstein, "The Pleasures and Pains of Information," *Science,* vol. 312 (May 5, 2006), pp. 704–6; Gregory S. Berns, "Neurobiological Substrates of Dread," *Science,* vol. 312 (May 5, 2006), pp. 754–58.

215 *At Stanford University:* Camelia M. Kuhnen and Brian Knutson, "The Neural Basis of Financial Risk Taking," *Neuron,* vol. 47 (Sept. 1, 2005), pp. 763–70; JZ e-mail interview with Knutson, June 19, 2006; Jennifer S. Lerner et al., "Heart Strings and Purse Strings," *PS,* vol. 15, no. 5 (2004), pp. 337–41; JZ e-mail interview with Lerner, June 20, 2006.

216 *When investors are disgusted:* JZ e-mail interviews with Wayne Wagner, former chairman, Plexus Group, now an independent consultant to ITG Solutions Network, April 7 and 11, 2006; Ian Domowitz and Benn Steil, "Automation, Trading Costs, and the Structure of the Securities Trading Industry," *Brookings-Wharton Papers on Financial Services* (Washington, D.C.: Brookings Institution, 1999), pp. 33–81; Donald B. Keim and Ananth Madhavan, "The Cost of Institutional Equity Trades," *FAJ* (July/Aug. 1998), pp. 50–69.

217 *Imagine that you buy:* Nina Hattiangadi, "Failing to Act," *International Journal of Aging and Human Development,* vol. 40, no. 3 (1995), pp. 175–85; Thomas Gilovich and Victoria Husted Medvec, "The Temporal Pattern to the Experience of Regret," *JPSP,* vol. 67, no. 3 (1994), pp. 357–65, and "The Experience of Regret," *PR,* vol. 102, no. 2 (1995), pp. 379–95; Thomas Gilovich et al., "Varieties of Regret," *PR,* vol. 105, no. 3 (1998), pp. 602–5; JZ telephone interview with Thomas Gilovich, July 17, 2006.

217 *As memories decay:* Ilana Ritov, "The Effect of Time on Pleasure with Chosen Outcomes," *JBDM,* vol. 19 (2006), pp. 177–90; Neal J. Roese and Amy Summerville, "What We Regret Most . . . and Why," *PSPB,* vol. 31, no. 9 (Sept. 2005), pp. 1273–85; Neal J. Roese, "Twisted Pair," *BHJDM,* pp. 258–73; Daniel Kahneman, "Varieties of Counterfactual Thinking," in Neal J. Roese and James M. Olson, eds., *What Might Have Been* (Mahwah, N.J.: Erlbaum, 1995), pp. 375–96; Neal Roese, *If Only* (New York: Broadway, 2005).

218 *As time goes by:* Suzanne O'Curry Fogel and Thomas Berry, "The Disposition Effect and Individual Investor Decisions," *JBF,* vol. 7, no. 2 (2006), pp. 107–16.

218 *It might not quite be true:* Daniel T. Gilbert et al., "Looking Forward to Looking Backward," *PS,* vol. 15, no. 5 (2004), pp. 346–50; JZ telephone interview with Thomas Gilovich, July 17, 2006.

219 *Face it:* JZ telephone interview with Dan Robertson, Feb. 17, 2005; fax to JZ from Robertson and Steve Schullo, Feb. 21, 2005.

220 *Have rules:* Terry Connolly and Marcel Zeelenberg, "Regret in Decision Making," *CDPS,* vol. 11, no. 6 (Dec. 2002), pp. 212–16; JZ e-mail interview with Thomas Gilovich, July 20, 2006.

221 *Seek the silver lining:* JZ telephone interview with Terrance Odean, July 12, 2005; Suzanne O'Curry Fogel and Thomas Berry, "The Disposition Effect and Individual Investor Decisions," *JBF,* vol. 7, no. 2 (2006), pp. 107–16; "Applying Behavioral Finance to Value Investing," presentation by Whitney Tilson (Nov. 2005), www.tilsonfunds.com/TilsonBehavioralFinance.pdf; Gretchen B. Chapman, "Similarity and Reluctance to Trade," *JBDM,* vol. 11 (1998), pp. 47–58.

222 *Cut your losses:* JZ, "How Losing Less Can Cost You More," *MM,* June 2005, p. 70.

223 *Put inertia:* JZ telephone interview with Michael Hadley, assistant counsel for pension regulation, Investment Company Institute, July 18, 2006.

224 *Change your frames:* Richard Zeckhauser, remarks at Oxford Programme on Investment Decision-Making, Saïd School of Business, Oxford, U.K. Oct. 22, 2004; JZ e-mail interview with Zeckhauser, July 20, 2006; Neal Roese, *If Only* (New York: Broadway, 2005), pp. 201–2.

225 *What you don't know:* JZ e-mail interview with Elke Weber, July 20, 2006; Vanguard data kindly provided by John Woerth, Vanguard public relations, July 24, 2006.

225 *Keep your balance:* Oded Braverman et al., "The (Bad?) Timing of Mutual Fund Investors," http://ssrn.com/abstract=795146; JZ e-mail interview with Shmuel Kandel and Avi Wohl, July 26, 2006; Pascal, *Pensées,* no. 139; Seth J. Masters, "Rebalancing," *JPM* (Spring 2003), pp. 52–57; Yesim Tokat, "Portfolio Rebalancing in Theory and Practice," Vanguard Investment Counseling & Research, report no. 31 (2006); Robert D. Arnott and Robert M. Lovell Jr., "Rebalancing," First Quadrant Corp. Monograph No. 3 (1992); Mark Riepe and Bill Swerbenski, "Rebalancing for Tax-Deferred Accounts," *JFP* (April 2006), pp. 40–44; Efficient Frontier, Sept. 1996, Jan. 1997, July 1997, www.efficientfrontier.com; JZ e-mails with William Bernstein, Jan. 23, 2004, and July 21, 2006; JZ telephone interview with Thomas Gilovich, July 17, 2006; http://therightmix.alliancebernstein.com/CmsObjectTRM/PDF/PressRelease_051102_INV.pdf; detailed survey results provided by Tiller LLC; JZ e-mail interview with John Ameriks, June 20, 2006.

10. HAPPINESS

228 *There is nothing, Sir:* James Boswell, *The Life of Samuel Johnson* (New York: Everyman's Library, 1992), p. 273.

229 *As the philosopher:* Ed Diener and Carol Diener, "Most People Are Happy," *PS,* vol. 7, no. 3 (May 1996), pp. 181–85; David G. Myers and Ed Diener, "Who Is Happy?" *PS,* vol. 6, no. 1 (Jan. 1995), pp. 10–19; David G. Myers, "The Funds, Friends, and Faith of Happy People," *AP,* vol. 55, no. 1 (Jan. 2000), pp. 56–67; PNC Advisors, "Wealth and Values Survey Findings" (2005), p. 5; David Futrelle, "Can Money Buy Happiness?" *MM,* Aug. 2006, pp. 127–31; Arthur Schopenhauer, *The Wisdom of Life* (Amherst, N.Y.: Prometheus Books, 1995), p. 45. "Money (That's What I Want)" was originally written by Berry Gordy Jr. and Janie Bradford and recorded by Barrett Strong.

229 *In an affluent society:* Robert Sapolsky, "Sick of Poverty," *SA* (Dec. 2005), pp. 93–99; Rosemarie Kobau et al., "Sad, Blue, or Depressed Days, Health Behaviors and Health-Related Quality of Life," *Health and Quality of Life Outcomes,* vol. 2, no. 40 (2004), www.hqlo.com/content/2/1/40; Maria J. Silveira et al., "Net Worth Predicts Symptom Burden at the End of Life," *Journal of Palliative Medicine,* vol. 8, no. 4 (2005), pp. 827–37; Andrew J. Tomarken et al., "Resting Frontal Brain Activity Linkages to Maternal Depression and Socio-economic Status Among Adolescents," *BP,* vol. 67 (2004), pp. 77–102; Peggy McDonough et al., "Income Dynamics and Adult Mortality in the United States, 1972 through 1989," *American Journal of Public Health,* vol. 87, no. 9 (Sept. 1997), pp. 1476–83; Janis L. Dickinson and Andrew McGowan, "Winter Resource Wealth Drives Delayed Dispersal and Family-Group Living in Western Bluebirds," *PRSLB,* vol. 272, no. 1579 (Nov. 22, 2005), pp. 2423–28; JZ e-mail interview with Dickinson, Nov. 9, 2005.

230 *But are the rich a lot happier:* Ed Diener et al., "Happiness of the Very Wealthy," *SIR,* vol. 16, no. 3 (April 1985), pp. 263–74; Ed Diener and Martin E. P. Seligman, "Beyond Money," *Psychological Science in the Public Interest,* vol. 5, no. 1 (2004), pp. 1–31; JZ e-mail interview with Ed Diener, Aug. 15, 2006.

231 *If you had passed through Pittsburgh:* George Loewenstein and Shane Frederick, "Predicting Reactions to Environmental Change," in Max Bazerman et al., *Psychological Perspectives on Ethics and the Environment* (San Francisco: New Lexington Press, 1997), pp. 52–72.

231 *In shock after the assassination:* Arthur M. Schlesinger Jr., *A Thousand Days* (New York: Mariner, 2002), p. 1028.

232 *Something strange happens:* Daniel Kahneman, "Objective Happiness," in *W-B,* pp. 3–25.

232 *Being is very different from becoming:* Philip Brickman et al., "Lottery Winners and Accident Victims," *JPSP,* vol. 36, no. 8 (1978), pp. 917–27; www.nefe.org/fple/windfallpage1.html; Beverly Keel, "A Dollar and a Dreamer," *New York,* Dec. 16, 2002, p. 18.

233 *Thus, although no one:* Richard Stensman, "Severely Mobility-Disabled People Assess the Quality of Their Lives," *Scandinavian Journal of Rehabilitation Medicine,* vol. 17, no. 2 (1985), pp. 87–99; Gale G. Whiteneck et al., "Mortality, Morbidity, and Psychosocial Outcomes of Persons Spinal Cord Injured More than 20 Years Ago," *Paraplegia,* vol. 30, no. 9 (Sept. 1992), pp. 617–30; Kenneth A. Gerhart et al., "Quality of Life Following Spinal Cord Injury," *Annals of Emergency Medicine,* vol. 23, no. 4 (April 1994), pp. 807–12; John R. Bach and Margaret C. Tilton, "Life Satisfaction and Well-Being Measures in Ventilator Assisted Individuals with Traumatic Tetraplegia," *APMR,* vol. 75, no. 6 (June 1994), pp. 626–32; Lawrence C. Vogel et al., "Long-Term Outcomes and Life Satisfaction of Adults Who Had Pediatric Spinal Cord Injuries," *APMR,* vol. 79, no. 12 (Dec. 1998), pp. 1496–1503; Marcel P. J. M. Dijkers, "Correlates of Life Satisfaction among Persons with Spinal Cord Injury," *APMR,* vol. 80, no. 8 (Aug. 1999), pp. 867–76; Karyl M. Hall et al., "Follow-up Study of Individuals with High Tetraplegia," *APMR,* vol. 80, no. 11 (Nov. 1999), pp. 1507–13.

234 *Jack Hurst:* JZ, "The Soul of an Investor," *MM,* March 2005, pp. 66–71.

234 *Even though our predictions:* JZ e-mail interview with Diener, Aug. 15, 2006.

235 *When we imagine the future:* Daniel T. Gilbert and Timothy D. Wilson, "Miswanting," in Joseph P. Forgas, ed., *Thinking and Feeling* (Cambridge, U.K.: Cambridge University Press, 2000), pp. 178–97; Timothy D. Wilson et al., "Focalism," *JPSP,* vol. 78, no. 5 (May 2000), pp. 821–36; Leaf Van Boven et al., "The Illusion of Courage in Social Predictions," *OBHDP,* vol. 96 (2005), pp. 130–41.

236 *We base our goals:* George Loewenstein and David Schkade, "Wouldn't It Be Nice?" in *W-B,* pp. 3–25; JZ telephone interview with Daniel Gilbert,

June 17, 2004; see also Gilbert's brilliant book, *Stumbling on Happiness* (New York: Knopf, 2006). "There are two tragedies": George Bernard Shaw, *Man and Superman,* 4:369 (Shaw borrowed, and polished, this line from Oscar Wilde).

237 *If it seems strange:* Derrick Wirtz et al., "What to Do on Spring Break?" *PS,* vol. 14, no. 5 (Sept. 2003), pp. 520–24; Terence R. Mitchell et al., "Temporal Adjustments in the Evaluation of Events," *JESP,* vol. 33 (1997), pp. 421–48; Robert I. Sutton, "Feelings about a Disneyland Visit," *Journal of Management Inquiry,* vol. 1, no. 4 (Dec. 1992), pp. 278–87; JZ e-mail interview with Sutton, Aug. 14, 2006; Dorothy Field, "Retrospective Reports by Healthy Intelligent Elderly People of Personal Events of Their Adult Lives," *International Journal of Behavioral Development,* vol. 4 (1981), pp. 77–97; Donald A. Redelmeier et al., "Memories of Colonoscopy," *Pain,* vol. 104 (2003), pp. 187–94; Daniel Kahneman et al., "Back to Bentham?" *QJE,* May 1997, pp. 375–404; Barbara L. Fredrickson, "Extracting Meaning from Past Affective Experiences," *C&E,* vol. 14, no. 4 (2000), pp. 577–606.

238 *Your memories, then:* Michael Ross, "Relation of Implicit Theories to the Construction of Personal Histories," *PR,* vol. 96, no. 2 (1989), pp. 341–57; Chu Kim-Prieto et al., "Integrating the Diverse Definitions of Happiness," *JOHS,* vol. 6 (2005), pp. 261–300; JZ e-mail interview with Justin Kruger, Aug. 24, 2006.

239 *We've seen that people:* Fritz Strack et al., "Priming and Communication," *European Journal of Social Psychology,* vol. 18 (1988), pp. 429–42; Norbert Schwarz et al., "Assimilation and Contrast Effects in Part-Whole Question Sequences," *Public Opinion Quarterly,* vol. 55 (1991), pp. 3–23; Norbert Schwarz and Fritz Strack, "Reports of Subjective Well-Being," in *W-B,* pp. 61–84; Daniel Kahneman et al., "A Survey Method for Characterizing Daily Life Experience," *Science,* vol. 306 (Dec. 3, 2004), pp. 1776–80; Shigehiro Oishi and Helen W. Sullivan, "The Predictive Value of Daily vs. Retrospective Well-Being Judgments in Relationship Stability," *JESP,* vol. 42 (2006), pp. 460–70.

240 *Measuring happiness, it seems:* Ed Diener and Martin E. P. Seligman, "Very Happy People," *PS,* vol. 13, no. 1 (Jan. 2002), pp. 81–84; David G. Myers, "The Funds, Friends, and Faith of Happy People," *AP,* vol. 55, no. 1 (Jan. 2000), pp. 56–67; William Pavot et al., "Extraversion and Happiness," *Personality and Individual Differences,* vol. 11, no. 12 (1990), pp. 1299–1306; Daniel Kahneman et al., "A Survey Method for Characterizing Daily Life Experience," *Science,* vol. 306 (Dec. 3, 2004), pp. 1776–80.

240 *We also know that when people are happy:* Richard J. Davidson and Nathan A. Fox, "Frontal Brain Asymmetry Predicts Infants' Response to Maternal Separation," *Journal of Abnormal Psychology,* vol. 98, no. 2 (1989), pp. 127–31; Richard J. Davidson, "Well-Being and Affective Style," *PTRSLB,* vol. 359 (2004), pp. 1395–1411; Robert E. Wheeler et al., "Frontal Brain

Asymmetry and Emotional Reactivity," *Psychophysiology,* vol. 30 (1993), pp. 82–89; Diego A. Pizzagalli et al., "Frontal Brain Asymmetry and Reward Responsiveness," *PS,* vol. 16, no. 10 (2005), pp. 805–13; Antoine Lutz et al., "Long-Term Meditators Self-Induce High-Amplitude Gamma Synchrony during Mental Practice," *PNAS,* vol. 101, no. 46 (Nov. 16, 2004), pp. 16369–73; Heather L. Urry et al., "Making a Life Worth Living," *PS,* vol. 15 (2004), no. 6, pp. 367–72.

241 *Knowing that money is a means:* Tim Kasser, *The High Price of Materialism* (Cambridge, Mass.: MIT Press, 2000), pp. 30–32; Tim Kasser et al., "The Relations of Maternal and Social Environments to Late Adolescents' Materialistic and Prosocial Values," *Developmental Psychology,* vol. 31, no. 6 (1995), pp. 907–14; Tim Kasser and Richard M. Ryan, "A Dark Side of the American Dream," *JPSP,* vol. 65, no. 2 (1993), pp. 410–22; Carol Nickerson et al., "Zeroing in on the Dark Side of the American Dream," *PS,* vol. 14, no. 6 (Nov. 2003), pp. 531–36; M. Joseph Sirgy, "Materialism and the Quality of Life," *SIR,* vol. 43 (1998), pp. 227–60; Ed Diener and Robert Biswas-Diener, "Will Money Increase Subjective Well-Being?" *SIR,* vol. 57 (2002), pp. 119–69; Daniel Kahneman et al., "Would You Be Happier If You Were Richer?" *Science,* vol. 312 (June 30, 2006), pp. 1908–10.

242 *To envy is human:* Robert M. Sapolsky, "The Influence of Social Hierarchy on Primate Health," *Science,* vol. 308 (April 29, 2005), pp. 648–52; Olivier Berton et al., "Essential Role of BDNF in the Mesolimbic Dopamine Pathway in Social Defeat Stress," *Science,* vol. 311 (Feb. 10, 2006), pp. 864–68; Sabrina S. Burmeister et al., "Rapid Behavioral and Genomic Responses to Social Opportunity," *PloS Biology,* vol. 3, no. 11 (Nov. 2005), pp. 1996–2004; Helen E. Fox et al., "Stress and Dominance in a Social Fish," *JN,* vol. 17, no. 16 (Aug. 15, 1997), pp. 6463–69; D. Caroline Blanchard et al., "Subordination Stress," *Behavioural Brain Research,* vol. 58 (1993), pp. 113–21; Robert O. Deaner et al., "Monkeys Pay Per View," *CB,* vol. 15 (March 29, 2005), pp. 543–48; Drake Morgan et al., "Social Dominance in Monkeys," *NNS,* vol. 5, no. 2 (Feb. 2002), pp. 169–74.

242 *Our modern minds:* Susanne Erk et al., "Cultural Objects Modulate Reward Circuitry," *NeuroReport,* vol. 13 (2002), pp. 2499–2503.

243 *Even if you think:* Michael R. Hagerty, "Social Comparison of Income in One's Community," *JPSP,* vol. 78 (2000), pp. 746–71; Alois Stutzer, "The Role of Income Aspirations in Individual Happiness," *JEBO,* vol. 54 (2004), pp. 89–109; Andrew E. Clark and Andrew J. Oswald, "Satisfaction and Comparison Income," *Journal of Public Economics,* vol. 61 (1996), pp. 359–81; Ed Diener et al., "Factors Predicting the Subjective Well-Being of Nations," *JPSP,* vol. 69 (1995), pp. 851–64; Michael Argyle, "Causes and Correlates of Happiness," in *W-B,* pp. 353–73; H. L. Mencken, "Masculum et Feminam Creavit Eos," in *A Mencken Chrestomathy* (New York: Knopf, 1978), p. 619; David Neumark and Andrew Postlewaite, "Relative Income

Concerns and the Rise in Married Women's Employment," *Journal of Public Economics,* vol. 70 (1998), pp. 157–83; Michael McBride, "Relative-Income Effects on Subjective Well-Being in the Cross-Section," *JEBO,* vol. 45 (2001), pp. 251–78 (the finding about a brother-in-law's income holds when both sisters are nonworking); "Affluent Americans and Their Money," survey for *MM* by RoperASW (2002), p. 46.

245 *Neuroscientist Richard Davidson:* Richard J. Davidson, "Affective Style, Psychopathology, and Resilience," *AP,* vol. 55, no. 11 (Nov. 2000), pp. 1196–1214; Melissa A. Rosenkranz et al., "Affective Style and in vivo Immune Response," *PNAS,* vol. 100, no. 19 (Sept. 16, 2003), pp. 11148–52; Richard J. Davidson et al., "Alterations in Brain and Immune Function Produced by Mindfulness Meditation," *Psychosomatic Medicine,* vol. 65 (2003), pp. 564–70; Carol D. Ryff et al., "Positive Health," *PTRSLB,* vol. 359 (2004), pp. 1383–94; Andrew Steptoe et al., "Positive Affect and Health-Related Neuroendocrine, Cardiovascular, and Inflammatory Processes," *PNAS,* vol. 102, no. 18 (May 3, 2005), pp. 6508–12; Erik J. Giltay et al., "Dispositional Optimism and All-Cause and Cardiovascular Mortality in a Prospective Cohort of Elderly Dutch Men and Women," *Archives of General Psychiatry,* vol. 61 (Nov. 2004), pp. 1126–35; Carol D. Ryff et al., "Psychological Well-Being and Ill-Being," *Psychotherapy and Psychosomatics,* vol. 75 (2006), pp. 85–95.

245 *In 1976:* Alice M. Isen, "Positive Affect and Decision Making," in Michael Lewis and Jeannette M. Haviland-Jones, *Handbook of Emotions* (New York: Guilford, 2004), pp. 417–35; Barbara L. Fredrickson, "The Broaden-and-Build Theory of Positive Emotions," *PTRSLB,* vol. 359 (2004), pp. 1367–77; Sonja Lyubomirsky et al., "The Benefits of Frequent Positive Affect," *PB,* vol. 131, no. 6 (2005), pp. 803–55; Ed Diener et al., "Dispositional Affect and Job Outcomes," *SIR,* vol. 59 (2002), pp. 229–59; Carol Graham et al., "Does Happiness Pay?" *JEBO,* vol. 55 (2004), pp. 319–42; Barry M. Staw et al., "Employee Positive Emotion and Favorable Outcomes at the Workplace," *Organization Science,* vol. 5 (1994), pp. 51–71; Andrew W. Lo, "Fear and Greed in Financial Markets," *AER,* vol. 95, no. 2 (2005), pp. 352–59.

246 *On a lovely morning:* JZ telephone interview with Barnett Helzberg, June 25, 2003; Bob Woodward, "How Mark Felt Became 'Deep Throat,'" *Washington Post,* June 2, 2005, p. A1; JZ telephone interview with Richard Wiseman, June 16, 2003; JZ, "Are You Lucky?" *MM,* Aug. 2003, p. 85; Richard Wiseman, "The Luck Factor," *Skeptical Inquirer,* May/June 2003, pp. 26–30, and *The Luck Factor* (New York: Miramax Books, 2003); www.luckfactor.co.uk.

247 *They look outward:* Robert K. Merton and Elinor Barber, *The Travels and Adventures of Serendipity* (Princeton, N.J.: Princeton University Press, 2004); http://members.aol.com/spencerlab/history/readdig.htm; www.raytheon.com/about/history/leadership/index.html; www.ieee-virtual-museum.org/collection/tech.php?id=2345891&lid=1.

249 *Would you rather get:* George Ainslie, *Picoeconomics* (Cambridge, U.K.: Cambridge University Press, 1992) and *Breakdown of Will* (Cambridge, U.K.: Cambridge University Press, 2000); Shane Frederick et al., "Time Discounting and Time Preference," in *TAD,* pp. 13–86; Richard Thaler and George Loewenstein, "Intertemporal Choice," in Richard H. Thaler, *The Winner's Curse* (Princeton, N.J.: Princeton University Press, 1992), pp. 92–106; Thomas C. Schelling, *Choice and Consequence* (Cambridge, Mass.: Harvard University Press, 1984), pp. 57–82.

250 *Getting rich quick:* Alex Kacelnik, "The Evolution of Patience," in *TAD,* pp. 115–38; Alex Kacelnik and Melissa Bateson, "Risky Theories," *American Zoologist,* vol. 36, no. 4 (Sept. 1996), pp. 402–34; Leonard Green and Joel Myerson, "Exponential versus Hyperbolic Discounting of Delayed Outcomes," *American Zoologist,* vol. 36, no. 4 (Sept. 1996), pp. 496–505; Joel Myerson and Leonard Green, "Discounting of Delayed Rewards," *JEAB,* vol. 64, no. 3 (Nov. 1995), pp. 263–76.

250 *One way or another: EBRI Issue Brief,* Jan. 2006, www.ebri.org/pdf/briefspdf/EBRI_IB_01-20061.pdf; Ronald T. Wilcox, "Bargain Hunting or Star Gazing," *JB,* vol. 76, no. 4 (2003), pp. 645–63; Edward S. O'Neal, "Mutual Fund Share Classes and Broker Incentives," *FAJ,* Sept.–Oct. 1999, pp. 76–87; Lawrence M. Ausubel, "The Failure of Competition in the Credit Card Market," *AER,* vol. 81, no. 1 (1991), pp. 50–81; Jonathan Clements, "Why It Pays to Delay," *WSJ,* Apr. 23, 2003, p. D1; www.tiaa-crefinstitute.org/research/papers/070102.html; Stefano Della Vigna and Ulrike Malmendier, "Contract Design and Self-Control," *QJE,* vol. 119, no. 2 (May 2004), pp. 353–402; John T. Warner and Saul Pleeter, "The Personal Discount Rate," *AER,* vol. 91, no. 1 (March 2001), pp. 33–53; Shlomo Benartzi and Richard H. Thaler, "Risk Aversion or Myopia?," *MS,* vol. 45, no. 3 (March 1999), pp. 364–81.

251 *A research team:* George Ainslie and John Monterosso, "A Marketplace in the Brain," *Science,* vol. 306 (Oct. 15, 2004), pp. 421–23; Samuel M. McClure et al., "Separate Neural Systems Value Immediate and Delayed Monetary Rewards," *Science,* vol. 306 (Oct. 15, 2004), pp. 503–7; JZ interview with Jonathan Cohen and Samuel McClure, Nov. 2004.

252 *Long ago, psychologist:* Walter Mischel et al., "Delay of Gratification in Children," *Science,* vol. 244 (May 26, 1989), pp. 933–38; Walter Mischel et al., "Sustaining Delay of Gratification over Time," *TAD,* pp. 175–200; John Ameriks et al., "Measuring Self-Control" (Feb. 2004), www.nber.org/papers/W10514; JZ, "Tie Me Down and Make Me Rich," *MM,* May 2004, p. 119; JZ telephone interview with Andrew Caplin, March 16, 2004.

253 *"I'll do it later":* James J. Choi et al., "Saving for Retirement on the Path of Least Resistance" (July 2004), http://post.economics.harvard.edu/faculty/laibson/papers/savingretirement.pdf; JZ telephone interviews with David Laibson, Nov. 11, 2005, and Joseph Ferrari, Nov. 15, 2005; JZ, "Do It Now," *MM,* Jan. 2006, pp. 80–81.

255 *Since old age seems:* Heather P. Lacey et al., "Hope I Die Before I Get Old," *JOHS,* vol. 7 (2006), pp. 167–82; JZ e-mail interview with Lacey, Aug. 15, 2006.

255 *Learning from your accumulated:* A. Rösler et al., "Effects of Arousing Emotional Scenes on the Distribution of Visuospatial Attention," *Journal of the Neurological Sciences* (2005), pp. 109–16; Mara Mather et al., "The Allure of the Alignable," *JEP,* vol. 134, no. 1 (2005), pp. 38–51; Susan Turk Charles et al., "Aging and Emotional Memory," *JEP,* vol. 132, no. 2 (2003), pp. 310–24; Mara Mather, "Why Memories May Become More Positive as People Age," in Bob Uttl et al., *Memory and Emotion* (Malden, Mass.: Blackwell, 2006), pp. 135–58.

256 *Neuroscientists believe:* Mara Mather et al., "Amygdala Responses to Emotionally Valenced Stimuli in Older and Younger Adults," *PS,* vol. 15, no. 4 (2004), pp. 259–63; Leanne M. Williams, "The Mellow Years?" *JN,* vol. 26, no. 24 (June 14, 2006), pp. 6422–30; Laura L. Carstensen et al., "Aging and the Intersection of Cognition, Motivation, and Emotion," in James E. Birren and K. Warner Schaie, eds., *Handbook of the Psychology of Aging* (San Diego: Academic Press, 2006), pp. 343–62. It is possible that the amygdala's ability to process negative emotions may decay with age, leaving positive processing intact; see Faith M. Gunning-Dixon et al., "Age-Related Differences in Brain Activation during Emotional Face Processing," *Neurobiology of Aging,* vol. 24 (2003), pp. 285–95.

256 *The aging brain:* Johnny Mercer and Harold Arlen, "Ac-Cent-Tchu-Ate the Positive" (1944).

257 *Once you learn the truth:* Hans-Ulrich Wittchen et al., "Lifetime Risk of Depression," *British Journal of Psychiatry,* vol. 26, supplement (Dec. 1994), pp. 16–22; Susan T. Charles and Laura L. Carstensen, "A Life Span View of Emotional Functioning in Adulthood and Old Age," *Advances in Cell Aging and Gerontology,* vol. 15 (2003), pp. 133–62; John F. Helliwell and Robert D. Putnam, "The Social Context of Well-Being," *PTRSLB,* vol. 359 (2004), pp. 1435–46; *BusinessWeek,* Aug. 13, 1979, pp. 54–59. I am grateful to William Bernstein for calling the "old fogies" line to my attention.

257 *All in all:* Robert Browning, "Rabbi Ben Ezra," lines 1–3, from *Dramatis Personae* (1855).

257 *The most powerful:* "You don't have to be rich to be happy" is a favorite phrase of my colleague Jean Chatzky of *Money* magazine (see her book *You Don't Have to Be Rich* [New York: Portfolio, 2003]); Fred B. Bryant et al., "Using the Past to Enhance the Present," *JOHS,* vol. 6 (2005), pp. 227–60.

258 *Turn off the tube:* M. Joseph Sirgy et al., "Does Television Viewership Play a Role in the Perception of Quality of Life?" *Journal of Advertising,* vol. 27, no. 1 (1998), pp. 125–42.

259 *Throw a party:* JZ telephone interview with Goldfine, Aug. 28, 2006.

259 *End on a high note:* Barbara L. Fredrickson, "Extracting Meaning from Past

Affective Experiences," *C&E,* vol. 14, no. 4 (2000), pp. 577–606; Dan Ariely and Ziv Carmon, "Gestalt Characteristics of Experiences," *JBDM,* vol. 13, no. 2 (2000), pp. 191–201.

259 *Go back to school:* Neal J. Roese and Amy Summerville, "What People Regret Most . . . and Why," *PSPB,* vol. 31, no. 9 (Sept. 2005), pp. 1273–85.

260 *Go for the goal:* Daniel Read et al., "Four Score and Seven Years from Now," *MS,* vol. 51, no. 9 (Sept. 2005), pp. 1326–35; Nava Ashraf et al., "Tying Odysseus to the Mast" (July 2005), http://ssrn.com/abstract=770387.

261 *Make your own luck:* JZ e-mail interview with Richard Wiseman, June 27, 2003; JZ, "Are You Lucky?," *MM,* Aug. 2003, p. 85. The article I read in the airport was Michael S. Gazzaniga, "The Split Brain Revisited," *SA,* July 1998, pp. 50–55. (See Chapter Four, "Pigeons, Rats, and Randomness," p. 57.)

262 *Do it now:* Roy F. Baumeister et al., "Ego Depletion," *JPSP,* vol. 74, no. 5 (1998), pp. 1252–65; Roy F. Baumeister and Kathleen D. Vohs, "Willpower, Choice, and Self-Control," in *TAD,* pp. 201–16; JZ telephone interview with Laibson, Nov. 11, 2005.

264 *Give yourself a happiness boost:* Martin E. P. Seligman et al., "Positive Psychology Progress," *AP,* vol. 60, no. 5 (2005), pp. 410–21; Martin E. P. Seligman et al., "A Balanced Psychology and a Full Life," *PTRSLB,* vol. 359 (2004), pp. 1379–81; www.edge.org/3rd_culture/seligman04/seligman_index.html; Kennon M. Sheldon and Sonja Lyubomirsky, "How to Increase and Sustain Positive Emotion," *Journal of Positive Psychology,* vol. 1 (2006), pp. 73–82; Martin E. P. Seligman, *Authentic Happiness* (New York: Free Press, 2002).

265 *Doing and being:* Chu Kim-Prieto et al., "Integrating the Diverse Definitions of Happiness," *JOHS,* vol. 6 (2005), pp. 261–300; Ed Diener and Shigehiro Oishi, "The Nonobvious Social Psychology of Happiness," *Psychological Inquiry,* vol. 16, no. 4 (2005), pp. 162–67; Leaf Van Boven and Thomas Gilovich, "To Do or to Have?," *JPSP,* vol. 85, no. 6 (2003), pp. 1193–1202; Sonja Lyubomirsky et al., "Pursuing Happiness," *Review of General Psychology,* vol. 9, no. 2 (2005), pp. 111–31; Kennon M. Sheldon and Sonja Lyubomirsky, "Achieving Sustainable Gains in Happiness," *JOHS,* vol. 7 (2006), pp. 55–86; Charles T. Munger in *Outstanding Investor Digest,* Sept. 24, 1998, p. 53; Richard M. Ryan and Edward L. Deci, "On Happiness and Human Potentials," *Annual Review of Psychology,* vol. 52 (2001), pp. 141–66.

Acknowledgments

My first debt of gratitude is to Greg Berns of Emory University, Paul Glimcher of New York University, Jordan Grafman of the National Institutes of Health, Scott Huettel of Duke University, Brian Knutson of Stanford University, and Read Montague of Baylor College of Medicine. Each of them welcomed my earliest inquiries and took me patiently under his wing, even though it was painfully obvious that I did not know my ventral tegmental area from my ventromedial prefrontal cortex. And they stuck with me through a steady bombardment of calls and e-mails.

The following scientists and educators freely shared the bounty of their learning and, in many cases, provided vital feedback on chapters in draft form: Antoine Bechara of the University of Southern California; Peter Bernstein of Peter L. Bernstein Inc.; William Bernstein, Efficient Frontier Advisors; Colin Camerer, California Institute of Technology; Jonathan D. Cohen, Princeton University; Antonio Damasio, University of Southern California; Ed Diener, University of Illinois; Thomas Gilovich, Cornell University; Robin Hogarth, Pompeu Fabra University; Carol Horner, Knight Center for Specialized Journalism, University of Maryland; Eric Johnson, Columbia University; Daniel Kahneman, Princeton University; David Laibson, Harvard University; Matthew Lieberman, UCLA; Taylor Larimore and Mel Lindauer, www.diehards.org; George Loewenstein, Carnegie Mellon University; Hiroyuki Nakahara, RIKEN Brain Science Institute; Terrance Odean, University of California, Berkeley; Taketoshi Ono, Toyama Medical and Pharmaceutical University; Camillo Padoa-Schioppa, Harvard Medical School; Massimo Piattelli-Palmarini, University of Arizona; Wolfram Schultz, University of Cambridge; Norbert Schwarz, University of Michigan; Paul Slovic, University of Oregon; Meir Statman, Santa Clara University; Elke Weber, Columbia University.

I would like to thank my agent, the peerless John W. Wright, for his inspiration, support, and friendship. My editor, Bob Bender, put his deft touch, decades of experience and learning, and a world of patience into this project.

I am grateful to Eric Schurenberg and Craig Matters of *Money* for generously giving me the time and space I needed.

I was propped up by readers, helpers, and friends like Ted Aronson, Kate Ashford, Carolyn Bigda, Jack Bogle, Jason Bram, Joan Caplin, Peter Carbonara, Jean Chatzky, Glenn Coleman, Roger Edelen, Eric Gelman, Amanda Gengler, William Green, Tara Kalwarski, Ed Klees, Denise Martin, Ellen McGirt, Peter Quinn, Janet Paskin, Pat Regnier, Bob Safian, Gary Schatsky, Mark Schweber, and Larry Siegel. My beloved friend Eric Schmuckler died before he could read this book; I wish we could have laughed together about whatever I got wrong.

The data demons at Aronson + Johnson + Ortiz, L.P. and Numeric Investors, L.L.P., provided stock returns and checked my math. Marian Hesseling and Lisa Muscolino kindly provided access to hundreds of articles published by Elsevier Science.

Finally, the work I put into writing this book was nothing compared to the work my wife put into enabling me to write it. Now more than ever, her love reminds me of David Ben-Gurion's words: "To be a realist, you must believe in miracles."

Index

Page numbers in *italics* refer to figures and illustrations.